Studies in Computational Intelligence

Volume 1036

Series Editor
Janusz Kacprzyk, Polish Academy of Sciences, Warsaw, Poland

The series "Studies in Computational Intelligence" (SCI) publishes new developments and advances in the various areas of computational intelligence—quickly and with a high quality. The intent is to cover the theory, applications, and design methods of computational intelligence, as embedded in the fields of engineering, computer science, physics and life sciences, as well as the methodologies behind them. The series contains monographs, lecture notes and edited volumes in computational intelligence spanning the areas of neural networks, connectionist systems, genetic algorithms, evolutionary computation, artificial intelligence, cellular automata, self-organizing systems, soft computing, fuzzy systems, and hybrid intelligent systems. Of particular value to both the contributors and the readership are the short publication timeframe and the world-wide distribution, which enable both wide and rapid dissemination of research output.

Indexed by SCOPUS, DBLP, WTI Frankfurt eG, zbMATH, SCImago.

All books published in the series are submitted for consideration in Web of Science.

More information about this series at https://link.springer.com/bookseries/7092

José Luis Verdegay · Julio Brito · Carlos Cruz
Editors

Computational Intelligence Methodologies Applied to Sustainable Development Goals

Springer

Editors
José Luis Verdegay
Department of Computer Science
and Artificial Intelligence
University of Granada
Granada, Spain

Julio Brito
Department of Engineering of Systems
University of La Laguna
San Cristóbal de La Laguna, Spain

Carlos Cruz
Department of Computer Science
and Artificial Intelligence
University of Granada
Granada, Spain

ISSN 1860-949X ISSN 1860-9503 (electronic)
Studies in Computational Intelligence
ISBN 978-3-030-97343-8 ISBN 978-3-030-97344-5 (eBook)
https://doi.org/10.1007/978-3-030-97344-5

This Springer imprint is published by the registered company Springer Nature Switzerland AG
The registered company address is: Gewerbestrasse 11, 6330 Cham, Switzerland

This book is dedicated to all those colleagues who wanted to contribute to it but were unable to do so because of COVID-19.

Preface

A few years ago, as one more stage in the close collaboration between the Group of Intelligent Computing of the University of La Laguna (Spain) and the Models of Decision and Optimization (MODO) Research Group of the University of Granada (Spain), both the teams agreed to constitute the Decision, Optimization and Sustainable Intelligent Systems (DOSIS) Working Group. Among the members of DOSIS are some of the authors in this book as well as all its three editors and, as it may be quite obvious, one of its main research lines is that one relative to the interface area between artificial intelligence (AI) and sustainable development goals (SDGs). The high interest for researching in that interface area is evident. There is no doubt that two of the leading issues that concern all of humanity today are those related to sustainable development and intelligent systems.

Every day new sectors appear in which the methodologies and techniques of AI are applied successfully, hoping that in the short term, they will affect the productive sector as a whole, equality and inclusion, environmental responsibility and many other areas. There are already studies that report the specific impacts produced by the application of AI in sustainable development. These impacts are generally positive, but there are also negative ones in some cases. However, to the best of our knowledge, there is no international referenced study that rigorously reports the extent to which AI could affect all aspects of sustainable development.

Thus arises a field of research and study, as pertinent as it is timely and critical, which focuses on analyzing how AI can influence the ability to comply in a timely manner with all the SDGs.

SDGs are a call for action by all countries to promote prosperity while protecting the planet. They recognize that ending poverty must go hand-in-hand with strategies that build economic growth and address a range of social needs including education, health, social protection and job opportunities, while tackling climate change and environmental protection. More important than ever, the goals provide a critical framework for COVID-19 recovery. As a whole, the SDGs are vital for a recovery that leads to a greener, more inclusive economies and more robust and resilient societies (https://www.un.org/sustainabledevelopment/).

On the other hand, although AI is umbrella for large variety of methodologies and techniques, it is patent that some of the current most successful AI systems are based on computational intelligence (CI). CI is the theory, design, application and development of biologically and linguistically motivated computational paradigms. Traditionally, the three main pillars of CI have been neural networks, fuzzy systems and evolutionary computation. However, in time, many nature-inspired computing paradigms have evolved. Thus, CI is an evolving field, and at present in addition to the three main constituents, it encompasses computing paradigms like ambient intelligence, artificial life, cultural learning, artificial endocrine networks, social reasoning and artificial hormone networks. CI plays a major role in developing successful intelligent systems, including games and cognitive developmental systems (https://cis.ieee.org/about/what-is-ci).

Encouraged by knowing potential applications of CI in the achievement of the 17 objectives and 169 goals recognized in the 2030 Agenda for Sustainable Development, in this book, we collect 18 contributions that are adjusted to that end. These are classified into four parts (a first cross-sectional and three more oriented to specific areas) that focus models on: CI and SDGs, health control, emissions control and transportation and distribution. As a whole, the 18 contributions approache problems from the three main areas of sustainable development: society, economy and environment, and they are authored by recognized scholars from 13 countries: Algeria, Brazil, Chile, Colombia, Cuba, Czechia, India, Iran, Mexico, Morocco, Poland, Spain and USA.

The first part opens with a paper by O. Castillo and P. Melin (from Mexico): "A Review on the Role of Computational Intelligence on Sustainability Development," in which an overview of the existing publications using CI techniques in applications to sustainability development is presented. Authors provide the up-to-date relevant statistics and analysis of the existing work in this area. In addition, they outline future possible trends for research on applying intelligent systems to problems in sustainability development. Then J. Kacprzyk, J. W. Owsinski, E. Szmidt and S. Zadrożny (from Poland) present the work entitled "Fuzzy Linguistic Summaries for Human Centric Analyses of Sustainable Development Goals (SDG) Related to Technological Innovations." Authors propose to use fuzzy logic-based linguistic data summaries for the assessment and evaluation of both the essence of SDGs and their fulfillment, as such a form is very human consistent because it uses natural language that is the only fully natural means of articulation and communication for the humans. The third contribution in this section is "Impact of Human-Computer Interaction in the Global Sustainable Development Goals" by J. Muñoz-Arteaga, H. Luna-García, C. Collazos and A. Granollers (from Colombia and Spain). The paper shows as the research area of human–computer interaction (HCI) plays a key role into the design and development of software applications and practices aimed to achieve the SDGs.

The second part is dedicated to the health control models and includes five chapters. In the first, "Investigating the Key Parameters Affecting Sustainable IoT-based Marketing," S. E. Najafi, H. Nozari and S. A. Edalatpanah (from Iran) show as perceived usefulness, ease of use, trust, social acceptance, enjoyment of use and controllability have been identified as effective variables on the use of Internet of things (IoT) in sustainable smart marketing. Then these key indicators are reviewed and prioritized, and a framework for expressing the positions of the impact of the IoT on sustainable marketing is presented. In the following chapter, R. Bello, M. M. García, Y. Caballero, A. Rosete and Y. Rodríguez (from Cuba), titled “Applications of Computational Intelligence in the Studies of COVID-19,” analyze how CI techniques have allowed developing different studies about COVID-19, creating several prediction models and formulating new knowledge about it. In “Quantification of African Swine Fever (ASF) Introduction Risk Using OWA Operators,” E. H. Cables, F. Moreno, M. T. Lamata and F. Gómez (from Colombia and Spain) introduce a novel methodology to quantify risk factors related to ASF introduction using expert knowledge. The methodology is based on operator weight average, in which weights are determined by using linguistic labels and a consensus mechanism.

The fourth chapter in this second part, by J. C. Figueroa–García, C. Franco and R. Neruda (from Colombia and Czechia), is entitled “An Optimization Model for Location-Allocation of Health Services Under Uncertainty,” and it considers the existence of fuzzy uncertainty in an optimization model for allocation of healthcare facilities to serve patients with different needs. The last chapter in this part is “Computational Intelligence and Tikhonov Regularization with Reduced Dimension Model: Applications in Health, Renewable Energy and Climate Heat Transfer Inverse Problems” by L. Correia da Silva Jardim, D. Campos Knupp, T. M. de Souza Monteiro de Barros, L. A. da Silva Abreu, C. Cruz-Corona and A. J. Silva Neto (from Brazil). Authors present a method to predict the optimal value of the Tikhonov’s regularization parameter by solving simplified versions of the inverse problems considered. The results obtained demonstrate the feasibility of the technique in three problems with potential practical applications in bioengineering, renewable energy and climate.

The third part, which deals with emissions control models, includes five chapters. Along the first one, “Intelligent Collaborative Freight Distribution to Reduce Greenhouse Gas Emissions: A Review,” C. Expósito-Izquierdo, A. Expósito-Márquez, B. Melián-Batista, J. A. Moreno-Pérez and J. M. Moreno-Vega (from Spain) review optimization models for horizontal collaborative freight transport that include environmental and economic criteria. Specifically, they consider models for planning delivery routes and for cross-docking. The next chapter, “Pressure Sensor Placement for Leak Location in Zones of a Water Distribution Network,” is authored by M. J. Ares-Milián, M. Quiñones-Grueiro, C. Verde, and O. Llanes-Santiago (from Cuba). Authors present a novel sensor placement approach for leak location in large-scale water distribution networks (WDN), and the satisfactory performance of the proposal is demonstrated under uncertain conditions and measurement noise even when the sensitivity of the sensors is limited. The third contribution in this part, by M. L. S. Araujo, Y. K. L. Kitagawa, D. M. Moreira and E. G. Sperandio Nascimento (from Brazil), is entitled “Forecasting Tropospheric Ozone Using Neural Networks and Wavelets: Case Study of a Tropical Coastal-Urban Area.” Authors present a study realized by using data from a tropical coast-urban site in Southeast Brazil, highly influenced by intense convective weather with complex terrain, and short-term forecasting of hourly ground-level ozone is performed using long short-term memory, a type of recurrent neural network, with the discrete wavelet transform. R. Coelho (from Brazil) authored the fourth work "Power Curve Estimation of Wind Farms with Imprecise Data by Fuzzy Quadratic Programming" where a parametric approach that solves quadratic programming problems under different kind of uncertainties in its data is shown. In order to demonstrate the efficiency of the approach, a fuzzy regression analysis problem is used. To close the part, the effect of the growth of carbon emissions in the function of GDP per capita, in countries of America and Europe, its behavior and the results they generate with regards to the SDG 8 and 13 are analyzed by A. Quintero, M. Gaitán-Angulo, M. I. Gómez-Caicedo, A. L. Mercado Suárez and R. E. Reyes-Gil (from Colombia) in "Effect of the Growth of Carbon Emissions as a Function of GDP per Capita and Its Relationship with the SDG in American and European Countries: An Application of Neural Networks."

Finally, the fourth part collects five papers on transportation and distribution models. In the first chapter, "A Fuzzy Approach for Organizational Transportation Aligned with the Sustainable Development Goals: Health Promotion, Inequality Reduction, and Responsible Consumption," E. Sánchez-Ansola, A. C. Pérez-Pérez and A. Rosete (from Cuba) introduce a new model for the School Bus Routing Problem (SBRP) that have a direct impact on three of the 3rd, 10th and 12th SDGs. The proposed model focuses on fuzzy modeling of maximum students' walking distance and bus capacity restrictions. The following chapter, by J. Ruiz-Meza, J. Brito and J. R. Montoya-Torres (from Spain and Colombia), is titled "A VND to Solve the Bi-objective Tourist Trip Design Problem, Including Reducing CO_2 with Fuzzy Constraints." Authors propose solving a bi-objective problem of planning tourist routes for multiple days, selecting the transport mode and multiple fuzzy constraints. To find the optimal planning they use two criteria: obtaining the most excellent satisfaction or benefit for the tourist and reducing CO_2 emissions. The third contribution in the section, "Time Variant Multi-objective Multi-item Solid Transportation Problem in Sustainable Development in Terms of Interval Valued Fuzzy Numbers," by T. Sifaoui and M. Aïder (from Algeria) presents a new procedure for converting the problem in the title to a deterministic form, and then, a hybrid method expressed with different membership function is used to solve the converted transport problem. H. El Raoui, D. Pelta, A. Rufián-Lizana, M. Oudani and A. El Hilali Alaoui (from Morocco and Spain) are the authors of "On the Generation of Alternative Solutions for a Perishable Food Distribution Problem." The work outlines the application of a modelling to generate alternatives approach that can generate a set of near optimal solutions, but maximally different from the best one, which is applied to solve a perishable food distribution problem by using as intermediate a general variable neighborhood search algorithm. Finally, the contribution by G. Maity and S. Kumar-Roy (from India) entitled "Solving Two-Stage Multi-objective Transportation Problem Using Goal Programming and Its Application to Sustainable Development" closes the last part to finish the book. Authors focus on a two-stage Multi-Objective Transportation Problem (MOTP) and solve it through goal programming. Considering goals corresponding to the objective functions of both stages, they solve two-stage MOTP and then derive Pareto optimal solution. Finally, a numerical example based on real scenario is presented to show the effectiveness of the study in sustainable development.

The variety and diversity of contributions collected in this book give a very good idea of the potential that CI has to contribute to the solution of the SDGs. But it is very important to emphasize that due to the inter and multi-disciplinary nature of the scientific, technological and social fields covered by the 2030 Agenda for Sustainable Development, this volume, and that is the main aim of the editors, want to be a sample, surely very limited, of the enormous work that lies ahead.

To finish, the editors want to state that there have been several authors who for different reasons but in all cases related to the COVID-19 have not finally been able to send their contributions to this book on time. We want to express our gratitude to all of them because we know the efforts they have made to participate in this editorial project. Finally, we also want to express that we are indebted to Prof. Janusz Kacprzyk, who since we presented our editorial idea to him has not stopped encouraging us to carry it out. To conclude, editors acknowledge the financial support provided by the projects PID2020-112754 GB-I00 (Spanish Ministry of Economy and Competitiveness and FEDER funds from the European Union) and B-TIC-640-UGR20 (Regional Govern of Andalusia).

Granada, Spain — José Luis Verdegay
Granada, Spain — Carlos Cruz
San Cristóbal de La Laguna, Spain — Julio Brito

Contents

Contributors

Ahmed El Hilali Alaoui Université Euro Méditerranéenne, Fez, Morocco

Mirella Lima Saraiva Araujo Manufacturing and Technology Integrated Campus (SENAI CIMATEC), Salvador, Bahia, Brazil

Marlon J. Ares-Milián Departamento de Automática y Computación, Universidad Tecnológica de la Habana José Antonio Echeverrıa, CUJAE, La Habana, Cuba

Méziane Aïder Faculty of Maths, LAROMAD, Bab Ezzouar, Algeria

Rafael Bello Universidad Central "Marta Abreu" de Las Villas, Santa Clara, Cuba

Julio Brito Departamento de Ingeniería Informática y de Sistemas, Instituto Universitario de Desarrollo Regional, Universidad de La Laguna, Santa Cruz de Tenerife, Spain

Yaile Caballero Universidad de Camagüey, Camaguey, Cuba

E. H. Cables Universidad Antonio Nariño, Bogotá, Bogota, Colombia

Oscar Castillo Tijuana Institute of Technology, Tijuana, BC, Mexico

Ricardo Coelho Departamento de Estatística e Matemática Aplicada, Centro de Ciências, Universidade Federal do Ceará, Fortaleza, CE, Brazil

César A. Collazos Universidad del Cauca, Cauca, Colombia

Carlos Cruz Corona Department of Computer Science and Artificial Intelligence, Granada University, Granada, Spain

Luiz Alberto da Silva Abreu LEMA—Patrícia Oliva Soares Laboratory of Experimentation and Numerical Simulation in Heat and Mass Transfer, Polytechnic Institute, Rio de Janeiro State University, Nova Friburgo, RJ, Brazil

Lucas Correia da Silva Jardim LEMA—Patrícia Oliva Soares Laboratory of Experimentation and Numerical Simulation in Heat and Mass Transfer, Polytechnic Institute, Rio de Janeiro State University, Nova Friburgo, RJ, Brazil

Taciano Magela de Souza Monteiro de Barros LEMA—Patrícia Oliva Soares Laboratory of Experimentation and Numerical Simulation in Heat and Mass Transfer, Polytechnic Institute, Rio de Janeiro State University, Nova Friburgo, RJ, Brazil

Seyyed Ahmad Edalatpanah Department of Applied Mathematics, Ayandegan Institute of Higher Education, Tonekabon, Iran

Hanane El Raoui Department of Computer Science and A.I., Universidad de Granada, Granada, Spain;
Modeling and Mathematical Structures Laboratory, Sidi Mohamed Ben Abdellah University, Fez, Morocco

Christopher Expósito-Izquierdo Universidad de La Laguna, San Cristóbal de La Laguna, Spain

Airam Expósito-Márquez Kaizten Analytics, San Cristóbal de La Laguna, Spain

Juan Carlos Figueroa–García Universidad Distrital Francisco José de Caldas, Bogotá, Colombia

Carlos Franco Universidad del Rosario, Bogotá, Colombia

Mercedes Gaitán-Angulo Fundación Universitaria Konrad Lorenz, Bogotá, Colombia

Huizilopoztli Luna García Universidad Autónoma de Zacatecas, Zacatecas, Mexico

María M. García Universidad Central "Marta Abreu" de Las Villas, Santa Clara, Cuba

Rosa Eugenia Reyes Gil Fundación Universitaria los Libertadores, Bogotá, Colombia

Antoni Granollers Universidad de Lleida, Lleida, Spain

Melva Inés Gómez-Caicedo Fundación Universitaria los Libertadores, Bogotá, Colombia

F. Gómez Universidad Nacional de Colombia, Bogota, Colombia

Janusz Kacprzyk Systems Research Institute, Polish Academy of Sciences, Warsaw, Poland;
Warsaw School of Information Technology, Warsaw, Poland

Yasmin Kaore Lago Kitagawa Department of Environmental Engineering, Federal University of Espírito Santo (UFES), Vitória, Espírito Santo, Brazil

Diego Campos Knupp LEMA—Patrícia Oliva Soares Laboratory of Experimentation and Numerical Simulation in Heat and Mass Transfer, Polytechnic Institute, Rio de Janeiro State University, Nova Friburgo, RJ, Brazil

M. T. Lamata Universidad de Granada, Granada, Spain

Orestes Llanes-Santiago Departamento de Automática y Computación, Universidad Tecnológica de la Habana José Antonio Echeverrı́a, CUJAE, La Habana, Cuba

Gurupada Maity Department of Applied Mathematics with Oceanology and Computer Programming, Vidyasagar University, Midnapore, West Bengal, India

Patricia Melin Tijuana Institute of Technology, Tijuana, BC, Mexico

Belén Melián-Batista Universidad de La Laguna, San Cristóbal de La Laguna, Spain

Jairo R. Montoya-Torres Facultad de Ingeniería, Universidad de La Sabana, Cundinamarca, Colombia

Davidson Martins Moreira Manufacturing and Technology Integrated Campus (SENAI CIMATEC), Salvador, Bahia, Brazil

José A. Moreno-Pérez Universidad de La Laguna, San Cristóbal de La Laguna, Spain

J. Marcos Moreno-Vega Universidad de La Laguna, San Cristóbal de La Laguna, Spain

F. Moreno Universidad Nacional de Colombia, Bogota, Colombia

Jaime Muñoz-Arteaga Universidad Autónoma de Aguascalientes, Aguascalientes, Mexico

Seyyed Esmaeil Najafi Department of Industrial Engineering, Science and Research Branch, Islamic Azad University, Tehran, Iran

Erick Giovani Sperandio Nascimento Manufacturing and Technology Integrated Campus (SENAI CIMATEC), Salvador, Bahia, Brazil

Roman Neruda Informatics Department, Charles University, Prague, Czech Republic

Antônio J. Silva Neto LEMA—Patrícia Oliva Soares Laboratory of Experimentation and Numerical Simulation in Heat and Mass Transfer, Polytechnic Institute, Rio de Janeiro State University, Nova Friburgo, RJ, Brazil

Hamed Nozari Department of Industrial Engineering, Central Tehran Branch, Islamic Azad University, Tehran, Iran

Mustapha Oudani TIC Lab, ESIN, International University of Rabat, Rabat, Morocco

Jan W. Owsinski Systems Research Institute, Polish Academy of Sciences, Warsaw, Poland

David A. Pelta Department of Computer Science and A.I., Universidad de Granada, Granada, Spain

Ana Camila Pérez-Pérez Department of Artificial Intelligence and Informatics Systems Infrastructure, Universidad Tecnológica de La Habana José Antonio Echeverría (CUJAE), La Habana, Cuba

Anderson Quintero Fundación Universitaria Konrad Lorenz, Bogotá, Colombia

Marcos Quiñones-Grueiro Institute for Software Integrated Systems, Vanderbilt University, Nashville, TN, USA

Yanela Rodríguez Universidad de Camagüey, Camaguey, Cuba

Alejandro Rosete Department of Artificial Intelligence and Informatics Systems Infrastructure, Universidad Tecnológica de La Habana José Antonio Echeverría (CUJAE), La Habana, Cuba

Sankar Kumar Roy Department of Applied Mathematics with Oceanology and Computer Programming, Vidyasagar University, Midnapore, West Bengal, India

Antonio Rufián-Lizana Dpto. Estadística e I.O., Universidad de Sevilla, Sevilla, Spain

José Ruiz-Meza Facultad de Ingeniería, Universidad de La Sabana, Cundinamarca, Colombia

Eduardo Sánchez-Ansola Department of Artificial Intelligence and Informatics Systems Infrastructure, Universidad Tecnológica de La Habana José Antonio Echeverría (CUJAE), La Habana, Cuba

Thiziri Sifaoui Faculty of Sciences, LAROMAD, UMM-Tizi-Ouzou, Algeria;
Faculty of Sciences and Technology, Amine Elokkal El Hadj Moussa Eghamouk University, Tamanghasset, Algeria

Álvaro Luis Mercado Suárez Fundación Universitaria los Libertadores, Bogotá, Colombia

Eulalia Szmidt Systems Research Institute, Polish Academy of Sciences, Warsaw, Poland;
Warsaw School of Information Technology, Warsaw, Poland

Cristina Verde Instituto de Ingeniería, Universidad Nacional Autónoma de México, Mexico City, Mexico

Sławomir Zadrożny Systems Research Institute, Polish Academy of Sciences, Warsaw, Poland;
Warsaw School of Information Technology, Warsaw, Poland

Computational Intelligence and SDG

A Review on the Role of Computational Intelligence on Sustainability Development

Oscar Castillo and Patricia Melin

Abstract This paper presents a review of the existing publications using computational intelligence techniques in applications to sustainability development. Computational intelligence is the area that deals with the design and development of intelligent systems for diverse scopes of application. Sustainable development can be viewed as a way to achieve human development goals while simultaneously maintaining the ability of natural systems to offer natural resources and ecosystem services for the benefit of the economy and society. In this regard, it is natural to think that the computational intelligence area, which includes models like neural networks, fuzzy systems, and metaheuristics, will significantly impact achieving the goals of sustainable development. This review paper reveals that there has been some work in this area. We will provide up to date relevant statistics and analysis of the current work. In addition, we will outline possible future trends for research on applying intelligent systems to problems in sustainability development.

Keywords Computational intelligence · Sustainability development · Intelligent systems

1 Introduction

Nowadays sustainable development has been receiving increasing attention from industry, government and academia [1–4]. Sustainable development can be viewed as a way for achieving human development goals, while simultaneously maintaining the ability of natural systems for offering the natural resources and ecosystem services for the benefit of economy and society [5–9]. More generally, it can also be put forward as the study of methods for designing, manufacturing/engineering, utilizing and discarding of computing devices in a way that sustains the ability to offer the natural resources and ecosystem services for the benefit of society [10–12]. For this reason, it is of upmost importance to consider this area in every aspect of human

O. Castillo (✉) · P. Melin
Tijuana Institute of Technology, 22379 Tijuana, BC, Mexico
e-mail: ocastillo@tectijuana.mx

J. L. Verdegay et al. (eds.), *Computational Intelligence Methodologies Applied to Sustainable Development Goals*, Studies in Computational Intelligence 1036,
https://doi.org/10.1007/978-3-030-97344-5_1

life [13–16]. At the same time, the use of computational intelligent techniques has become also very important for everyday life, ranging from domestic appliances to autonomous systems and automated medical diagnosis. In this form, we come to realize that the interaction of both sustainable development and computational intelligence would become very important in the near future [14–20].

Computational Intelligence is a relatively new area of Computer Science that deals with the development of intelligent systems based on relatively new intelligent models, like fuzzy systems, neural networks, swarm intelligence, genetic algorithms and others [21–24]. Fuzzy logic was originally proposed by Zadeh and generalizes traditional Boolean logic by allowing truth values to be between 0 and 1. The main goal of fuzzy logic is to represent and handle the intrinsic uncertainty in real-world problems by using fuzzy systems. Neural networks are computer models that simulate the human brain by representing real neural networks with mathematical models (as artificial networks) and by using an iterative learning algorithm. Neural networks can learn from real-world data to form supervised or unsupervised neural systems that can be applied in many diverse real-world situations ranging from classification and pattern recognition to time series forecasting. Genetic algorithms are computer methods that simulate real natural evolution and the principle of survival of the fittest for achieving an efficient search for solutions to optimization problems. Genetic algorithms and related evolutionary techniques have also found many real applications, like in scheduling, planning, manufacturing and others. Swarm intelligence consists of techniques based on collective intelligence, like particle swarm intelligence, ant colony optimization, the grey wolf optimizer, firefly algorithm, flower pollination, bee colony optimization and others. These recent techniques have been applied in general optimization problems, as well as in optimization of fuzzy and neural models [25–30]. In addition, the prudent combination of several of the previous techniques can be utilized for constructing powerful hybrid intelligent systems, which can solve many real-world problems in different areas of applications, such as intelligent control, pattern recognition, robotics, diagnosis, classification and others [31–40].

The contribution of this paper is the proposal of a review of the existing works in the literature, where both sustainable development and computational intelligence have been used together in real-world problems, so that the potential advantages for the combination of these areas can be evaluated and possible future works can be envisioned. In addition, based on the scientometric analysis of the existing works, another contribution of the paper was the proposal of future trends (in the short, medium and long terms) for possible research works based on the combination of sustainable development and computational intelligence.

The rest of the article is organized in the following fashion. Section 2 offers an overview of existing works in the literature on intelligent systems applied to sustainable development problems. Section 3 presents a scientometric analysis of sustainable development and computational intelligence techniques. Section 4 outlines possible future trends of research on the intersection of computational intelligence and sustainable development. Finally, Sect. 5 summarizes the conclusions and outlines interesting lines of research.

2 Overview of Existing Works

In this section we describe a sample of the papers that have been published in sustainable development and computational intelligence. This sample of papers provides a general view of the work that has been done in these areas.

In Chen et al. [17], the authors presented a bi-level distributed decision approach based on swarm intelligence techniques. Current existing works have shown that more benefits regarding energy cost saving, environmental sustainability and reliable power supply can be achieved by clustering buildings together so that free exchange of information and energy can be achieved. In this case, both centralized and distributed decision approaches have been developed in recent years, to enable efficient operation among buildings in the cluster. However, most of the existing methods are only applicable for small scale building clusters and/or the privacy of the stakeholders is not well protected. To efficiently address these research gaps, in this work a bi-level distributed decision approach based on swarm intelligence was proposed. A particle swarm optimizer was applied at the system level to coordinate the operations among buildings, and a mixed integer programming model was put forward for each building to simultaneously calculate operation decisions for its energy systems. The performance of the proposed decision method in terms of accuracy, scalability, and robustness was evaluated using several clusters of buildings with the number of buildings ranging from 2 to 256. It was shown that the proposed approach is computationally efficient, as the computational complexity is linear (O(n)), where n is the number of buildings in the cluster.

In Ekici et al. [29], the authors presented a systematic review and summary of performative computational architecture using swarm and evolutionary optimization techniques. The classification for one hundred types of studies was illustrated, which includes different sub-categories of performative computational architecture, such as sustainability cost, functionality, and structure. The main findings, including optimization and all the types of parameters, are presented by focusing on different aspects of buildings. In addition, utilization of form-finding parameters of all reviewed studies and the distributions for each performance objectives are also presented. Moreover, the utilization of swarm and evolutionary optimization algorithms in the reviewed works is summarized. The future trends in publications, published years, problem scales, and building functions, were also analyzed. Finally, future trends for research works were outlined by focusing on different aspects of computational architecture in accordance to the evidence that was obtained based on the review process.

In Garg et al. [38], a new evolutionary approach for modeling the milling process with energy conservation in manufacturing operation was presented. From the point of view of conservation of energy, the concept of modeling the energy consumption as a crucial element of environmental sustainability in any manufacturing industry remains an important study focus for climate change experts around the world. Among the existing manufacturing operations, machining is still widely performed.

Extensive studies by researchers reveal that the focus has been on modeling and optimization of the manufacturing facets (e.g. surface roughness, tool wear rate, dimensional accuracy) of the machining operations by using computational intelligence methods, such as artificial neural networks. Alternatively, an evolutionary approach based on multi-gene genetic programming can also be applied, but its effective performance depends on its fitness function. This study proposes a new multi-gene genetic programming approach based on orthogonal functions and compares its performance to that of the normal multi-gene genetic programming in modeling energy consumption of the milling process. The existing hidden relationships between the energy consumption and the input process parameters were discovered by performing sensitivity and parametric analysis. From these relationships, an optimum set of input settings was obtained, which was able to conserve a greater amount of energy from these operations. It was found that the cutting speed has the highest impact on the milling process followed by feed rate and depth of cut.

In Naserifar et al. [40], the authors have offered a hybrid computational approach including an efficient polarization at the electrode–electrolyte interface, which combined with machine learning, was able to predict new, highly improved performance of realistic electrocatalysts. This approach enables in silico design aimed at highly improving the performance along with other applications. The portfolio for energy and environmental sustainability demands development of new electrocatalysts with greatly improved performance. This in turn needs developing a profound knowledge of the fundamental mechanisms underlying electrocatalysis processes to develop in silico combinatorial calculations to find the best possible new potential solutions for experimental synthesis and characterization. The hybrid approach was able to efficiently cope with the problem of finding the new potential solutions needed.

In Pinter et al. [41], the authors presented an approach for real estate prices prediction with the help of artificial intelligence. The development of efficient models for real estate prices prediction is very important for urban development and several issues in this regard were presented. In this case, due to the significant uncertainties and dynamic variables involved in the process, modeling real estate has been considered as complex systems. In this work, a new machine learning approach is proposed to handle the complexity of real estate modeling. Call detail records can provide excellent opportunities for in-depth investigation of the mobility characterization. This work explored the potential of call detail records in predicting the real estate prices with the help of artificial intelligence (AI). A set of very important mobility entropy factors, including dweller entropy, dweller gyration, workers' entropy, worker gyration, dwellers' work distance, and workers' home distance, are used as input variables. The prediction model was developed using a multi-layered neural network perceptron trained with the collective intelligence algorithm of particle swarm optimization (PSO). Model performance was evaluated using mean square error (MSE), sustainability index (SI), and Willmott's index (WI). The proposed model exhibited promising results discovering that the entropy of the workers and the dwellers' work distances directly influence the real estate price.

3 Scientometric Analysis

In this section, the Scientometric analysis for the existing published papers in the Sustainability Development and Computational Intelligence areas, performed for this work is presented. We will present a sequence of plots and tables that show the analysis of the data (published papers) from different points of view or facets used in the analysis. We can say that this analysis helps us understand the nature of publications and research that has been done in the areas of Sustainability Development and Computational Intelligence. First, we start with Fig. 1 showing a plot of citations per year since 2010. Another important measure of the impact of the publications in these areas is the h index, which can be defined as the highest value that represents the number of times a particular paper has been cited. In this case, in Fig. 2 we show the plot for the h index, which has a value of 13 for the areas of Sustainability

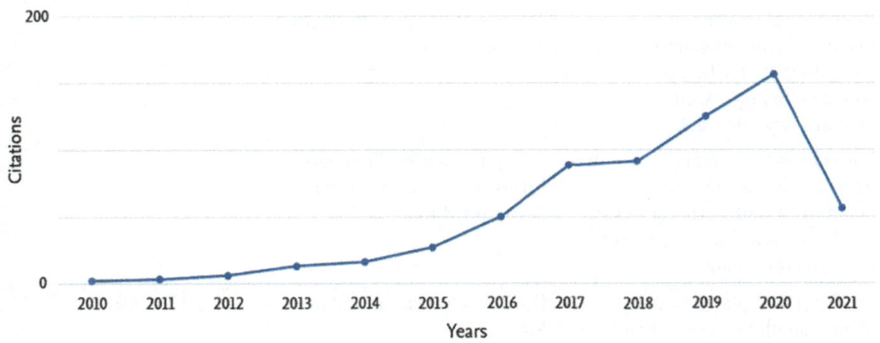

Fig. 1 Plot of citations per year for articles published since 2010

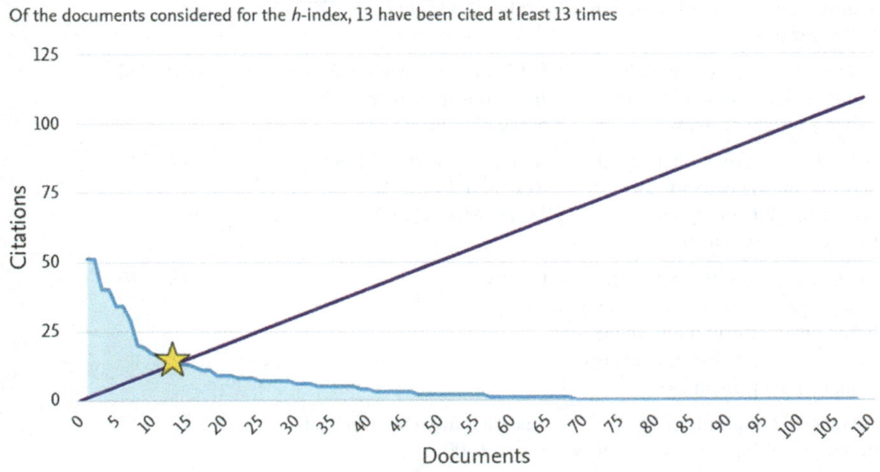

Fig. 2 Plot illustrating h index values for the papers published since 2010

Development and Computational Intelligence. The value of 13 means that there are 13 papers with a least 13 references or more. Table 1 shows the list of the top eight most cited papers in Sustainability Development and Computational Intelligence.

We describe in Fig. 3 the area distribution (in percentage) of the papers that have been published in sustainability development and computational intelligence. The area with the higher percentage is Computer Science with a 36.6%.

We illustrate in Fig. 4 a plot of the documents per year, where we can notice an increasing trend.

We show in Fig. 5 a plot with bars to illustrate the number of publications of the top authors in sustainability development and computational intelligence.

Table 1 List of the top 8 most cited papers on sustainable development and computational intelligence

Title	Published in	Year	Total citation
Energy conservation in manufacturing operations: modelling the milling process by a new complexity-based evolutionary approach	Journal of Cleaner Production 108, pp. 34–45	2015	51
Integration of sustainability issues in strategic transportation planning: a multi-criteria model for the assessment of transport infrastructure plans	Computer-Aided Civil and Infrastructure Engineering 25(6), pp. 440–451	2010	51
From parallel plants to smart plants: intelligent control and management for plant growth	IEEE/CAA Journal of Automatica Sinica 4(2), 7,894,126, pp. 161–166	2017	40
Evaluating electric bus operation for a real-world BRT Public Transportation using simulation optimization	IEEE Transactions on Intelligent Transportation Systems 17(10), 7,415,957, pp. 2777–2786	2016	40
A real-time service-oriented framework to support sustainable cyber-physical systems	IEEE International Conference on Industrial Informatics (INDIN) 5,549,473, pp. 15–21	2010	34
A hidden Markov model-based acoustic cicada detector for crowdsourced smartphone biodiversity monitoring	Journal of Artificial Intelligence Research 51, pp. 805–827	2014	20
Synthesis of an intelligent rural village microgrid control strategy based on smartgrid multi-agent modelling and transactive energy management principles	Energy 147, pp. 263–278	2018	19
Advances in multiple criteria decision making for sustainability: modeling and applications	Sustainability (Switzerland) 10(5), 1600	2018	17

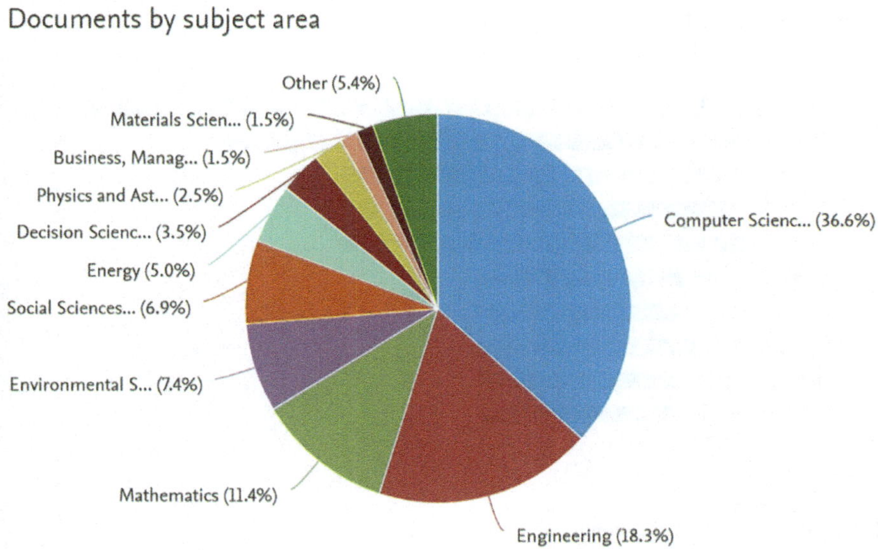

Fig. 3 Distribution of papers according to their area of application

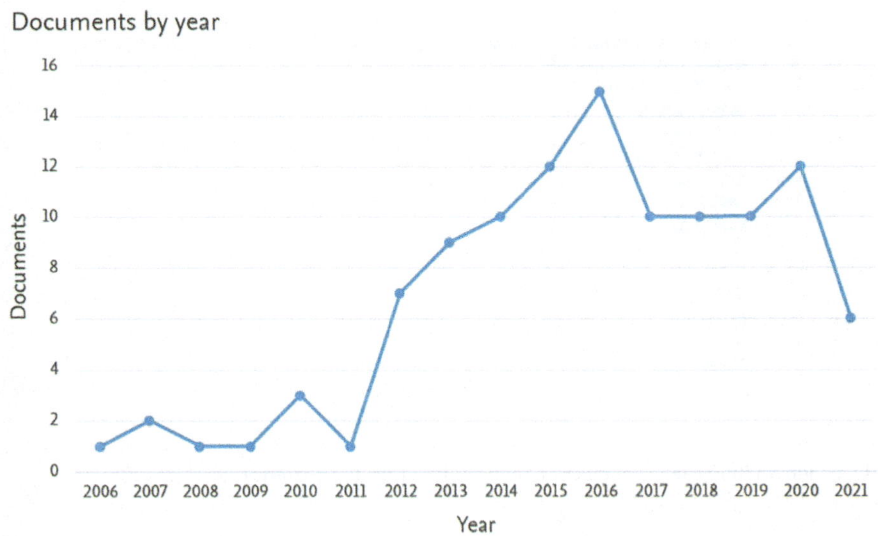

Fig. 4 Plot of the published documents per year according to Scopus

We show in Fig. 6 a plot with bars to illustrate the number of publications of the top countries in sustainability development and computational intelligence.

We show in Fig. 7 the distribution of the published documents according to their type. In this case, almost 50% of the documents are conference papers and at the moment only 30% are journal papers.

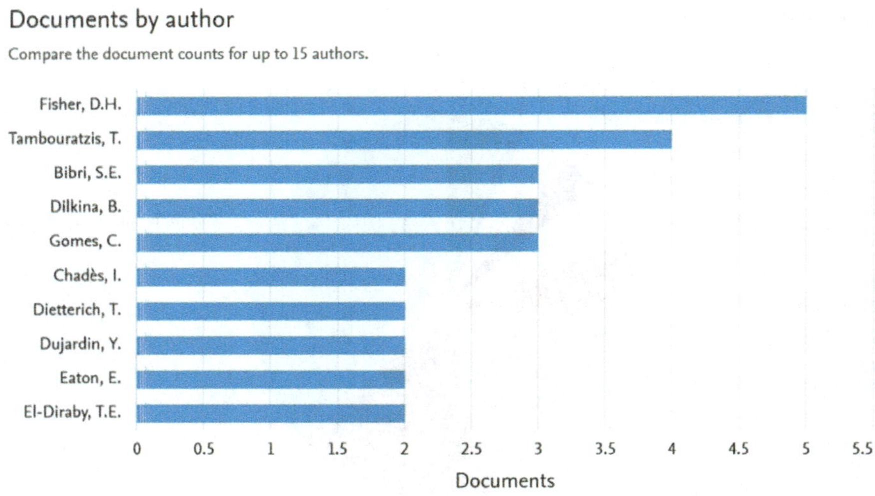

Fig. 5 Plot to illustrate the number of publications of the top authors

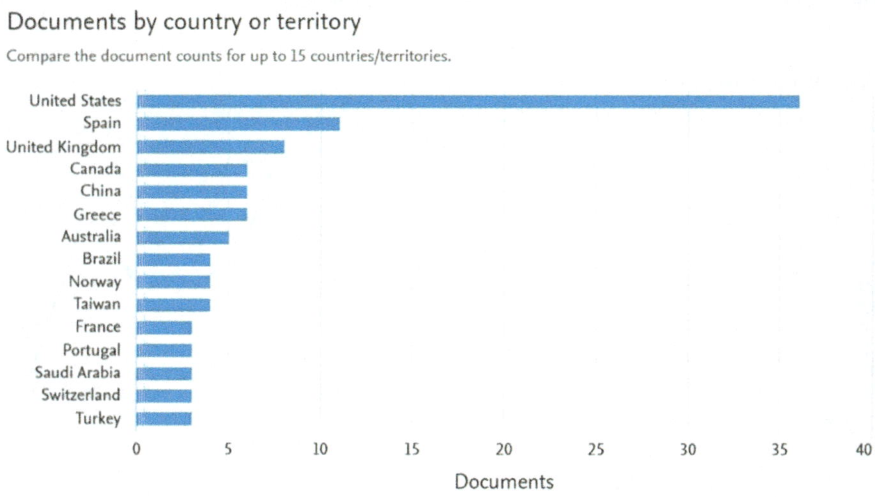

Fig. 6 Plot to illustrate the number of publications of the top countries

We show in Fig. 8 a plot with bars to illustrate the number of publications according to funding sponsors in sustainability development and computational intelligence.

In Fig. 9 we show a cluster visualization for the density of the main authors based on the papers that have been published from 2006 to 2021. This Figure shows authors that are close together working in research collaborations associated to sustainable development and computational intelligence. We have to say that names with larger fonts represent authors with more published papers.

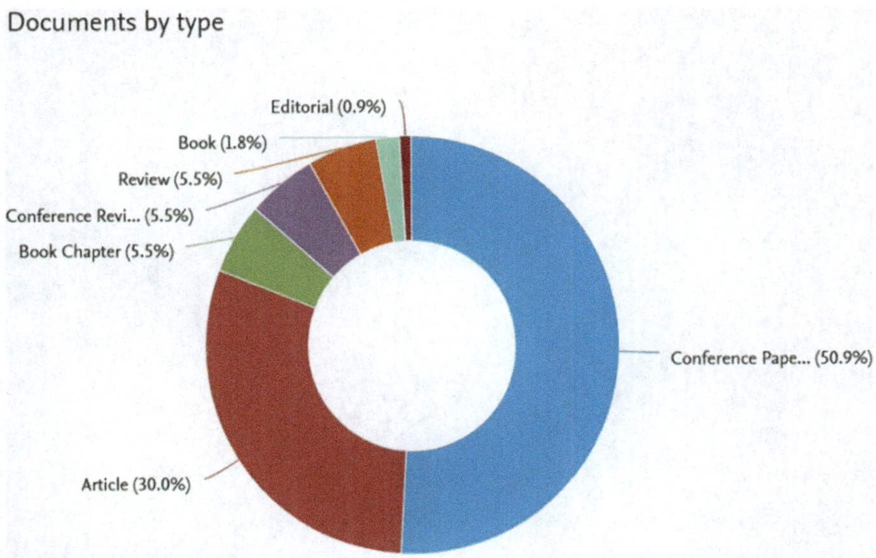

Fig. 7 Distribution of published documents according to their type

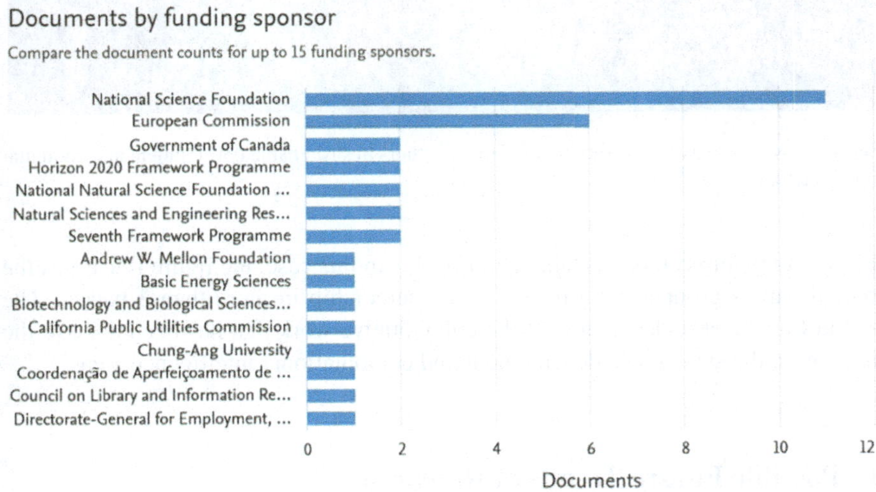

Fig. 8 Plot to illustrate the number of publications of the top funding sponsors

In Fig. 10 we show a cluster visualization for the density of the main authors based on the citations that have been made to their papers from 2006 to 2021. We have to say that names with larger fonts represent authors with more citations for their published papers.

Co-authorship is essential for any Scientometric study and for the case of the interaction of sustainable development and computational intelligence techniques

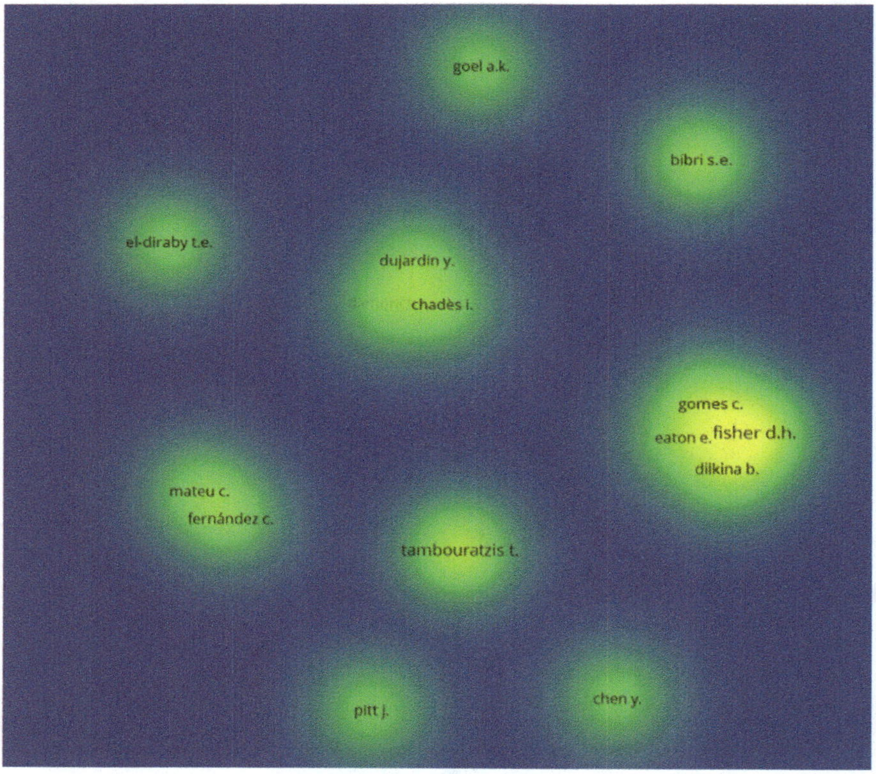

Fig. 9 Cluster density visualization of the main authors in sustainable development and computational intelligence

the co-authorship can be visualized in Fig. 11. In this case, the figure illustrates the journal papers grouped with respect to co-authorship in more than 3 papers. The plot in Fig. 11 provides a view of the collaborative work that has been done to the moment in the sustainable development and computational intelligence areas.

4 Possible Future Trends of Research

Based on the papers described on the previous sections and the performed scientometric analysis, we can say that to the moment, computational models like fuzzy logic, neural networks, genetic algorithms and swarm intelligence techniques have been used in sustainable development applications. We still feel that more papers will come in future years with these same intelligent techniques, as there many key problems in sustainable development that can be tackled with these techniques that have not been considered to the moment. However, there are still other kinds of models coming from the computational intelligence area that can be applied to

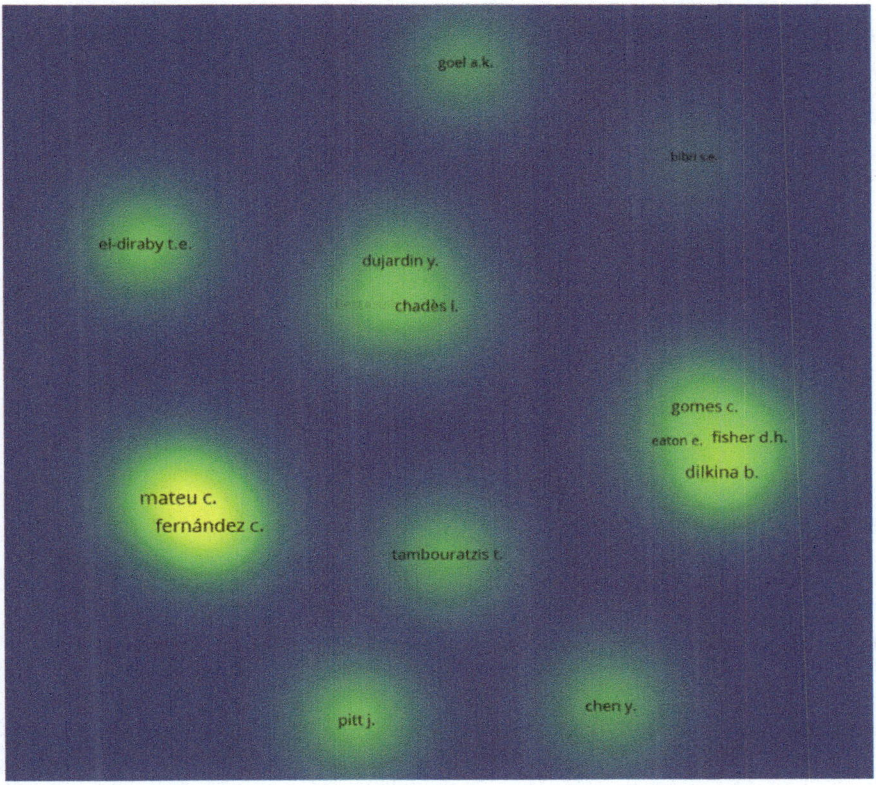

Fig. 10 Cluster density visualization of the main authors by their citations in sustainable development and computational intelligence

sustainable development problems. For example, some swarm intelligent and nature-inspired methods like, particle swarm optimization (PSO), ant colony optimization (ACO), grey wolf optimization (GWO), firefly algorithm (FA) and other similar methods [42–48], have not been yet applied in sustainable development problems. Also, other extensions of the original fuzzy logic (that is now called type-1) are still to be considered, like general type-2 fuzzy systems, intuitionistic fuzzy logic, fractional fuzzy systems, hesitant fuzzy systems and others [49–52], have not been yet applied in sustainable development. Finally, new neural models, such as modular neural networks and convolutional neural models, as well as type-2 fuzzy neural hybrid systems are also to be considered in sustainable development problems [53–58]. Finally, we have to say that there is a wide number of opportunities of future research work on applying computational intelligence techniques in the sustainable development area and we envision this will continue to be a fruitful area of research in the short, medium and the long terms. Finally, we have to say that the computational intelligence area is also evolving into newer directions, like the use of quantum

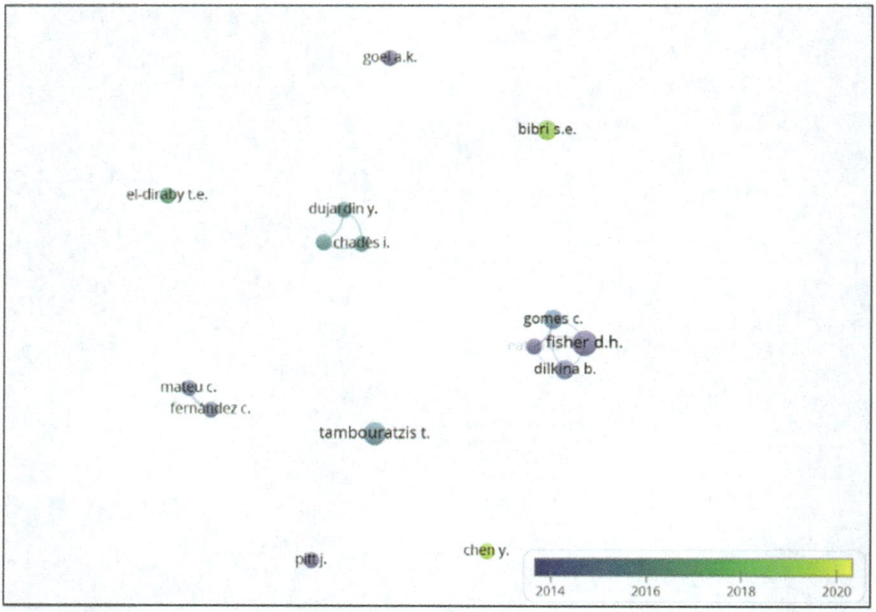

Fig. 11 Co-authorship pattern for sustainable development and computational intelligence

computing, nanotechnology and other newer technologies that may come to be developed in conjunction with intelligent techniques and models. So, we expect that as the computational intelligence area evolves with new theory and concepts, newer methods will be used for solving in a better way the existing problems in sustainable development. It is also possible, that newer problems will arise in sustainable development that may motivate the development of new theories and concepts in the computational intelligence areas.

5 Conclusion

A relatively large volume of literature on sustainable development computing with computational intelligence techniques for a period of 2006–2020 recorded in Scopus of Elsevier has been studied and analyzed with scientometrics. Various scientometric tools have been employed in this article to understand the patterns of growth in this field. Based on the results and findings, the following observations have been made. A total of 255 papers were found for a period of 15 years. Till 2009, the growth had a slow pace. After 2010, this research area has shown an escalating growth in the number of research publications (except for 2017 and 2018). Significant growth in the literature has been noticed in the last ten years. More than 80% of the growth has been observed after 2010 (Fig. 6). We have described in more detail a representative sample of these papers, which illustrate the types of works that have been made in this area. The maximum number of articles have been published from United States of America (USA), followed by Spain, followed by United Kingdom, Canada, China, and Greece. The growth rate has not been so clear before the year 2010. However, after 2010 there has been a consistent growth of the output of published articles. This study can provide valuable information for the academicians, research scholars, and people interested in scientometric analytical views of sustainable development and computational intelligence techniques. As future work, we imagine the researchers doing the application of generalized type-2 fuzzy logic, modular neural networks, evolutionary and swarm intelligence techniques or hybrid combinations of these methods, in sustainable development for solving real-world problems in many areas.

References

1. Adao, T., Soares, A., Padua, L., Guimardes, N., Pinho, T., Sousa, J.J., Morais, R., Peres, E.: Mysense-Webgis: a graphical map layering-based decision support tool for agriculture. Int Geosci Remote Sens Symp (IGARSS) 4195 (2020)
2. Ahmed, Q., Anifowose, F.A., Khan, F.: System availability enhancement using computational intelligence-based decision tree predictive model. Proc. Inst. Mechan. Eng. Part O: J. Risk. Reliab. **229**(6), 612–626 (2015)
3. Alptekin, S.E., Alptekin, G.I.: A fuzzy quality function deployment approach for differentiating cloud products. Int. J. Comput. Intel. Syst. **11**(1), 1041–1055 (2018)
4. An, B.: Game theoretic analysis of security and sustainability. In: IJCAI International Joint Conference on Artificial Intelligence, p. 5111 (2017)
5. Arabameri, A., Saha, S., Roy, J., Tiefenbacher, J.P., Cerda, A., Biggs, T., Pradhan, B., Thi Ngo, P.T., Collins, A.L.: A novel ensemble computational intelligence approach for the spatial prediction of land subsidence susceptibility. Sci. Total Environ. **726** (2020)
6. Barrientos, F., Moral, A., Rodríguez, J., Martínez, C., Campo, F., Carnerero, R., Parra, M., Benítez, J.M., Sainz, G.: Knowledge-based minimization of railway infrastructures environmental impact. Transport. Res. Procedia **840** (2016)
7. Basak, A., Mengshoel, O., Hosein, S., Martin, R., Jayakumaran, J., Morga, M.G., Aghav, I.: Identifying contributing factors of occupant thermal discomfort in a smart building. In: AAAI Workshop—Technical Report, pp. 219 (2016)
8. Beyer, B., Geldermann, J., Lauven, L.: Agent-based model of the German heating market: simulations concerning the use of wood pellets and the sustainability of the market. In: International Conference on the European Energy Market, EEM (2017)
9. Bibri, S.E.: Novel intelligence functions for data–driven smart sustainable urbanism. In: Utilizing Complexity Sciences in Fashioning Powerful Forms of Simulations Models (2019)
10. Bibri, S.E.: Smart Sustainable Urbanism: Paradigmatic, Scientific, Scholarly, Epistemic, and Discursive Shifts in Light of Big Data Science and Analytics (2019)
11. Bibri, S.E.: The Leading Smart Sustainable Paradigm of Urbanism and Big Data Computing: A Topical Literature Review (2019)
12. Bondi, E.: Bridging the gap between high-level reasoning in strategic agent coordination and low-level agent development: Doctoral consortium. In: Proceedings of the International Joint Conference on Autonomous Agents and Multiagent Systems, AAMAS, p. 2402
13. Buitrago, P.A., Nystrom, N.A.: Neocortex and Bridges-2: A High Performance AI+HPC Ecosystem for Science, Discovery, and Societal Good (2021)
14. Cascales, M.S.G., Lozano, J.M.S., Arredondo, A.D.M., Corona, C.C.: Soft computing applications for renewable energy and energy efficiency. In: Soft Computing Applications for Renewable Energy and Energy Efficiency, pp. 1–408 (2014)

15. Chang, N., Kumar, R., Yen, G., Wang, C.: Guest Editorial: Special Issue on 'Cyber-Innovated Environmental Sensing, Monitoring, and Modeling for Sustainability'. IEEE Syst. J. **10**(3), 1236–1238 (2016)
16. Chen, G., He, Y., Yang, T.: An ISMP approach for promoting design innovation capability and its interaction with personal characters. IEEE Access **8**, 161304–161316 (2020)
17. Chen, Y., Hu, M.: A swarm intelligence based distributed decision approach for transactive operation of networked building clusters. Energy Build. **169**, 172–184 (2018)
18. Chui, K.T., Alhalabi, W., Pang, S.S.H., de Pablos, P.O., Liu, R.W., Zhao, M.: Disease diagnosis in smart healthcare: innovation, technologies and applications. Sustainability (Switzerland) **9**(12) (2017)
19. Chvátalová, Z., Hřebíček, J., Bartulec, T.: Sustainability performance indicators construction with using neural networks in Maple. In: Proceedings of the 26th International Business Information Management Association Conference—Innovation Management and Sustainable Economic Competitive Advantage: From Regional Development to Global Growth, IBIMA 2015, p. 2035 (2015)
20. Davis, J., Edgar, T., Porter, J., Bernaden, J., Sarli, M.: Smart manufacturing, manufacturing intelligence and demand-dynamic performance. Comput. Chem. Eng. **47**, 145–156 (2012)
21. De Gracia, A., Fernández, C., Castell, A., Mateu, C., Cabeza, L.F.: Control of a PCM ventilated facade using reinforcement learning techniques. Energy Build. **106**, 234–242 (2015)
22. De Melo, J.G., De Souza Farias, F., Kato, O.R.: Assessment of the sustainability of agroecosystems in the Amazon region using neural artificial networks. IEEE Lat. Am. Trans. **14**(8), 3804–3810 (2016)
23. Du, R., Lu, Z., Pandit, A., Kuang, D., Crittenden, J., Park, H.: Toward social media opinion mining for sustainability research. AAAI Workshop—Technical Report, p. 21 (2015)
24. Dujardin, Y., Dietterich, T., Chadès, I.: Three new algorithms to Solve N-POMDPs. In: 31st AAAI Conference on Artificial Intelligence, AAAI 2017, p. 4495 (2017)
25. Dujardin, Y., Dietterich, T., Chadès, I.: α-min: A compact approximate solver for finite-horizon POMDPs. In: IJCAI International Joint Conference on Artificial Intelligence, p. 2582 (2015)
26. Dursun, P., Kaya, T.: Fuzzy multiple criteria sustainability assessment in forest management based on an integrated AHP-TOPSIS methodology. In: Computational Intelligence Foundations and Applications—Proceedings of the 9th International FLINS Conference, FLINS 2010, p. 438 (2010)
27. Eaton, E., Gomes, C., Williams, B.: Computational sustainability. AI Mag. **35**(2), 3–7 (2014)
28. Echeverry, A.X.H., Montoya-Torres, J.R., Richards, D., Neira, N.O.: Computational Intelligence to Support Cooperative Seaport Decision-Making in Environmental and Ecological Sustainability (2015)
29. Ekici, B., Cubukcuoglu, C., Turrin, M., Sariyildiz, I.S.: Performative computational architecture using swarm and evolutionary optimisation: a review. Build. Environ. **147**, 356–371 (2019)
30. Emerson, A., Henderson, N., Rowe, J., Min, W., Lee, S., Minogue, J., Lester, J.: Investigating Visitor Engagement in Interactive Science Museum Exhibits with Multimodal Bayesian Hierarchical Models (2020)
31. Fernandes, A.S., Bacciu, D., Jarman, I.H., Etchells, T.A., Fonseca, J.M., Lisboa, P.J.G.: p-Health in breast oncology: a framework for predictive and participatory e-systems. Proc. Int. Conf. Dev. eSyst. Eng. DeSE **2009**, 123 (2009)
32. Fernández, C., Manyà, F., Mateu, C., Sole-Mauri, F.: Modeling energy consumption in automated vacuum waste collection systems. Environ. Model. Softw. **56**, 63–73 (2014)
33. Fisher, D.H.: A selected summary of AI for computational sustainability. In: 31st AAAI Conference on Artificial Intelligence, AAAI 2017, pp. 4852 (2017)
34. Fisher, D.H.: Recent advances in AI for computational sustainability. IEEE Intell. Syst. **31**(4), 56–61 (2016)
35. Fisher, D.H., Dilkina, B., Eaton, E., Gomes, C.: Incorporating computational sustainability into AI education through a freely-available, collectively-composed supplementary lab text. In: Proceedings of the National Conference on Artificial Intelligence, p. 2369 (2012)

36. Fuchino, T., Batres, R., Shimada, Y.: A Knowledge-Based Approach for the Analysis of Abnormal Situations (2007)
37. Ganapathi Subramanian, S., Crowley, M.: Combining MCTS and A3C for Prediction of Spatially Spreading Processes in Forest Wildfire Settings (2018)
38. Garg, A., Lam, J.S.L., Gao, L.: Energy conservation in manufacturing operations: modelling the milling process by a new complexity-based evolutionary approach. J. Clean. Prod. **108**, 34–45 (2015)
39. Geem, Z.W., Chung, S.Y., Kim, J.: Improved optimization for wastewater treatment and reuse system using computational intelligence. Complexity (2018)
40. Naserifar, S., Chen, Y., Kwon, S., Xiao, H., Goddard, W.A.: Artificial intelligence and QM/MM with a polarizable reactive force field for next-generation electrocatalysts. Matter **4**(1), 195–216 (2021)
41. Pinter, G., Mosavi, A., Felde, I.: Artificial intelligence for modeling real estate price using call detail records and hybrid machine learning approach. Entropy **22**(12), 1–14 (2020)
42. Olivas, F., Valdez, F., Castillo, O., Melin, P.: Dynamic parameter adaptation in particle swarm optimization using interval type-2 fuzzy logic. Soft Comput. **20**(3), 1057–1070 (2016)
43. Olivas, F., Valdez, F., Castillo, O., Gonzalez, C.I., Martinez, G., Melin, P.: Ant colony optimization with dynamic parameter adaptation based on interval type-2 fuzzy logic systems. Appl. Soft Comput. **53**, 74–87 (2017)
44. Sanchez, D., Melin, P., Castillo, O.: Optimization of modular granular neural networks using a firefly algorithm for human recognition. Eng. Appl. of AI **64**, 172–186 (2017)
45. González, B., Valdez, F., Melin, P., Prado-Arechiga, G.: Fuzzy logic in the gravitational search algorithm for the optimization of modular neural networks in pattern recognition. Expert Syst. Appl. **42**(14), 5839–5847 (2015)
46. González, B., Valdez, F., Melin, P., Prado-Arechiga, G.: Fuzzy logic in the gravitational search algorithm enhanced using fuzzy logic with dynamic alpha parameter value adaptation for the optimization of modular neural networks in echocardiogram recognition. Appl. Soft Comput. **37**, 245–254 (2015)
47. Miramontes, I., Guzman, J., Melin, P., Prado-Arechiga, G.: Optimal design of interval type-2 fuzzy heart rate level classification systems using the bird swarm algorithm. Algorithms **11**(12), 206 (2018)
48. Gonzalez, C.I., Melin, P., Castro, J.R., Castillo, O., Mendoza, O.: Optimization of interval type-2 fuzzy systems for image edge detection. Appl. Soft Comput. **47**, 631–643 (2016)
49. Castillo, O., Castro, J.R., Melin, P., Rodriguez-Diaz, A.: Application of interval type-2 fuzzy neural networks in non-linear identification and time series prediction. Soft. Comput. **18**(6), 1213–1224 (2014)
50. Melin, P., Gonzalez, C.I., Castro, J.R., et al.: Edge-detection method for image processing based on generalized type-2 fuzzy logic. IEEE Trans. Fuzzy Syst. **22**(6), 1515–1525 (2014)
51. Castillo, O., Melin, P.: A review on interval type-2 fuzzy logic applications in intelligent control. Inf. Sci. **279**, 615–631 (2014)
52. Gaxiola, F., Melin, P., Valdez, F., Castro, J.R., Castillo, O.: Optimization of type-2 fuzzy weights in backpropagation learning for neural networks using GAs and PSO. Appl. Soft Comput. **38**, 860–871 (2016)
53. Melin, P., Castillo, O.: Intelligent control of complex electrochemical systems with a neuro-fuzzy-genetic approach. IEEE Trans. Industr. Electron. **48**(5), 951–955 (2001)
54. Sanchez, M.A., Castillo, O., Castro, J.R., Melin, P.: Fuzzy granular gravitational clustering algorithm for multivariate data. Inf. Sci. **279**, 498–511 (2014)
55. Sánchez, D., Melin, P.: Optimization of modular granular neural networks using hierarchical genetic algorithms for human recognition using the ear biometric measure. Eng. Appl. Artif. Intell. **27**, 41–56 (2014)
56. Sanchez, M.A., Castro, J.R., Castillo, O., Mendoza, O., Rodriguez-Diaz, A., Melin, P.: Fuzzy higher type information granules from an uncertainty measurement. Granul. Comput. **2**(2), 95–103 (2017)

57. Melin, P., Miramontes, I., Prado-Arechiga, G.: A hybrid model based on modular neural networks and fuzzy systems for classification of blood pressure and hypertension risk diagnosis. Expert Syst. Appl. **107**, 146–164 (2018)
58. Guzmán, J.C., Miramontes, I., Melin, P., Prado-Arechiga, G.: Optimal genetic design of type-1 and interval type-2 fuzzy systems for blood pressure level classification. Axioms **8**(1), 8 (2019)

Fuzzy Linguistic Summaries for Human Centric Analyses of Sustainable Development Goals (SDG) Related to Technological Innovations

Janusz Kacprzyk, Jan W. Owsinski, Eulalia Szmidt, and Sławomir Zadrożny

Abstract We are concerned with 17 Sustainable Development Goals (SDGs) the fulfillment of which is crucial for both the whole world and particular nations. The assessment of attainment of targets included in these SDGs involves human judgments, preferences, intentions, etc. which are human specific and can often be expressed in natural language possibly with a provision for the handling of imprecision. We propose to use fuzzy logic based linguistic data summaries for the assessment and evaluation of both the essence of SDGs and their fulfillment, and such a form is very human consistent because it uses natural language that is the only fully natural means of articulation and communication for the humans. We show the application of linguistic summaries for the analyses of innovation and innovativeness which is explicitly involved in the ninth sustainable development goal (SDG 9). However, as innovations are considered as relevant for all SDGs, then our approach can be used for the analyses of other SDGs too.

Keywords Sustainability · Sustainable development · Sustainable development goals · Transformation of sustainable development goals · Innovativeness · Innovation · Fuzzy logic · Linguistic summary

---

J. Kacprzyk (✉) · J. W. Owsinski · E. Szmidt · S. Zadrożny
Systems Research Institute, Polish Academy of Sciences, ul. Newelska 6, 01-447 Warsaw, Poland
e-mail: kacprzyk@ibspan.waw.pl

J. W. Owsinski
e-mail: owsinski@ibspan.waw.pl

E. Szmidt
e-mail: szmidt@ibspan.waw.pl

S. Zadrożny
e-mail: zadrozny@ibspan.waw.pl

J. Kacprzyk · E. Szmidt · S. Zadrożny
Warsaw School of Information Technology, ul. Newelska 6, 01-447 Warsaw, Poland

J. L. Verdegay et al. (eds.), *Computational Intelligence Methodologies Applied to Sustainable Development Goals*, Studies in Computational Intelligence 1036,
https://doi.org/10.1007/978-3-030-97344-5_2

1 Introduction

Sustainable development is still one of the most crucial concepts concerning the way the world should develop and operate. The concept is usually reported to appear for the first time in 1987 in the so-called Brundtland Report [47], called after Gro Harlem Brundtland, the former Prime Minister of Norway, who was a Chairperson of the World Commission on Environment and Development (WCED) that had prepared that report, entitled "Our common future". Though the report did not give a universally acceptable definition of sustainable development, it clearly mentioned its very essence which can be stated as that the sustainable development is the type of development which is aimed at meeting the present day needs and expectations of the human kind and individuals but, at the same time, does not jeopardize the meeting of the same and other needs and expectations of the future generations.

This report was a subject of a debate at the UN General Assembly in 1989 who decided to hold a UN Conference on Environment and Development. Then, the very essence of the report had been assumed by practically all countries and organizations as the way to follow. In 2015 the UN General Assembly issued a UN Resolution which had been popularly termed Agenda 2030 (UN General Assembly) that more precisely specified steps that should be taken to achieve the goals that are important to attain the sustainable development.

In this context, a special role for our purposes is played by the so-called Post-2015 Development Agenda that has been set up as the future global development framework to follow the previously set up Millennium Development Goals which ended in 2015. A crucial part of the Post-2015 Development Agenda was the formulation of 17 so-called Sustainable Development Goals (SDG) that subsumed in a brief and comprehensible form what was really relevant for the individual, social, economic, and technological development. We will now briefly list and explain these SDGs.

The 17 Sustainable Development Goals (SDGs) listed in the above mentioned reports and widely assumed throughout the world, and meant to be fulfilled until 2030, are as follows:

- Goal 1: "No poverty"

This goal, SDG 1, is aimed at ending poverty in all forms in all regions and countries and in all social classes. From a technical point of view, SDG 1 has 7 targets and 13 indicators to measure its progress, and 5 outcome targets, i.e. the eradication of extreme poverty, reduction of all poverty by 50%, implementation of social protection systems, equal rights to ownership, basic services, technology and economic resources, and resilience to environmental, economic and social disasters.

- Goal 2: Zero hunger

This goal, SDG 2, states as its objectives ending hunger, providing healthy food to everybody, attaining adequate food security and improved nutrition, and promoting sustainable agriculture. SDG 2 has 8 targets and 14 indicators to measure progress.

The 5 outcome targets are: end of hunger and improved access to food; end of malnutrition, higher agricultural productivity, sustainable food production and resilient agriculture, genetic diversity of seeds, cultivated plants and farmed and domesticated animals, more investments, research and technology. Moreover, three means of achieving targets are set as to limit trade restrictions and distortions in world food and agriculture markets and their derivatives.

- Goal 3: Good health and well-being

SDG 3 aims at ensuring healthy lives and promote well-being for all people at all ages. SDG 3 has 13 targets and 28 indicators to measure progress, with the first 9 targets being the outcome targets: to reduce maternal mortality, end all preventable deaths of children under 5 years of age, fight communicable diseases, ensure reduction of mortality from non-communicable diseases and promote mental health, reduce road injuries and deaths, provide a universal access to sexual and reproductive care, family planning and education, provide a universal health coverage, and reduce mortality and illnesses caused by hazardous chemicals and pollution. The four means to attain the SDG 3 targets are: to implement tobacco control, to support research, development and universal access to affordable vaccines and medicines, to increase health financing and support health care in developing countries, and to improve early warning systems for global health risks.

- Goal 4: Quality education

This goal, called SDG 4, aims at ensuring inclusive and equitable quality education and promote lifelong learning opportunities for all. SDG 4 has 10 targets measured by 11 indicators, with 7 outcome-oriented targets: free primary and secondary education, equal access to quality pre-primary education, affordable technical, vocational and higher education, increased number of people with relevant skills for financial success, elimination of all discrimination in education, universal literacy and numeracy, and education for sustainable development and global citizenship.

- Goal 5: Gender equality

SDG 5 aimes at attaining gender equality and empower all women and girls. SDG 5 has 5 aims to grant women and girls equal rights, opportunities to live free without discrimination, including workplace discrimination, or any violence.

- Goal 6: Clean water and sanitation

SDG 6 aims at ensuring the availability and sustainable management of water and sanitation for all, has 8 targets measured by 11 indicators, with 6 outcome-oriented targets to: provide safe and affordable drinking water and fresh water supply; end open defecation and provide access to sanitation and hygiene, improve water quality, wastewater treatment and safe reuse, increase the efficiency of water use and freshwater supplies, protect and restore water-related ecosystems, etc. The means of achieving targets are to expand water and sanitation support to developing countries, and to support local involvement in water and sanitation management.

- Goal 7: Affordable and clean energy

This goal, SDG 7, aims at ensuring access to affordable, reliable, sustainable and modern energy for all, and it has five targets among which three are outcome targets: universal access to modern energy, increase of the global percentage of renewable energy, as well as improvement in energy efficiency, and the two remaining targets are to promote access to research, technology and investments in clean energy; and expand and upgrade energy services for developing countries.

- Goal 8: Decent work and economic growth

SDG 8 aims at promoting sustained, inclusive and sustainable economic growth with a full and productive employment and decent work for all. It has 12 targets with 10 being the outcome targets: to attain the sustainable economic growth, diversify, innovate and upgrade for economic productivity, promote policies to support job creation and growing enterprises, improve resource efficiency in consumption and production, reach full employment and decent work with equal pay, promote youth employment, education and training, end modern slavery, trafficking, and child labour, protect labour rights, promote safe working environments, promote beneficial and sustainable tourism, and provide an universal access to banking, insurance and financial services. The remaining two targets are the means of achieving, namely: to increase aid for trade support, and to develop a global youth employment strategy.

- Goal 9: Resilient infrastructure, inclusive and sustainable industrialization, and fostering of innovation

The SDG 9 goal has 8 targets, and progress is measured by 12 indicators, with the first 5 targets being the outcome targets: to develop sustainable, resilient and inclusive infrastructures, to promote inclusive and sustainable industrialization, to increase access to financial services and markets, to upgrade all industries and infrastructures for sustainability, to enhance research and upgrade industrial technologies, and the remaining three targets are the means-related targets: to facilitate sustainable infrastructure development for developing countries, to support domestic technology development and industrial diversification, and to grant an universal access to information and communications technology.

- Goal 10: Reduced inequality

SDG 10 aims at reducing income inequality within and among countries. It has 10 targets, of which seven are the outcome targets: to reduce income inequalities, promote universal social, economic and political inclusion, ensure equal opportunities and end discrimination, adopt fiscal and social policies that promote equality; improved regulation of global financial markets and institutions, to attain an enhanced representation for developing countries in financial institutions, and to implement responsible and well-managed migration policies. The other three targets are means-oriented: special and differentiated treatment for developing countries, development assistance and investment in least developed countries, reduction of transaction costs for migrant remittances.

- Goal 11: Sustainable cities and communities

The SDG 11 goal aims at making cities and human settlements inclusive, safe, resilient, and sustainable. It has 10 targets to be achieved, measured by 15 indicators, with seven outcome targets: safe and affordable housing, affordable and sustainable transportation, inclusive and sustainable urbanization, protection of the world's cultural and natural heritage, reduction of adverse effects of natural disasters, reduction of the environmental impacts of cities, provision of access to safe and inclusive green and public spaces. The 3 means of achieving targets are: strong national and regional development planning, adequate policies for inclusion, resource efficiency and disaster risk reduction, support for least developed countries in sustainable and resilient construction.

- Goal 12: Responsible consumption and production

This goal, SDG 12, aims at ensuring sustainable consumption and production patterns. It has 11 targets: to implement the 10-Year Framework of Programs on Sustainable Consumption and Production Patterns, achieve the sustainable management and efficient use of natural resources, reduce by a half the per capita global food waste at the retail and consumer levels, reduce food losses along production and supply chains, attaining of environmentally sound management of chemicals and all wastes throughout their life cycle, reduction of waste generation through prevention, reduction, recycling and reuse, encouraging of companies to adopt sustainable practices, promotion of sustainable public procurement practice, providing the relevant information and awareness for sustainable development to all. The three means of achieving targets are: to support developing countries to strengthen their scientific and technological capacity, to develop and implement tools to monitor sustainable development impacts, to remove market distortions, like fossil fuel subsidies, that encourage wasteful consumption.

- Goal 13: Climate action

SDG 13 aims at taking an urgent action to combat climate change and its impacts by regulating emissions and promoting developments in renewable energy. It has five targets, with three output targets: to strengthen resilience and adaptive capacity to climate-related disasters, to integrate climate change measures into policies and planning, to build knowledge and capacity to meet climate change; and two means of achieving targets: to implement the UN Framework Convention on Climate Change, and to promote mechanisms to raise capacity for planning and management.

- Goal 14: Life below water

SDG 14 aims at conserving and sustainable use of the oceans, seas and marine resources. It has seven outcome targets: to reduce marine pollution, protect and restore ecosystems, reduce ocean acidification, implement sustainable fishing, conserve coastal and marine areas, end subsidies contributing to overfishing. The last 3 targets are the means of achieving targets: to increase and improve scientific knowledge, research and technology concerning oceans, support small scale fishers, implement and enforce international sea law.

- Goal 15: Life on land

SDG 15 aims at protecting, restoring and promoting sustainable use of terrestrial ecosystems, sustainable forestry, combatting desertification, stopping and reversing land degradation and loss. It has nine outcome targets: to conserve and restore terrestrial and freshwater ecosystems, to end deforestation and restore degraded forests, to end desertification and restore degraded land, to ensure the conservation of mountain ecosystems, to protect biodiversity and natural habitats, to protect access to genetic resources and fair sharing of the benefits, to eliminate poaching and trafficking of protected species, to prevent invasive alien species in land and water ecosystems, and to integrate ecosystem related issues and biodiversity in governmental planning. The three means of achieving targets are: to increase financial resources to conserve and sustainably use ecosystems and biodiversity, to finance and incentivize sustainable forest management and use, to combat global poaching and trafficking.

- Goal 16: Peace, justice and strong institutions

This goal, called SDG 16, aims at promoting peaceful and inclusive societies for sustainable development, providing access to justice for all, and building effective, accountable and inclusive institutions at all levels. It has 10 outcome targets: to reduce violence, to protect children from abuse, exploitation, trafficking and violence, to promote the rule of law and ensure equal access to justice, to combat organized crime and illicit financial and arms flows, to reduce corruption and bribery, to develop effective, accountable and transparent institutions, to ensure responsive, inclusive and representative decision making. to strengthen the participation in global governance, to provide universal legal identity, and to ensure public access to information and protect fundamental freedoms. The 2 means of achieving targets are: to strengthen national institutions to prevent violence and combat crime and terrorism, and to promote and enforce non-discriminatory laws and policies.

- Goal 17: Partnership for the goals

This goal, SDG 17, aims at strengthening the means of implementation and revitalizing the global partnership for sustainable development. It has 19 outcome targets and 24 indicators. An increasing international cooperation is considered vital to attain all 16 former SDGs, but SDG 17 is included to make sure that countries and organizations cooperate and not compete.

We have therefore briefly described the very essence of the 17 SDGs. We did not mention ways of monitoring their fulfillment, updating, etc., and above all how to operationalize their fulfillment. This all is out of t scope of this paper. Moreover, we did not mention some crtiticism of the SDGs, for instance their too wide a scope, attempt to attain too many targets too soon, etc.

One should also mention that there are some concepts that reduce the number of sustainable development goals. For instance, drawing on the so called "The World in 2050" initiative, Sachs et al. [44] propose the so called Sustainable Development

Goals Transformations which concern in a more integrated way the goals to be attained, namely now concerning:

- education, gender and inequality;
- health, well-being and demography;
- energy decarbonization and sustainable industry;
- sustainable food, land, water and oceans;
- sustainable cities and communities; and
- digital revolution for sustainable development;

so that these transformations include 17 SDGs already mentioned and take into account strong interrelations between these SDGs, while each transformation points out grand issues that are challenging and can be operationalized as they somehow reflect, for instance, the governmental structure.

In this paper we will not deal with this and other versions of the SDGs because, from the point of view of our work, i.e. the role of innovation and its evaluation, this is not a primary concern, even though a deeper insight into the interrelations between the goals, targets, measures etc. might in fact be of very pronounced importance, potentially leading to more coherent and effective plans and strategies (see, e.g., [42]).

2 Sustainable Development Goals (SDGs) and Technology, and Innovation

Our interest in this paper is related to the role of science, technology and—in particular—innovations in the context of sustainable development in general, and the SDGs in particular. It is quite obvious that research, development, deployment, and widespread diffusion of socially and environmentally sound technologies in this context is also closely linked to other key aspects, notably—for our purposes—innovations. Technological progress, hence innovations, is essential to economic growth and development, and the more advanced the technology available, the more quickly the local and global economy can improve and serve the society. Science and technology, the driving forces of economic and social development, are fundamental to a proper, effective and efficient use of natural resources so that they are essential for environmental sustainability. The use of traditional tools and techniques is not sufficient in the rapidly changing present world, and innovative solutions have to be in the foreground of virtually all development projects. The social aspect of innovation and its entire process, involving needs and their satisfaction, side effects and systemic consequences, is another crucial element of true sustainability.

From a broader point of view, innovation concerns all 17 SDGs. However, from the point of view in which SDG innovation can be viewed to be explicitly concerned, one can consider Goal 9, i.e. SDG 9, which aims at setting up a resilient infrastructure, promoting inclusive and sustainable industrialization and fostering innovation.

The study of innovation is a very popular and important subject of interest by both researchers and scholars, but also economic, business and political leaders because of its acute importance for the society, organizations, countries, etc. The importance of innovation has clearly implied a huge research effort, and many new concepts related to the supporting, monitoring, controlling, operationalizing, etc. of innovations have been proposed. A comprehensive and wide survey of various issues and perspectives related to the role of innovation for reaching targets imposed on the SDGs can be found in [38].

In this paper we are concerned with a general assessment and evaluation of technological innovation which is mostly related to SDG 9, even if innovation is relevant for all SDGs. We will consider innovation in a general sense and will not consider its various forms like, for instance, sustaining, incremental, radical or disruptive, or due to any other classification that is performed for other purposes.

Our interest is in the development of tools and techniques that make it possible to briefly assess the nature and extent of innovation. We advocate a human consistent form of such evaluations, which are based on the use of natural language that is the only fully natural means of articulation and communication of the human being. In this paper we just show a possible use of a new method for a human consistent evaluation of innovation by using fuzzy logic based linguistic summaries which make it possible to obtain short, natural language like descriptions of relations between some innovation indicators even if the data set used is very large. We will extend our approach to the use of linguistic data summaries for the analysis of innovations proposed by Baczko et al. and [32], and tailor it to the context of the analysis od SDGs.

It is not just a side remark that the methods here forwarded might, in fact, be applied to any of the SDGs listed in the Introduction, given the complex nature of these constructs, the multiplicity of targets, measures, undertakings etc. This complexity, coupled with unavoidable diversity of circumstances, errors, subjective judgments and so on, calls for a possibly simplifying, yet well-founded and easily communicable methodology that could help in dealing not so much with details of the entire scheme, but with its integrative facets, such as the SDGs.

This approach is perfectly in line with the main purpose of the volume, that is to show how methods of computational intelligence, specifically fuzzy logic, can be useful for a human centric, natural language based summarization of data related to innovation. Such assessments can then be employed for various purposes that can help operationalize the attainment of targets set for the SDG 9 and also other SDGs.

Essentially, the works by Baczko et al. and [32] are concerned with the analyses of innovativeness of companies, branches, industries, etc. from the point of view of a national innovation system, which involves some constituent systems for the monitoring, evaluation and supporting of innovativeness and innovations. Innovation at the national level is there meant in the sense of, for instance, Archibugi et al. [4], many papers in Llerena and Mat [34], Malerba and Brusoni [36] or Malerba and Cantner [37], and Baczko [5–8]. For the regional context, the analyses in the sense of Baraczyk et al. [11] or Howells [13], and in a sectorial context, in the sense of Malerba [35], are justified. Some publicly available national, regional and sectorial indicators of innovativeness developed along the lines of the Frascati Manual [40] and the Oslo Manual [41] are used. In the examples showed in this paper, corporate data from 2004 to 2006 on selected Polish companies are used, gathered via questionnaires containing both quantitative and qualitative data as well as public statistics, patents granted, stock data, firms reports, data concerning signed contracts, and experts judgments. Then, use of these historic data has been motivated by data protection issues notably in the case of commercial companies.

3 Assessment and Evaluation of Innovativeness

Innovativeness assessment and evaluation of an innovation system is a well known problem (cf. [39]) in which a crucial role of companies, entrepreneurs and workers in the development and implementation of innovativeness is emphasized. The system is meant to attain some goals like making it possible to analyze the structure, and effectiveness and efficiency of business expenditures for R&D and to identify factors that have impact on the level of innovativeness of a company, which is crucial. This all should help manage insufficient innovativeness and undertake proper action, including proper allocation of public and private expenditures in science, technology and innovation. This should also help identify most promising companies, branches, regions, etc. Notice that the above aspects are crucial for the operationalization of the analysis and management of technological innovations that is the essence of SDG 9.

4 Verbalization of Data Analysis Results via Linguistic Summaries

In this paper we propose the use of linguistic (data) summaries proposed by Yager [49], and in their constructive and implementable form by Kacprzyk and Yager [17], and Kacprzyk et al. [19], and then implemented in Kacprzyk and Zadrożny, [24, 26].

Suppose that:

- V is a quality (attribute) of interest, e.g. "salary" in a database of workers,
- $Y = \{y_1, \ldots, y_n\}$ is a set of objects (records) that manifest quality V, e.g. the set of workers; hence $V(y_i)$ are values of quality V for object y_i,
- $D = \{V(y_1), \ldots, V(y_n)\}$ is a set of data (the "database" on question)

A linguistic summary of a data set (data base) is constitutes by:

- a summarizer S, i.e., a linguistic term (e.g. "young"),
- a quantity in agreement Q (e.g. "most"),
- truth T—e.g. 0.7,
- a qualifier K (optionally), i.e. another linguistic term (e.g. "well-earning"), determining a fuzzy subset of Y,

and can be exemplified by: "T(most of employees are young) = 0.7".

The summarizer S (and qualifier K) is a simple linguistic expression such as, e.g., "young", semantically represented as a fuzzy set in $\{1, 2, \ldots, 90\}$, or a confluence of attribute values as, e.g., "young and well paid", and even a non-trivial, human-consistent summarizers (concepts) as, e.g.: "productive workers", involving complicated combinations of attributes via ANDing and/or ORing, taking k out of n, *most*, etc. of the attribute values.

The calculation of truth (validity) of the linguistic summary is equated with the calculation of the truth value (from [0,1]) of a linguistically quantified statement (e.g., "most of the employees are young") that can be done using [53] calculus of linguistically quantified propositions [56] or [50] OWA operators (cf. [51]), see, in particular, Kacprzyk et al. [18] for a recent comprehensive exposition.

Thus, a linguistic summary is a linguistically quantified proposition, written

$$Qy\text{'s are } S \tag{1}$$

where Q is a (relative) linguistic quantifier (e.g., "most"), $Y = \{y\}$ is a set of objects (e.g., employees), and S is some property (e.g., "well paid"), as, e.g., in "most of the employees are well-paid".

By adding another property K, we obtain

$$QKy's \text{ are } S \tag{2}$$

e.g., "most (Q) of the highly qualified (K) employees (y's) are well paid (S)".

We seek the truth values of such statements which are equal to: if S and K are fuzzy sets in Y, and a (proportional, nondecreasing) linguistic quantifier Q is a fuzzy set in [0,1] exemplified by

$$\mu_Q(x) = \begin{cases} 1 & \text{for } x \geq 0.8 \\ 2x - 0.6 & \text{for } 0.3 < x < 0.8 \\ 0 & \text{for } x \leq 0.3 \end{cases} \tag{3}$$

then, due to Zadeh [53]

$$\text{truth}(Qy's \text{ are } S) = \mu_Q\left[\tfrac{1}{n}\sum_{i=1}^{n} \mu_S(y_i)\right] \tag{4}$$

$$\text{truth}(QKy's \text{ are } S) = \mu_Q\left[\sum_{i=1}^{n} (\mu_K(y_i) \wedge \mu_S(y_i)) / \sum_{i=1}^{n} \mu_K(y_i)\right] \tag{5}$$

One can use other criteria (degrees of imprecision, covering, appropriateness, etc.) for the evaluation of the linguistic summaries, cf. Kacprzyk and Zadrożny [30]; cf. also Szmidt et al. [45].

The above linguistic summaries can be extended to the dynamic context, notably to the linguistic summarization of time series as proposed in a series of our papers [15, 16], Kacprzyk and Zadrożny [30] but this will not be considered here.

A crucial problem is the generation of linguistic summaries and, again, this problem will not be considered here in detail due to a lack of space. For instance, Kacprzyk and Zadrożny's approach, based on protoforms of linguistic summaries and Kacprzyk and Zadrożny's FQUERY for Access fuzzy querying package can be used. To be more specific, the following linguistic terms, which form a dictionary of the FQUERY for Access system, are releveant for linguistic summaries construction:

- fuzzy values, e.g. "low" in "profitability is low",
- fuzzy relations, e.g. "much greater than" in "income is much greater than spending", and
- linguistic quantifiers, e.g. "most" in "most conditions to be met",

and these terms are elicited by an interface.

In our setting, the fuzzy queries directly correspond to the linguistic summaries, and therefore the derivation of a linguistic summary may proceed interactively within the fuzzy querying system making it possible to interactively define the summarizers (indication of attributes and their combinations). The possible queries are:

- simple as, e.g., "salary is high"
- compound as, e.g., "salary is low AND age is old"
- compound with quantifier, as, e.g., "most of {salary is high, age is young, …, training is well above average}".

An usual granulation of the set of possible values is used by using some reasonable number (7 ± 2!) of values exemplified by: very low, low, medium, high, very high, and most, almost all, …, etc. for the linguistic quantifiers.

The matching degree, $md(\cdot, \cdot)$, for the query "Q of N conditions are satisfied" for record t is equal to

$$md(Q, \mathrm{condition_i,t}) = \mu_{\mathrm{Q}}\left[\tau\left(\sum_i md(\mathrm{condition_i,t})\right)\right] \tag{6}$$

and, with additional importance values of the particular conditions, we have as in (5), and then

$$\begin{aligned} &md(QK, \mathrm{condition_i, t}) = \\ &= \mu_Q\left[\tau\left(\sum_i (md(\mathrm{condition_i,t}) \wedge \mu_K(\mathrm{condition_i})\right)/\right. \\ &\left.\left(\sum_i \mu_K(\mathrm{condition_i})\right)\right] \end{aligned} \tag{7}$$

To derive a linguistic summary:

- the user derives a set of linguistic summaries of interest (relevance) via the fuzzy querying add-on as briefly shown above,
- the system retrieves consecutive records from the database and calculates the truth of each summary by matching the records with the summaries, and
- the most truthful linguistic summary is found.

This generation procedure can be extended by the use of Kacprzyk and Zadrożny's [26, 28, 29] protoforms defined as an abstract prototype of a linguistically quantified proposition, that is, given as (1) and (2) which are the most abstract protoforms, and their less abstract forms are obtained via instantiation of their particular elements, i.e., quantifier, summarizer and qualifier. For details, we refer the reader to the literature, notably Kacprzyk and Zadrożny [29]. Moreover, for the use of association rules in the context of linguistic summaries and their generation, cf. Kacprzyk and Zadrożny [24].

5 Examples of Linguistic Summaries for the Assessment of Innovation

In this section we will briefly show some examples of linguistic summaries used for a human centric and comprehensive description and evaluation of the innovativeness of companies. This will best show the potential of this techniques for the analysis of innovativeness in the context of SDG 9 too.

For simplicity, we assume a minimal set of linguistic terms: "low", "medium" and "high", defined by appropriate membership functions of fuzzy sets defined in specified domains, and the linguistic quantifier "most" which is widely used.

We use for our purposes the attributes of companies shown in Table 1. Due to some data protection restrictions, data from the period of several years ago are used.

As a result, many interesting and insightful linguistic summaries on what happens in the companies involved are obtained which can be exemplified by:

> Most companies having high net revenues from sales and equivalent in 2004 had high total assets in 2004.

which can be of relevance both for the company itself but also—which is more important for our purposes, from the point of view of a more general analysis of innovativeness, for instance, sectorial.

An interesting summary on dependencies between assessments of experts evaluating innovativeness of the companies can be like:

> Most of the companies having at least a few points ("successes") for their RTD related activities in 2006 had also some points for that in 2005.

Table 1 Attributes of companies used for the linguistic summarization

Attribute name	Description
totalAssets2006	Total assets of the company in 2006
totalAssets2005	Total assets of the company in 2005
totalAssets2004	Total assets of the company in 2004
equityCapital2006	Equity of the company in 2006
equityCapital2005	Equity of the company in 2005
equityCapital2004	Equity of the company in 2004
netSales2006	Net sales of the company in 2006
netSales2005	Net sales of the company in 2005
netSales2004	Net sales of the company in 2004
grossProfit2006	Gross profit of the company in 2006
grossProfit2005	Gross profit of the company in 2005
grossProfit2004	Gross profit of the company in 2004
netProfit2006	Net profit of the company in 2006
netProfit2005	Net profit of the company in 2005
netProfit2004	Net profit of the company in 2004
salesDynamics20062005	Sales dynamics in 2005–2006
salesDynamics20052004	Sales dynamics in 2004–2005
ROA2005	Return on assets (ROA) of the company in 2005
ROADynamics20062005	ROA dynamics in 2005–2006
PointsRTDTotal2006	Total points (given by experts) related to RTD related activities in 2006
PointsRTDTotal2005	Total points (given by experts) related to RTD related activities in 2005
PointsRTDTotal20052006	Total points (given by experts) related to RTD related activities in 2005–2006
PointsPatentsTotal2006	Total points (given by experts) related to patents obtained in 2006
PointsTotal	Total points given by experts

which points out a continuity in the RTD related activities of the companies; however, in the case considered, the summary shows that the number of companies that get those points is not growing in the period analyzed.

A similar relation, now involving also patents, is:

> A majority of companies having some points related to patents in 2006 AND some points for their RTD related activities in 2005 had also some points for RTD related activities in 2006.

which states that the companies that are in RTD in 2005 do not necessarily continue to do so in 2006 but those who get some patents in 2006 usually also have RTD related activities in 2006.

These are just examples of many linguistic summaries obtained by using the Baczko et al. and Kacprzyk et al. [32] approach and system for the evaluation of innovations and innovativeness of companies. It can be seen that the form of these summaries is very human consistent and comprehensible, mainly because of the use of natural language which is the only fully natural means of articulation and communication for the humans.

It is easy to see that the approach presented above can be readily extended for covering issues related to SDG 9, that is "Resilient infrastructure, inclusive and sustainable industrialization, and fostering of innovation", notably what concerns the targets of: upgrading all industries and infrastructures for sustainability, enhancing research and upgrading industrial technologies, and supporting domestic technology development and industrial diversification. These linguistic summaries can clearly help the analysts to better assess to which extent the SDG 9 is fulfilled. However, what we have been emphasized many times in literature, and also mentioned in this paper, innovations and innovativeness are an inherent part of all 17 SDGs so that an extension and modification of the approach proposed above can readily be done.

It is also quite obvious that the methodology here outlined could be effectively applied in both the analytic discourse and the actual assessments, concerning all of the SDGs. The complexity of the assessment and monitoring tasks could be well harnessed by the smart use of similar linguistic summaries as those, presented here, especially when the queries are adequately formulated, given the potential multiplicity of such summaries. On the other hand, if the applications were based on well-defined ultimate objectives, formulation of such summeries could become also automatic, this being one of the future research challenges.

6 Concluding Remarks

We have been concerned with 17 Sustainable Development Goals (SDGs) which are crucial for both the whole world and particular nations. Their fulfillment is a sine qua non condition for the world, not only in terms of present day prosperity and quality of life but maybe above all for keeping those values for the next generations. The operationalization of the fulfillment of goals represented in SDGs is a difficult problem that requires a concerted and coordinated action of all stakeholders at all local and global levels. While considering and then choosing the best, or even good enough, procedures, the point of departure is the availability of proper tools and techniques that can be used to assess and evaluate particular options.

This assessment and evaluation concerns highly complicated aspects that concern human judgments, references, intentions, etc. which are often not clear-cut and changeable. They should be comprehensible to the humans who need not be familiar with formal tools and techniques.

We have proposed therefore the use of fuzzy logic based linguistic summaries that can present in a simple and comprehensible way basic relations among various aspects involved in SDG 9. The main reasons are that the fuzzy linguistic summaries are short, relatively simple natural language like descriptions, which are clear and comprehensible per se, and also because the use of fuzzy logic makes it possible to represent and handle imprecision of meaning which is an inherent feature of natural language.

The approach presented in this paper has concerned SDG 9, which is explicitly related to technological innovations, but—as emphasized by many authors (cf. [38])—practically all SDGs are explicitly and implicitly related to innovation, and all of them represent complex systems of targets, measures, objectives, undertakings etc., so that our approach can be extended to cover all SDGs so that a full fledged system can be developed.

References

1. Agrawal, R., Srikant, R.: Fast algorithms for mining association rules. In: Proceedings of the 20th International Conference on Very Large Databases, Santiago, Chile, Morgan Kaufmann, San Mateo, CA, USA 1994, pp. 487–499 (1994)
2. Anadon, L.D., Chan, G., Harley, A.G., Matus, K., Moon, S., Murthy, S.L., Clark, W.C.: Making technological innovation work for sustainable development. Proc. Natl. Acad. Sci. U S A **113**(35), 9682–9690 (2016)
3. Anderson, M.W., Teisl, M.F., Noblet, C.L.: Whose values count: is a theory of social choice for sustainability science possible? Sustain. Sci. **11**, 373–383 (2016)
4. Archibugi, D., Howells, J., Michie, J.: Innovation systems in a global economy. Technol. Anal. Strateg. Manage. **11**, 527–539 (1999)
5. Baczko, T., et al.: The Report on Innovativeness of Polish Economy in 2007. Institute of Economics, Polish Academy of Sciences (in Polish), Warsaw (2007)
6. Baczko, T.: Integrated micro indicators of innovativeness-new market and public policy institutional solution. In: Jakubowska, P., Kukliński, A., Żuber, P. (eds.) The Future of European Regions, pp. 326–335. Ministry of Regional Development, Warsaw (2007)
7. Baczko, T., et al.: Standortbedingungen in Ostdeutschland und Polen aus Sicht der Unternehmen. Wochenbericht des DIW Berlin **9**, 91–97 (2008)
8. Baczko, T., et al.: The Report on Innovativeness of Polish Economy in 2008. Institute of Economics, Polish Academy of Sciences, Warsaw (in Polish) (2009).
9. Baczko, T. (ed.): The Future Science and Technology and Innovation Indicators and the Challenges Implied. Institute of Economics, Polish Academy of Sciences, Warsaw (2009b)
10. Baczko, T., Kacprzyk, J., Zadrożny, S.: Towards knowledge driven individual integrated indicators of innovativeness. In: Józefczyk, J., Orski, D. (eds.) Knowledge-Based Intelligent System Advancements: Systemic and Cybernetic Approaches, pp. 129–140. IGI Global, Hershey (2011)
11. Baraczyk, H., Cook, P., Heidenreich, R. (eds.): Regional Innovation Systems. University of London Press, London (1996)
12. Bonvoisin, N.: Decision Making for Sustainable Development: How Assessment Can Help. UNECE Annual Report Economic Essays 2009-6, UNECE (2009)
13. Howells, J.: Innovation and regional economic development: a matter of perspective? Res. Policy **34**, 1220–1234 (2005)
14. Kacprzyk, J., Pasi, G., Vojtaš, P., Zadrożny, S.: Fuzzy querying: issues and perspective. Kybernetika **36**, 605–616 (2000)
15. Kacprzyk, J., Wilbik, A., Zadrożny, S.: Linguistic summarization of time series using a fuzzy quantifier driven aggregation. Fuzzy Sets Syst. **159**, 1485–1499 (2008)

16. Kacprzyk, J., Wilbik, A., Zadrożny, S.: An approach to the linguistic summarization of time series using a fuzzy quantifier driven aggregation. Int. J. Intell. Syst. **25**, 411–439 (2010)
17. Kacprzyk, J., Yager, R.R.: Linguistic summaries of data using fuzzy logic. Int. J. Gen Syst **30**, 133–154 (2001)
18. Kacprzyk, J., Yager, R.R., Merigo, J.M.: Towards human-centric aggregation via ordered weighted aggregation operators and linguistic data summaries: a new perspective on Zadeh's inspirations. IEEE Comput. Intell. Mag. **14**(1), 16–30 (2019)
19. Kacprzyk, J., Yager, R.R., Zadrożny, S.: A fuzzy logic based approach to linguistic summaries of databases. Int. J. Appl. Math. Comput. Sci. **10**, 813–834 (2000)
20. Kacprzyk, J., Zadrożny, S.: FQUERY for access: fuzzy querying for a Windows-based DBMS. In: Bosc, P., Kacprzyk, J. (eds.) Fuzziness in Database Management Systems, pp. 415–433. Physica-Verlag, Heidelberg (1995)
21. Kacprzyk, J., Zadrożny, S.: Data mining via linguistic summaries of data: an interactive approach. In: Yamakawa, T., Matsumoto, G. (eds.) Methodologies for the Conception, Design and Application of Soft Computing—Proceedings of IIZUKA'98, Iizuka, Japan, pp. 668–671 (1998)
22. Kacprzyk, J., Zadrożny, S.: The paradigm of computing with words in intelligent database querying. In: Zadeh, L.A., Kacprzyk, J. (eds.) Computing with Words in Information/Intelligent Systems. Part 2. Foundations, pp. 382–398. Physica-Verlag (Springer-Verlag), Heidelberg and New York (1999)
23. Kacprzyk, J., Zadrożny, S.: On a fuzzy querying and data mining interface. Kybernetika **36**, 657–670 (2000)
24. Kacprzyk, J., Zadrożny, S.: Computing with words in intelligent database querying: standalone and Internet-based applications. Inf. Sci. **34**, 71–109 (2001)
25. Kacprzyk, J., Zadrożny, S.: Fuzzy linguistic summaries via association rules. In: Kandel, A., Last, M., Bunke, H. (eds.) Data Mining and Computational Intelligence, pp. 115–139. Physica-Verlag (Springer-Verlag), Heidelberg, New York (2001)
26. Kacprzyk, J., Zadrożny, S.: Linguistic database summaries and their protoforms: towards natural language based knowledge discovery tools. Inf. Sci. **173**(4), 281–304 (2005)
27. Kacprzyk, J., Zadrożny, S.: Protoforms of linguistic database summaries as a human consistent tool for using natural language in data mining. Int. J. Softw. Sci. Comput. Intel. **1**(1), 100–111 (2009)
28. Kacprzyk, J., Zadrożny, S.: Modern data-driven decision support systems: the role of computing with words and computational linguistics. Int. J. General Syst. **39**(4), 379–393 (2010)
29. Kacprzyk, J., Zadrożny, S.: Computing with words is an implementable paradigm: fuzzy queries, linguistic data summaries, and natural-language generation. IEEE Trans. Fuzzy Syst. **18**, 461–472 (2010)
30. Kacprzyk, J., Zadrożny, S.: Linguistic summarization of the contents of Web server logs via the Ordered Weighted Averaging (OWA) operators. Fuzzy Sets Syst. **285**, 182–198 (2016)
31. Kacprzyk, J., Zadrożny, S.: Fuzzy logic-based linguistic summaries of time series: a powerful tool for discovering knowledge on time varying processes and systems under imprecision. WIREs Data Min. Knowl. Discov. **6**, 37–46 (2016)
32. Kacprzyk, J., Zadrożny, S., Baczko, T.: Towards a human consistent analysis of innovativeness via linguistic data summaries and their protoforms. In: Proto, A.N., Squillante, M., Kacprzyk, J. (eds.) Advanced Dynamic Modeling of Economic and Social Systems, pp. 91–107. Springer, SCI 448 (2013)
33. Krozer, Y., Nentjes, A.: An essay on innovations for sustainable development. Environ. Sci. **3**(3), 163–174 (2006)
34. Llerena, P., Matt, M. (eds.): Innovation Policy in a Knowledge-Based Economy Theory and Practice. Springer, Heidelberg (2004)
35. Malerba, F.: Sectoral Systems of Innovation. Cambridge University Press, Cambridge (2004)
36. Malerba, F., Brusoni, S.: Perspectives on Innovation. Cambridge University Press, Cambridge (2007)

37. Malerba, F., Cantner, U.: Innovation, Industrial Dynamics and Structural Transformation. Springer, Heidelberg (2007)
38. Maier, D., Maier, A., Aşchilean, I., Anastasiu, L., Gavriş, O.: The relationship between Innovation and sustainability: a bibliometric review of the literature. Sustainability **12**(10), 1–20 (2020)
39. McCraw, Th.K.: Prophet of Innovation: Joseph Schumpeter and Creative Destruction. Harvard University Press (2007)
40. OECD: The measurement of scientific and technological activities. In: Frascati Manual 2002. Proposed Standard Practice for Surveys on Research and Experimental Development. OECD Publishing (2003)
41. OECD: Oslo Manual: Guidelines for Collecting and Interpreting Innovation Data, 3rd edn. OECD Publishing (2005)
42. Owsiński, J.W.: Is there any 'law of requisite variety' in construction of indices for complex systems? Soc. Indic. Res. **136**(3), 1125–1137 (2017). https://doi.org/10.1007/s11205-016-1545-5
43. Pastoors, S., Scholz, U., Becker, J.H., van Dun, R.: Towards Sustainable Innovation. A Five Step Approach to Sustainable Change. Tectum Nomos, Baden-Baden (2017)
44. Sachs, J.D., Schmidt-Traub, G., Mazzucato, M., Messner, D., Nakicenowicz, N., Rockström, J.: Six transformations to achieve the sustainable development goals. Nat. Sustain. **2**, 805–814 (2019)
45. Szmidt, E., Kacprzyk, J., Bujnowski, P.: How to measure the amount of knowledge conveyed by Atanassov's intuitionistic fuzzy sets. Inf. Sci. **257**, 276–285 (2014). https://doi.org/10.1016/j.ins.2012.12.046
46. UN General Assembly, Transforming Our World: The 2030 Agenda for Sustainable Development, 21 October 2015, A/RES/70/1, available at: https://www.refworld.org/docid/57b6e3e44.html
47. World Commission on Environment and Development: Our Common Future. Oxford University Press, Oxford (1987)
48. Wu, J., Guo, S., Huang, H., Liu, W., Xiang, Y.: Information and communications technologies for sustainable development goals: state-of-the-art, needs and perspectives. IEEE Commun. Surv. Tutorials **20** (2018)
49. Yager, R.R.: A new approach to the summarization of data. Inf. Sci. **28**, 69–86 (1982)
50. Yager, R.R.: On ordered weighted averaging operators in multicriteria decision making. IEEE Trans. Syst. Man Cybern. **SMC-18**, 183–190 (1988)
51. Yager, R.R., Kacprzyk, J. (eds.): The Ordered Weighted Averaging Operators: Theory and Applications. Kluwer, Boston (1997)
52. Yager, R., Kacprzyk, J., Beliakov, G. (eds.): Recent Developments in the Ordered Weighted Averaging Operators: Theory and Practice. Springer, Heidelberg (2011)
53. Zadeh, L.A.: A computational approach to fuzzy quantifiers in natural languages. Comput. Math. Appl. **9**, 149–184 (1983)
54. Zadeh, L.A.: Syllogistic reasoning in fuzzy logic and its application to usuality and reasoning with dispositions. IEEE Trans. Syst. Man Cybern. **SMC-15**, 754–763 (1985)
55. Zadeh, L.A.: A Prototype-Centered Approach to Adding Deduction Capabilities to Search Engines—The Concept of a Protoform, in BISC Seminar, 2002. University of California, Berkeley (2002)
56. Zadeh, L.A., Kacprzyk, J. (eds.): Computing with Words in Information/Intelligent Systems, 1. Foundations, 2. Applications. Physica-Verlag (Springer-Verlag), Heidelberg and New York (1999)
57. Zadrożny, S., Kacprzyk, J.: Summarizing the contents of web server logs: a fuzzy linguistic approach. In: Proceedings of 2007 IEEE Conference on Fuzzy Systems, London, UK, 1860–1865 (2007)

Impact of Human-Computer Interaction in the Global Sustainable Development Goals

Jaime Muñoz-Arteaga, Huizilopoztli Luna García, César A. Collazos, and Antoni Granollers

Abstract Sustainability and sustainable development are extremely critical aspects in modern society. Sustainability is the term employed for the practice of ensuring that goods and services are produced in ways that do not use resources that cannot be replaced. This concept can be extended to any environmental, political, and socio-economical system, becoming a global concern that has been highlighted in several different re-search agendas. According to the general assembly of the United Nations defined in 2015 a universal call to action, defining 17 Sustainable Development Goals (SDGs) addressed not only to governments but also to businesses and civil society. The agenda goes to the heart of tackling a number of interrelated global issues such as poverty, inequality, hunger, and environmental degradation. In this scenario, the research area of Human-Computer Interaction (HCI) plays a key role into the design and development of software applications and practices aimed to achieve the SDGs. This paper depicts a vision about how HCI could be integrated in a better way through the inclusion of human values aspects in the design of interactive applications) lines before and after the abstract.

Keywords Human-computer interaction · Interactive applications

J. Muñoz-Arteaga (✉)
Universidad Autónoma de Aguascalientes, Aguascalientes, Mexico
e-mail: jaime.munoz@edu.uaa.mx

H. L. García
Universidad Autónoma de Zacatecas, Zacatecas, Mexico
e-mail: hlugar@uaz.edu.mx

C. A. Collazos
Universidad del Cauca, Cauca, Colombia
e-mail: ccollazo@unicauca.edu.co

A. Granollers
Universidad de Lleida, Lleida, Spain
e-mail: toni.granollers@udl.cat

J. L. Verdegay et al. (eds.), *Computational Intelligence Methodologies Applied to Sustainable Development Goals*, Studies in Computational Intelligence 1036,
https://doi.org/10.1007/978-3-030-97344-5_3

1 Introduction

Information technology (IT) has become an integral part of modern society, affecting nearly every aspect of our lives, including education, medicine, government, business, entertainment, and social interactions. Innovations in IT have been fueled by a continuous and extraordinary increase in computer development. Along with the technological advances and the rapid flow of information, people continue creating more and more innovative technology products to make life more convenience and comfortable [1]. From the invention of wheels to the application of computers and software in everyday life, technological innovations have always been game changers in human development [2]. However, huge industrial processes bring significant changes in the environment, the earth's resources are getting to dry up, the natural environment and ozonosphere have been destroyed, and leading to serious threats to the survival of the Earth's species [3].

In a society that is technologically dominated and mediated day by day, computing has a large role to play in the future of societal sustainability. This is not only due to computing's impact on the global ecosystem (e.g., due to power consumption, e-waste, etc.) but also due to the patterns of human-computing behavior that trace out the path of much that takes place in industrial economies today [4]. De-spite their few weaknesses, modern Information and Communication Technologies (ICTs) such as computers, smartphones and numerous software transformed modern societies and human interactions thereby bringing about unprecedented progress in the history of humankind.

There has been a recognition of the need to integrate sustainability goals into the process of user-centered design along with analyzing the role of computing within a broader ecological scope. Human-Computer Interaction (HCI) is now central to the success of many if not most interactive systems. This discipline should be responsible for how the things we design are used (e.g. "responsible design"). This is a tough requirement as it would not just mean a responsibility for design and development but would also encompass deployment and end-of-life responsibilities [5], Creating knowledge and understanding through science equips allow us to find solutions to today's acute economic, social, and environmental challenges and to achieving sustainable development and greener societies. In this paper, we present ours vies about how HCI could be important into the achievement of Sustainable Development Goals (SDGs) [6] through the embracement of human values in the design of interactive applications.

The next section depicts a brief background of SDG and HCI. Then section three describes some related works about how HCI has been addressed in SDG. The conceptual model of this works is presented in section four. Section five proposes the use of well-known HCI methodologies aligned to proposal for this work. Finally, we present some conclusions and further work.

2 Background

IT can be a crucial enabler in helping to achieve the SDGs, particularly in today's developing countries where closing the development gap requires substantial effort, innovation, and investment. IT is an existing and widely deployed that can be mobilized to step up the pace and scale of transformation. The central role ICT can play in delivering innovative, integrated, cross sectoral sustainable development outcomes has been highlighted by the UNESCO Broadband Commission for Sustainable Development [7].

2.1 SDG: Sustainable Development Goals

The new Sustainable Development Goals, or SDGs set out a shared global schedule for human development based on prosperity, social inclusion, and environmental sustainability. The SDGs include several bold objectives to be achieved by the year 2030, including universal coverage in health, education, poverty eradication, and modern energy services. These global goals are even more ambitious than the Millennium Development Goals (MDGs) Meanwhile MDGs helped to halve extreme poverty, the SDGs aim to end it completely. The MDGs focused on the poorest countries, the SDGs engage all nations in a shared, universal agenda [8]. The MDGs prioritized prosperity and inclusion: this time, environmental sustainability is also addressed as a fundamental pillar of global wellbeing. These goals are achievable, but they require a breakthrough in both the speed and degree of progress.

The United Nations 2030 Sustainable Development Goals (SDGs) were adopted in 2015 [6] by 193 countries in the agenda "Transforming our world". The 17 goals are depicted in Fig. 1.

Each of the objectives is briefly described below:

1. End poverty in all its forms everywhere.
2. End hunger, achieve food security and improved nutrition and promote sustainable agriculture.
3. Ensure healthy lives and promote well-being for all at all ages.
4. Ensure inclusive and equitable quality education and promote lifelong learning opportunities for all.
5. Achieve gender equality and empower all women and girls.
6. Ensure availability and sustainable management of water and sanitation for all.
7. Ensure access to affordable, reliable, sustainable, and modern energy for all.
8. Promote sustained, inclusive, and sustainable economic growth, full and productive employment, and decent work for all.
9. Build resilient infrastructure, promote inclusive and sustainable industrialization, and foster innovation.
10. Reduce inequality within and among countries.
11. Make cities and human settlements inclusive, safe, resilient, and sustainable.

Fig. 1 Sustainable Development Goals, available in: https://sustainabledevelopment.un.org/sdgs

12. Ensure sustainable consumption and production patterns.
13. Take urgent action to combat climate change and its impacts.
14. Conserve and sustainably use the oceans, seas and marine resources for sustainable development.
15. Protect, restore and promote sustainable use of terrestrial ecosystems, sustainably manage forests, combat desertification, and halt and reverse land degradation and halt biodiversity loss.
16. Promote peaceful and inclusive societies for sustainable development, provide access to justice for all and build effective, accountable and inclusive institutions at all levels.
17. Strengthen the means of implementation and revitalize the global partnership for sustainable development.

In short, the ultimate goal of the SDGs is to promote a new worldview and provide the beginnings of a plan to end poverty without imposing significant costs on Earth's life support systems [9].

2.2 *Human Computer Interaction*

Human-Computer Interaction (HCI) is a multidisciplinary field of study (see Fig. 2) focusing on the design of computer technology and, in particular, the interaction between humans (the users) and computers. While initially concerned with computers, HCI has since expanded to cover almost all forms of information technology design [10].

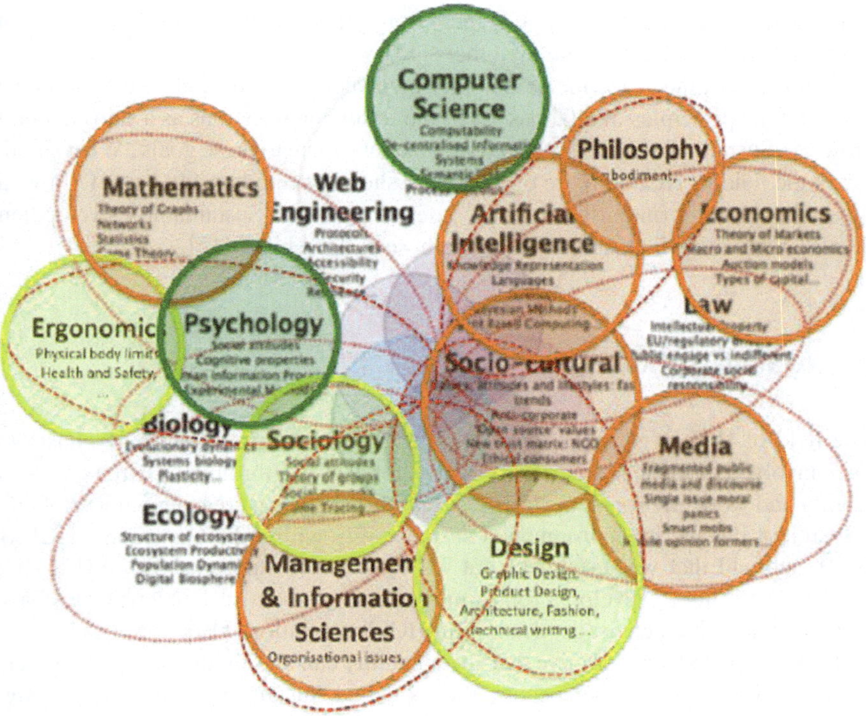

Fig. 2 Heat map for HCI showing its multidisciplinary by Hooper and Dix [11]

The growth of the HCI has traversed across disciplines and technological aspects encompassing elements that are both tangible and intangible. It is hard to exclude HCI from domains of Informatics, Internet of Everywhere (IoE), Quality of Life and others, more so with a diverse range of users and possibilities [12]. Incorporating HCI and various application domains have resulted in a variety of user requirements, needs and experience, converting to everchanging conditions. In line with these transformations, sustainable development goals (SDGs) have been chosen as the theme for this work. Sustainability is a growing concern in a world where climate change threatens to inflict massive environmental and human damage in the coming decades. As climate change becomes a global issue, researchers and professionals from many pursuits are beginning to rally around the cause of bringing about an economically and environmentally sustainable future. Computer science and the related fields of Human-Computer Interaction have both a moral obligation and profound opportunity to contribute to environmental sustainability [13].

In order to contribute to solving the problems related to the SDGs, the HCI academic community has generated groups with the purpose of addressing issues that are central to the SDGs, for example, the HCI for development community (HCI4D), explored how HCI can be useful in global development contexts; the Sustainable HCI Community (SHCI) has focused on sustainability concerns as it relates to HCI, the

Crisis Informatics Community (CI) increasingly investigates crises resulting from environmental instabilities, among others [14].

In this sense, the academic community involved in the development of solutions based on the principles of HCI has asked the following questions as a starting point. How can Sustainable HCI be inspired by, and contribute to these goals? What should we in the field of HCI do more of, and what should we perhaps do less of? In what areas should we form partnerships in order to reach the Sustainable Development Goals? And with whom should we form these partnerships? [15].

3 Related Works

The field of HCI in general and the NordiCHI community in particular, has long been involved in social sustainability issues, for example through research on accessibility and work environments [16, 17]. Furthermore, an increasing number of HCI researchers engage in environmental sustainability within the Sustainable HCI (S-HCI), subfield that was established at 7th NordiCHI conference [15, 18] S-HCI, has been working to develop a robust understanding about sustainability as a basic element from which envisage meaningful HCI interventions [19].

Blevis depicted a perspective that sustainability can and should be a central focus of interaction design, a perspective that is called Sustainable Interaction Design (SID). As a starting point for a perspective of sustainability, design is defined as an act of choosing among or informing choices of future ways of being [20].

Mankoff et al. [21] raise awareness of the most serious consequences of climate change can be averted, but only if fundamental changes are made in the CHI community and start a conversation about the possibilities and responsibilities to address issues of sustainability [21]. Lately, it has been a recognition of the need to integrate SDG into the process of User Centred Design (UCD) along with analyzing the role of computing within a broader ecological scope [22]. Pathak et al. [23] designed an interactive and user-friendly apps, where climate data for regions in New Jersey are considered, predicting the output of precipitation, humidity and temperature [23].

The Sustainable HCI perspective is challenging to engage with, as it is difficult to strike the right balance. Some work tends to address sustainability concerns only superficially while other work delves so deeply into ecology that the HCI questions are lost, as we noted previously [15, 18]. Nevertheless, strong work on this topic manages to avoid these pitfalls and advances our understanding of design criteria that have often been ignored in the past [4, 24]. Preist et al. [24] analyzed the unsustainable cornucopian paradigm and its often implicit implications for design. Authors like Ekbia and Nardi [26] have identified political economy generally and inequality specifically as often ignored sources of concern that demand a shift in thinking in HCI.

Fig. 3 Proposed model

4 Proposal

With the intention of doing our bit in this challenge, we pretend to design and implement interactive applications considering a Responsible Research and Innovation, which is defined as a transparent, interactive process by which societal actors and innovators become mutually responsive to each other with a view to the (ethical) acceptability, sustainability and societal desirability of the innovation process and its marketable products (in order to allow a proper embedding of scientific and technological advances in our society) [27]. It becomes necessary to consider three important aspects: Business in terms of viability, Technology in terms of feasibility and, Human Values in terms of usability and User eXperience (UX). Only through the intersection of these three elements, we could pretend to obtain responsible design innovations.

The model we have designed to obtain innovative designs and so support some of the GSD include the combination of three methodologies: User Centered Design; Value Sensitive Design; Design Thinking (see Fig. 3).

Value Sensitive Design: is a concept that advocates the consideration of human principles and standards when planning technology [28]. VSD brings human values to the forefront of the technical design process, providing technologists, designers, business leaders, and others involved in developing technology with strategies for identifying and incorporating human values into the design and development processes. VSD aims to consider the values not only of the users but all others impacted by the technologies regardless of whether those individuals will ever actually use the technologies. As such, VSD advises designers and developers to consider the values of direct stakeholders as well as indirect stakeholders. In addition to being a concept, VSD is also a methodology for developing technologies in a way that considers and addresses the values of all stakeholders. It advocates that developers start with a conceptual phase, in which they begin to identify these direct and indirect stakeholders and how the planned technologies might impact their values; an empirical phase in which the developers confirm, expand and tweak the ideas developed in the first phase through investigations using surveys, interviews and other methods; and then a technical phase in which the information gathered is used to shape the final product. The idea of VSD is to develop technologies that respect the human values, such as such as the values of human dignity, justice, welfare, human rights, privacy, trust, informed consent, respect for intellectual property rights, universal usability, environmental sustainability, moral responsibility, accountability, honesty, and democracy. All those impacted by them by identifying and addressing those values early in the development process, thereby limiting or eliminating potential problems once the technologies have been deployed.

User Centered Design: also known as Human Centered Design, is an approach to interactive systems development that aims to make systems usable and useful by focusing on the users, their needs and requirements, and by applying human factors/ergonomics, and usability knowledge and techniques. This approach enhances effectiveness and efficiency, improves human well-being, user satisfaction, accessibility and sustainability; and counteracts possible ad-verse effects of use on human health, safety and performance (ISO 9241-210:2019) [29]. It is considered as a process in which needs, wants and limitations of end-users of a product are given extensive attention at each stage of the design process [30]. On this approximation, people are more than users, and is a Holistic (not just a disembodied task or set of the cognitive process), Social (not standalone organisms), and Complex (human systems are more complex than technical systems) process.
Design Thinking: is a Human-Centered approach to innovation "anchored in understanding customer's needs, rapid prototyping, and generating creative ideas" that will transform the way you develop products, services, processes, and organizations. It encourages organizations to focus on the people they are creating for, which leads to better products, services, and internal processes. It has 5 stages: Empathize, Define, Ideate, Prototype; Test [31].

5 HCI Methodologies

HCI has already identified some methodological approaches to be combined in order to deal with the sustainability issues such as Design Thinking Lean Startup Agile methodologies (Source Interaction-design [32]). It combines three flexible models to demonstrate a reduced risk of failing (see Fig. 4). All three methodologies use a feedback loop to ensure the requirements of end user. It further enables developers to improve product/service features based on a continuous feedback loop. Each methodology includes a list of stages leading customers from problem to its solution. These include:

- Design Thinking: Emphasize—Define—Ideate
- Lean Startup: Measure—(multiple iterations)—Build—Lean
- Agile: Sprint review—shippable increment—sprint execution—sprint planning—product backlog

According to Fig. 4, the Lean UX methodology takes the best of "agile development" or Agile UX, but it is not the same. It is worth clarifying, as many confuse both concepts. Agile focuses more on the efficiency with which the team works when developing a digital product/service, but the Lean methodology, apart from taking these principles, seeks to make the work process faster, to achieve the goals in less time. The lean methodology applied to digital products and services is due to its effectiveness, this method has been taken to the design and development of digital products, such as web sites; products focused on user experience (UX) to what is called Lean UX methodology. It is important to mention that Lean UX eliminates

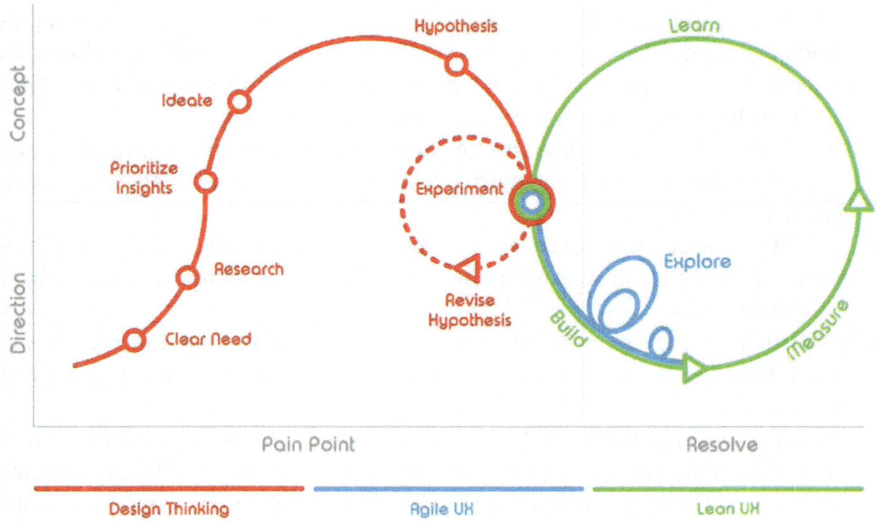

Fig. 4 Design thinking lean startup agile methodologies (Interaction-Design, 2021)

any waste that does not provide something useful to the customer and relies on learning to improve its products. In short, Lean UX is a method that allows you to create products that meet all user requirements, reducing the possibility of failure, speeding up your production time, and achieving the desired objectives in a shorter time.

6 Discussion

It is not possible to keep track of everything the goals are supposed to achieve without massive IT support. There are plenty of opportunities for the HCI community to contribute also for supporting the goals both indirectly as an enabling infrastructure for communication and directly with applications. Some examples given by United Nations [6]:

- Adopt measures to ensure the proper functioning of food commodity markets and their derivatives and facilitate timely access to market information, including on food reserves, in order to help limit extreme food price volatility.
- Achieve universal health coverage, including financial risk protection, access to quality essential healthcare services, and access to safe, effective, quality, and affordable essential medicines and vaccines for all.
- By 2030, ensure that all learners acquire the knowledge and skills needed to promote sustainable development, including, among others, through education for sustainable development and sustainable lifestyles, human rights, gender equality,

promotion of a culture of peace and non-violence, global citizenship, and appreciation of cultural diversity and culture's contribution to sustainable development.

- Enhance the use of enabling technology, in particular information and communications technology, to promote the empowerment of women.
- Significantly increase access to information and communications technology and strive to provide universal and affordable access to the Internet in the least developed countries by 2020.
- By 2030, enhance inclusive and sustainable urbanization and capacity for participatory, integrated, and sustainable human settlement planning and management in all countries [4].
- By 2030, halve per capita global food waste at the retail and consumer levels and reduce food losses along production and supply chains, including post-harvest losses.
- Fully operationalize the technology bank and science, technology, and innovation capacity-building mechanism for least developed countries by 2017 and enhance the use of enabling technology, in particular information and communications technology.
- By 2020, enhance capacity-building support to developing countries, including for least developed countries and small island developing states, to significantly increase the availability of high-quality, timely and reliable data disaggregated by income, gender, age, race, ethnicity, migratory status, disability, geographic location and other characteristics relevant in national contexts.

On a general level we need to horizons and encourage more interdisciplinary work/cooperation. Ethical challenges, "Implications for design" section/angle, why then could not there be an "Implications for the Sustainable Development Goals".

7 Conclusions and Further Work

Social and technological changes make human Computer interaction fundamental to the success of most computer systems at this time. We argue that with the integration of methodologies such as Value Sensitive Design, User Centered Design and Design Thinking -all together- it is possible to develop current and future technology that contributes to minimizing the problems identified in the 17 Sustainable Development Goals.

In this article we have placed the 17 SDGs as a central part to answer questions such as What kind of technological systems must be developed to move towards a technologically sustainable society? and How can this technology contribute to solving the identified SDGs?

As further work we pretend to validate our proposal (integration of Value Sensitive Design, User Centered Design and Design Thinking) in the development process of Interactive systems and propose a set of guidelines in order to design applications centered on human values, involving different Sustainable Development Goals.

References

1. Aker, J.C., Ghosh, I., Burrell, J.: The promise (and pitfalls) of ICT for agriculture initiatives. Agric. Econ. **47**(S1), 35–48 (2016)
2. Harper, R., Rodden, T., Rogers Y., Sellen, A.: Being Human: Human-Computer Interaction in the Year 2020. Microsoft Research (2008)
3. World Health Organization: Preventing Disease Through Healthy Environments: Towards an Estimate of the Environmental, Burden of Disease, pp. 1–1106 (2019)
4. Babsa, A., Adekola, F., Ayodele, D.: Study of metals dissolution from a brand of mobile phone waste. Metalurgija—J. Metall. **16**, 269–277 (2010)
5. DiSalvo, C., Sengers P., BrynjarsdÂ'ottir H.: Mapping the land-scape of sustainable hci. In: Proceedings of the SIGCHI Conference on Human Factors in Computing Systems, pp. 1975–1984 (2010)
6. United Nations: Transforming our world: the 2030 Agenda for Sustainable Development, A/70/L.1 (2015)
7. UNESCO: Broadband Commission, Means of Transformation, ITU/UNESCO (2014)
8. Weitz, N. Persson A., Nilsson, M., Tenggren, S.: Sustainable development goals for Sweden: In: Sights on Setting a National Agenda, In Stockholm Environment Institute (2015)
9. Gaffney, O.: Sustainable development goals improving human and planetary wellbeing. IGBP's Glob. Change Mag. **1**, 20–23 (2014)
10. Dix, A., Finlay, J., Abowd, G., Beale, R.: Human-Computer Interaction. 3rd edn. Pearson (2004)
11. Hooper, C., Dix, A.: Web science and human-computer interaction: forming a mutually supportive relationship. Interactions **20**, 52–57 (2013)
12. Jacko, J.: Human–Computer Interaction Handbook, 3rd edn. CRC Press (2012)
13. Jackson, S: Sustainability in Computer Science. In: Human-Computer Interaction, and Interaction Design. CMC Senior Theses, p. 1329 (2016)
14. Kumar, N., Kamath, V., Gamage, D., Sible, A., Sturm, C., Rivera, C., Sabie, D., Moinuddin, M., Moreno-Rocha, M.: HCI Across Borders and Sustainable Development Goals, CHI 2020. Honolulu, HI, USA (2020)
15. Eriksson, E., Pargman, D., Bates, O., Normark, M., Gulliksen, J., Anneroth, M., Berndtsson, J.: HCI and UN's sustainable development goals: responsabilities. Barriers Oppor. NordiCHI **2016**, 1–2 (2016)
16. Bodker, S.: When second wave HCI meets third wave challenges. In: Proceedings of NordiCHI'06, New York, NY, USA, pp. 1–8. ACM (2006)
17. Grudin, J.: NordiCHI 2006: learning from a regional conference. Interactions **14**, 52–53 (2006)
18. Moore H.: Global prosperity and sustainable development goals. J. Int. Dev. **27** (2015)
19. Kowles, B., Bates, O.: Hakanson, M., Gamage D.: This changes sustainable HCI. In: Proceedings of the 2018 CHI Conference on Human Factors in Computing Systems (2018)
20. Blevis, E.: Sustainable interaction design: invention & disposal, renewal & reuse CHI '07. In: Proceedings of the SIGCHI Conference on Human Factors in Computing Systems, pp. 503–512 (2007)
21. Mankoff, J.C., Blevis, E., Borning, A., Friedman, B. Fussell, S. R., Hasbrouck, J., Woodruff, A., Sengers P.: Environmental sustainability and interaction. In: CHI'07 Extended Abstracts on Human Factors in Computing Systems, pp. 2121–2124. ACM (2007)
22. Raghavan, B., Pargman, D.: Means and Ends in Human-Computer Interaction: Sustainability through Disintermediation, CHI 2017. Denver, CO, USA (2017)
23. Pathak, D., Varde, A., Alo., C., Oteng F.: Ubiquitous access for local water management through HCI based app development. In: 2019 IEEE 10th Annual Ubiquitous Computing, Electronics & Mobile Communication Conference (UEMCON), New York City, pp. 0227–0233. NY, USA (2019)
24. Baumer, E.P., Silberman, M.: When the implication is not to design (technology). In: Proceedings of the SIGCHI Conference on Human Factors in Computing Systems, pp. 271–2274. ACM (2011)

25. Preist, C. Schien, D., Blevis E.: Understanding and mitigating the effects of Preist, 2016. Device and cloud service design decisions on the environmental footprint of digital infrastructure. In: Proceedings of the 2016 CHI Conference on Human Factors in Computing Systems, pp. 1324–1337. ACM (2016)
26. Ekbia H., Nardi, B.: Social inequality and HCI: the view from political economy. In: Proceedings of the 2016 CHI Conference on Human Factors in Computing Systems, pp. 4997–5002. ACM (2016)
27. von Schomberg, R.: Prospects for technology assessment in a framework of responsible research and innovation. In: Dusseldorp, M., Beecroft, R. (eds.) Technikfolgen abschÃ'tzen lehren. VS Verlag fÃijr Sozialwissenschaften (2012)
28. Friedman, B., Hendry, D.: Value Sensitive Design: Shaping Technology with Moral Imagination. MIT Press (2019)
29. ISO 9241-210:2019, Ergonomics of human-system interaction. In: Human-Centered Design for interactive systems (2019)
30. Ferris, T.: User-centered design: an integrated approach. IEEE Trans. Profess. Commun. **47**, 75–77 (2004)
31. Cross, N.: A Brief History of the Design Thinking Research Symposium Series, Design Studies, vol. 57, pp. 160–164. ACM (2018)
32. Interaction-design: A Simple Introduction to Lean UX, Available online at: https://www.interaction-design.org/literature/article/a-simple-introduction-to-lean-ux, Ac-Cited 1 May 2021

Health Control Models

Investigating the Key Parameters Affecting Sustainable IoT-Based Marketing

Seyyed Esmaeil Najafi, Hamed Nozari, and Seyyed Ahmad Edalatpanah

Abstract Over the past four decades, the world has witnessed an interesting new trend. Technologies are made just to make the world a better place. All industries have developed clean energy such as wind and solar. Today, all organizations strive to create a better world. In this regard, the vision of the World Economic Forum's Community Economics and Digital Society Initiative is to include our digital future as reliable and sustainable. IoT is the main focus in this regard. The Internet of Things and the capabilities it provides to managers are an effective tool for achieving the organization's sustainable competitive advantage. Based on studies, perceived usefulness, ease of use, trust, social acceptance, enjoyment of use and controllability have been identified as effective variables on the use of IoT in sustainable smart marketing. In this study, these key indicators were reviewed and prioritized. Finally, a framework for expressing the positions of the impact of the Internet of Things on sustainable marketing is presented.

Keywords Marketing · Sustainable marketing · Internet of things (IoT) · Sustainable development · IoT based marketing

1 Introduction

The Internet of Things has brought about change in all aspects of daily life, including business. Meanwhile, the rapid development of mobile communications and computing has provided a better opportunity for this change and has created a

S. E. Najafi
Department of Industrial Engineering, Science and Research Branch, Islamic Azad University, Tehran, Iran

H. Nozari
Department of Industrial Engineering, Central Tehran Branch, Islamic Azad University, Tehran, Iran

S. A. Edalatpanah (✉)
Department of Applied Mathematics, Ayandegan Institute of Higher Education, Tonekabon, Iran
e-mail: s.a.edalatpanah@aihe.ac.ir

J. L. Verdegay et al. (eds.), *Computational Intelligence Methodologies Applied to Sustainable Development Goals*, Studies in Computational Intelligence 1036,
https://doi.org/10.1007/978-3-030-97344-5_4

new generation of marketing under the smart interactive marketing paradigm. This new generation of marketing has long been felt because customers in shopping are not just looking to meet their basic needs and are interested in applying their feedback on purchased products. In this type of marketing, products are directly connected to the Internet, and instead of reporting on their products and experiences regarding the performance and capabilities of a product they use in a social network, it is the products themselves that they do this directly [1].

The IoT is able to seamlessly integrate a large number of heterogeneous and diverse end systems by providing free access to select subset of data to improve a set of digital services. The Internet of Things is used as an umbrella-like keyword that covers various aspects of the Internet and the Web in the physical world. This cover is done by using a wide range of distributed devices and identifying the sensor, sensor or other capabilities used and embedded. The Internet of Things envisions a vision for the future in which physical and digital beings are interconnected using appropriate communication and information technologies to provide a whole new set of applications and services [2]. The IoT is generally characterized by real-world small things, widespread distribution, limited storage, and processing capacity, with respect to object concerns IoT, such as reliability, performance, security, and privacy. Technologies that are able to communicate via the Internet [3]. IoT technology is currently used in various fields, such as supply chain management, urban design, library management, retail tracking, stock control, digital logic, optimal transportation, home automation, mobile payment, housekeeping, Privacy and healthcare. IoT technology offers many improvements among other industries and their benefits to customers are significant. For example, users may use IoT-based technologies used in smart refrigerators to automatically control food and beverage consumption and re-order goods. A set of models has been created to express customer acceptance of new technology [4]. It is a very cognitive method based on the technology acceptance model, and has created a wide stream of studies to examine tendencies, customer use, and actual use. For IoT practitioners, the more practical question is: What partnerships increase the customer's understanding that IoT is simple, useful, fun, and secure. Therefore, this study is based on the domains of honesty, social impact, benefit, and behavior control from individual perspectives, social context and technology, to provide an integrated framework that the effects of these variables on customer acceptance of technology [5]. This study allows us to paint a more realistic and broader picture of IoT incentives in IoT technology than the comprehensive approach. The knowledge that this study provides is useful for attendees in IoT technology. A more complete understanding of the impact of IoT technologies on their acceptance of technology can help attendees gain more knowledge about IoT. How to entice customers to use IoT users more consistently, and entice non-users to get started. In the following, we will examine sustainable marketing and the effects of the Internet of Things on it, as well as the key indicators influencing the use of this technology on marketing sustainability.

2 Sustainable Marketing

Today, the category of marketing in sustainable development is usually considered not as part of the problem but as part of the strategic solution. Due to the forecast of world population growth and increasing demand for food and consumer goods due to the existing ecological and social constraints, the supply of these needs cannot be met through current and current methods of production and marketing, so marketing systems to achieve development Sustainable need to adopt new business strategies. There are several definitions of the concept of sustainability, some of which have been proposed along with side ideas such as sustainable marketing [6]. In general, sustainable development means achieving social and economic progress without affecting the natural balance of the environment. Therefore, the concept of sustainable development should always consider the three issues of economic growth, environmental protection and respect for human and social rights and try to promote it. The main concepts of sustainable development refer to the needs and limitations. Given the unfair distribution of wealth between rich and poor countries, sustainable development focuses on meeting the minimum basic needs [7].

The possibility of using the concept of marketing depends on the use of a socio-economic environment. According to the Board of Directors of the American Marketing Association: Marketing is an activity, a set of institutions and processes to create, communicate and exchange offers that are valuable to customers, clients, partners and the community at large. Obviously, studying the concept of social marketing helps us a lot to understand and achieve the goals of sustainable marketing, because after the formation of social marketing, the concept of sustainable marketing has appeared in macro marketing [8]. Because the mission of macro marketing is to examine marketing processes, activities, and outcomes that examine cultural, social, and political interactions. The concept of social responsibility of organizations is in fact the commitment of the organization to maintain and improve the welfare of society in the present and future conditions, which is itself formed through various social activities. Therefore, business with the approach of social responsibility has a broad concept of sustainable development, with the difference that the social responsibility of the organization limits the social issues of society to issues related to the performance of the organization and operates only in this area. Obviously, the two concepts of social marketing and socially responsible marketing are formed [9]. Although the performance of both is to improve the social environment, but its goals are different because in social marketing, unlike social marketing, special attention is paid to the ethical aspect of the issue. The main principles of sustainable marketing are shown in Fig. 1.

Green marketing also tries to market products that do not harm the environment. Accordingly, green marketing encompasses a wide range of activities such as plant breeding, changing the production process, changes in product packaging and even changes in the way products are advertised. Green marketing pays attention to the compatibility of the product or service with the environment and follows all the stages of production, distribution, promotion, packaging and consumption of products in

Fig. 1 Principles of sustainable marketing

a way that responds to ecological issues. Manufacturers, wholesalers, retailers and service companies also always have the opportunity to participate in green marketing. Today, green marketing is also considered as a factor of competitive advantage [10]. Although green marketing is a forward-looking approach and necessarily requires sustainability, in order to achieve this, it seems necessary to integrate it with political and governmental affairs so that consumers also have the opportunity to act in accordance with the goals set in the long run. Time can improve their standard of living. Achieving sustainability in this category requires deep thinking and reflection in this field, and the long recession is one of the obstacles to achieving this goal, and most people are not familiar with the correct concept of sustainability in marketing and cannot work with these limitations. Understand the improvement of your life. In fact, the meaning of sustainability in marketing has an implicit relationship with its realization in the long run. Therefore, sustainable marketing can be considered as an advanced stage of green marketing [11].

In fact, sustainable marketing is a holistic approach that aims to meet the demands and needs of the customer and at the same time pays attention to environmental and social issues. In this way, it provides responsible profitability [12]. In another definition: Sustainable marketing is the process of design, implementation, control and development, pricing, promotion and distribution of products in a way that firstly meets the needs of customers, secondly achieves organizational goals and thirdly this process is compatible with the ecosystem. Undoubtedly, sustainable marketing can achieve these three goals, the two primary goals are in line with organizational goals and the third goal is in line with sustainable development. Based on this, the holistic approach of sustainable marketing is proven. Of course, implementing this approach is not an easy task and can only be done by formulating an appropriate strategy with an approach to paying attention to social responsibility. In this way, a superior competitive advantage can be achieved because it causes continuity of communication and is a kind of acceptance of the common marketing philosophy by accepting ecological and social constraints. Sustainable marketing can play an important role in sustainable development as well as job creation in agriculture and other industries, as well as increasing income and using healthy products that promote community health and increase life expectancy in the community. On the other hand, consumers will strive to achieve sustainability by consuming sustainable products [13]. The use of sustainable marketing also ensures the protection of soil and water resources and non-renewable resources.

One of the problems in the field of sustainable marketing is that this issue has been discussed more theoretically than practically, and this has led to several misunderstandings about its general and main concept. Sustainable marketing is in fact a concept on the demand for sustainable development because sustainable development in the form of feedback always affects sustainable marketing [14]. Today, it seems necessary for organizations to consider ways to achieve sustainability in their planning, and this is one of the conditions for the success of organizations. It is important to mention that development management without considering marketing will certainly not have the necessary efficiency and it is necessary that marketing as an active and commercial economic process in the first step to use successful forecasting methods to be able to relative trends in accordance with development. Identify stable. In fact, the performance of sustainable marketing is similar to a magic wand because it is expected to provide a final and decisive solution in the event of instability.

3 Sustainable IoT Based Marketing

3.1 IoT Based Marketing

The main tool to achieve marketing goals is communication, and just as communication in the world in recent decades due to the evolution of information technology has undergone extensive changes, marketing has also benefited from these changes and has undergone changes. Of course, it should be noted that the existing theories in marketing have the ability to be used in new communications, because the basic needs and desires of human beings have not changed and only how they are met has changed [15]. Over the years, Internet technology has revolutionized the way marketing is done. The Internet of Things has also brought about fundamental changes in Internet marketing and traditional marketing due to some of its special features. Due to the great potential of the Internet of Things, IoT can be used in various fields as a new model of modern and wireless telecommunications. Using the Internet of Things, customers can receive a software from the shopping center server as soon as they enter the shopping center and thus become a member of the shopping center intelligent interactive system [16]. Today, with the advent of social media, users' experiences with the use of a product have become very important, and these experiences are integrated with the information obtained from searches performed through Internet search engines and in the user's decision to The choice and purchase of a product affects. In addition, new inferred information can be re-shared with other users through tools such as mass media and social networks, influencing other users' choice to buy a product. In IoT-based interactive marketing, products connect directly to the Internet, and instead of customers reporting on their products and experiences with the performance and capabilities of a product they use on a social network, this It is the products themselves that do this directly [17]. IoT provides the Internet to all parties involved in the transaction, including the main parties to the transaction, such as customers, product manufacturers, and sub-parties involved, such as product user activity, such as user opinions and experiences. The benefits of IoT-based intelligent interactive marketing include:

- ***Increasing marketing influence***

The Internet of Things will take marketing to a new level, no part of life, no product, no service and no non-marketable solution.

- ***Interoperability and ability to provide information simultaneously***

IoT-based intelligent and interactive marketing can provide the customer with the information they need at the same time, 24 h a day, seven days a week, which means more interaction, better service and faster response.

- ***Smart customer relationship management***

When it comes to trust and customer relationship management, the Internet of Things comes first. The IoT will be able to do more than just collect data and organize customer information.

- ***Social networks forecasting***

Years ago, when Facebook and Twitter were first introduced to the public, marketers did not even imagine that one-day social media would be one of their main goals. But now, we all understand this great change well. Today, 74% of brand marketers, with only 6 h per week for social media marketing, are faced with huge traffic and flood of audiences in the virtual environment of the Internet.

3.2 Components of IoT-Based Sustainable Marketing

Now, by summarizing the theories and findings of the mentioned experimental researches, the theoretical framework of the research can be presented. Theoretical framework of the research is a conceptual model based on theoretical relationships between factors and variables affecting the research topic which is taken from the research literature. In this study, key indicators affecting sustainable IoT-based marketing have been extracted using literature reviews and opinions of experts active in the field of marketing. For this purpose, 10 active experts in marketing who were fully acquainted with the field of IoT and information technology were selected [18]. These experts were selected on the recommendation of senior managers of organizations and based on positive work records. The key indicators of marketing sustainability are shown in Table 1. As can be seen, these indicators are classified into two general environmental and social sections.

In this study, in order to rank key indicators, a fuzzy ranking method based on hierarchical analysis has been used. This method was first proposed in 2003 by Mikhailov [19]. In this method, first, the integration matrix of fuzzy pairwise comparisons based on triangular numbers and linguistic criteria is obtained (Table 1). These values are then used using fuzzy nonlinear mathematical modeling (Eq. 1) to rank the criteria.

$$\begin{aligned}
&\max\ \lambda \\
&s.t:\ (m_{ij} - l_{ij})\lambda w_j - w_i + l_{ij} w_j \le 0 \\
&\quad (u_{ij} - m_{ij})\lambda w_j + w_i - u_{ij} w_j \le 0 \\
&\quad \sum_{k=1}^{n} w_k = 1 \\
&w_k > 0, k = 1, 2, \ldots, n;\ \ i = 1, 2, \ldots, n-1;\ \ j = 2, 3, \ldots, n \\
&j > i
\end{aligned} \tag{1}$$

In Eq. 1, the values obtained from fuzzy proximal numbers obtained from the pairwise comparison matrix are related and then the generated model must be solved. Since the generated model is nonlinear, so using simple program methods Mathematical finesse cannot be solved. Therefore, software such as GAMS or LINGO is used to solve the created model. In the present study, LINGO software was used to solve the models based on the pairwise comparison matrix.

The main stages of review and ranking are divided into two general sections. In the first stage, by producing fuzzy questionnaires, the opinions of experts are extracted and integrated (averaging). Then, using the nonlinear mathematical model presented in the research, the most practical sections are ranked (Tables 2 and 3).

Table 1 Components of IoT-based sustainable marketing

Main criterion	Sub-index	Symbol
Environmental	Consumer environmental knowledge	W1
	Green transportation and distribution	W2
	Green ad	W3
	Environmental costs	W4
Social	Consumer happiness	W5
	Social costs	W6

Table 2 Fuzzy pairwise comparisons based on integration of expert opinions for options based on environmental criteria

	W1			W2			W3			W4		
W1	–	–	–	–	–	–	–	–	–	–	–	–
W2	4.5	4.6	5.1	–	–	–	–	–	–	–	–	–
W3	3.2	3.8	4.1	3.2	4.1	5.1	–	–	–	–	–	–
W4	3.5	3.9	4.2	1.2	2.3	3.5	2.8	4.1	6.1	–	–	–

Table 3 Fuzzy pairwise comparisons based on integration of expert opinions for options based on Social criteria

	W5			W6		
W5	–	–	–	–	–	–
W6	3.1	4.1	6.3	–	–	–

Table 4 Weight of each option based on environmental criteria

	Symbol	Weight
Consumer environmental knowledge	W1	0.228862
Green transportation and distribution	W2	0.395375
Green ad	W3	0.148203
Environmental costs	W4	0.225893

Table 5 Weight of each option based on Social criteria

	Symbol	Weight
Consumer happiness	W5	0.749757
Social costs	W6	0.250243

By placing the data obtained from pairwise comparison tables obtained from the integration of expert opinions, we have formed fuzzy nonlinear mathematical models and solved the resulting models using LINGO software. The weight and rank of each option were obtained, as shown in Tables 4 and 5.

By examining the weights obtained from the mathematical modeling solution, one can understand the importance of each of the options.

3.3 Managing IoT-Based Sustainable Marketing

Traditional and sustainable approaches in marketing have a similar approach to decision making and it is considered as a process of planning, implementation, control, which is influenced by various factors. Although the approach of the two approaches to decision making is almost similar, the results of the decision in these two approaches may be fundamentally different from each other. The Internet of Things provides a consistent source of big data (such as location, time, speed, acceleration, temperature, mood, walking steps, and heart rate) [20]. The availability of this big data paves the way for the development of generalizable and predictive behavioral models that have already been provided in any form apart from the thick descriptions written by researchers.

Access to audience information and analysis of this information will be the most important tool to formulate our marketing strategies correctly and purposefully. This information will be used to analyze and predict the behavior of the audience so that at the most appropriate moment, according to the needs of the audience, we can send our message that we have already prepared. Using this technology, as mentioned in the previous section, it is possible to create many social and environmental impacts on the market space, which in turn can increase sustainability.

Figure 2 provides a framework that can be used to see the elements that affect IoT-based marketing.

A marketing platform is a tool through which a sustainable marketing strategy is implemented in order to achieve strategies and goals related to pollution avoidance and resource acquisition. This is because decisions about this marketing platform create inevitable cycles of resource use and waste generation, and therefore determine any environmental impact (positive or negative) attributed to consumption.

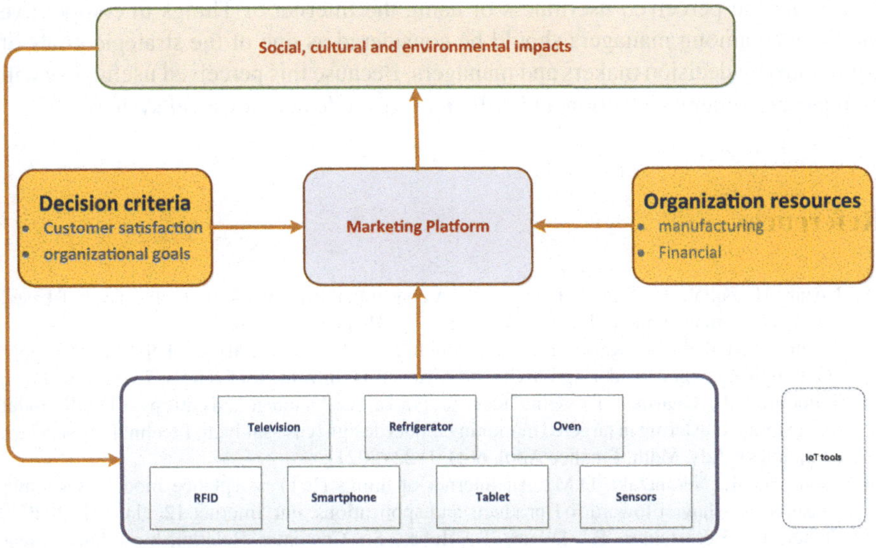

Fig. 2 A framework for managing IoT-based sustainable marketing

4 Conclusion

Smart marketing has made great strides in its evolution. In recent years, competitive intelligence has become one of the most important management concepts and has become ingrained in the culture of leading companies. Increasing competitive intelligence enables organizations to analyze environmental information faster and more accurately, and to store the results in a useful way and make them available to decision makers when appropriate. This accelerates the flow of information and knowledge in the context of the organization and dramatically improves the effectiveness of the collective thinking and decision-making process. In this way, executives with a better understanding of competitors and the challenging environment of competition can rely on specific tools and techniques, information related to the capabilities, strengths and weaknesses and intentions of competitors to review and select the appropriate strategy and create Sustainable development is always one step ahead of competitors. A marketing platform is a tool through which a sustainable marketing strategy is implemented in order to achieve strategies and goals related to pollution avoidance and resource acquisition. This is because decisions about this marketing platform create inevitable cycles of resource use and waste generation, and therefore determine any environmental impact (positive or negative) attributed to consumption. Therefore, it should always be considered that the use of this technology makes things faster. The use of this technology will also lead to better work performance. Increasing the perceived usefulness of using the Internet of Things in competitive intelligence among managers should be considered as one of the strategic goals of companies by decision makers and managers. Because this perceived usefulness will increase customer satisfaction and will create a stable and powerful system.

References

1. Nozari, H., Fallah, M., Szmelter-Jarosz, A.: A conceptual framework of green smart IoT-based supply chain management. Int. J. Res. Indus. Eng. **10**(1), 22–34 (2021)
2. Nozari, H., Fallah, M., Kazemipoor, H., Najafi, S.E.: Big data analysis of IoT-based supply chain management considering FMCG industries. Bus. Inform. **15**(1 (eng)), 78–96 (2021)
3. Teimouri, H.B., Gharib, J., Hossein Zadeh, A., Pouya, A.: An integrated entropy/VIKOR model for customer clustering in targeted marketing model design (case study: IoT technology services companies). Adv. Math. Finance Appl. **6**(4), 1–22 (2021)
4. Tsourela, M., Nerantzaki, D.M.: An internet of things (IoT) acceptance model. Assessing consumer's behavior toward IoT products and applications. Fut. Internet **12**(11), 191 (2020)
5. Ploder, C., Bernsteiner, R., Dilger, T., Huber, S.: Customer Relationship Management Improvement Using IoT Data (2021)
6. Munnia, A., Nicotra, M., Romano, M.: Big Data, Predictive Marketing and Churn Management in the IoT Era. In: The Internet of Things Entrepreneurial Ecosystems, pp. 75–93. Palgrave Pivot, Cham (2020)
7. Pimpan, S., Ruaguttamanun, C., Wongkhae, K.: Sustainable marketing strategy and marketing success of ISO 14001 certified manufacturing businesses in Thailand. J. Humanit. Soc. Sci. Thonburi Univ. **15**(3), 37–48 (2021)

8. Mohammadian, A., Mirbagheri, F., Ghorbani, A.: Prioritizing IOT implications for innovation in marketing mix by considering technological, legal and market factors in Iran. J. Executive Manage. Univ. Mazandaran **12**(23) (2020)
9. Moradi, M.: Importance of Internet of Things (IoT) in Marketing Research and Its Ethical and Data Privacy Challenges (2021)
10. Sivathanu, B.: Adoption of internet of things (IOT) based wearables for healthcare of older adults–a behavioural reasoning theory (BRT) approach. J. Enabl. Technol. (2018)
11. Peterson, M., Minton, E.A., Liu, R.L., Bartholomew, D.E.: Sustainable marketing and consumer support for sustainable businsses. Sustain. Product. Consump. **27**, 157–168 (2021)
12. Ray, S.: Implication of predictive maintenance for industrial marketing: a case study from the air compressor industry. In: Innovations in the Industrial Internet of Things (IIoT) and Smart Factory, pp. 231–243. IGI Global (2021)
13. Suny, M.F.I., Fahim, M.M.R., Rahman, M., Newaz, N.T., Akhund, T.M.N.U.: IoT past, present, and future a literary survey. In: Information and Communication Technology for Competitive Strategies (ICTCS 2020), pp. 393–402. Springer, Singapore (2021)
14. Syniuk, A.: Sustainable Apparel Marketing and Consumption: An Analysis of Patagonia's Worn Wear Video Marketing Campaign (2021)
15. Janardhana, K., Mahaboobjan, R., Vijayaragavan, M., Vinoth, K., Kanagavalli, N.: Study Analysis of IoT in FMCG Smart Tags and their Significance in Counterfeit Prevention (2021)
16. Jung, I.: An IoT-based smart parking management system. Int. J. Comput. Vis. Rob. **10**(2), 122–132 (2020)
17. Rozalena, A., Sulaeman, M., Mulyati, S., Gunawan, H.: Business communication skill model based on internet of thing (IoT). J. Phys. Conf. Ser. **1477**(7), 072010 (2020)
18. Tudor, A.I.M., Chițu, I.B., Dovleac, L., Brătucu, G.: IoT technologies as instruments for SMEs' innovation and sustainable growth. Sustainability **13**(11), 6357 (2021)
19. Mikhailov, L.: Deriving priorities from fuzzy pairwise comparison judgements. Fuzzy Sets Syst. **134**(3), 365–385 (2003)
20. Ekren, B.Y., Mangla, S.K., Turhanlar, E.E., Kazancoglu, Y., Li, G.: Lateral inventory share-based models for IoT-enabled E-commerce sustainable food supply networks. Comput. Oper. Res. **130**, 105237 (2021)

Applications of Computational Intelligence in the Studies of Covid-19

Rafael Bello, María M. García, Yaile Caballero, Alejandro Rosete, and Yanela Rodríguez

Abstract Health is an extremely important aspect in the United Nations Sustainable Development Goals because there is not possible progress for humankind without it. The Covid19 pandemic has evidenced to what extent society can be affected in all its facets when suddenly a phenomenon affects human's health. In this chapter, it is analyzed how computational intelligence techniques have allowed developing different studies about this disease, creating several prediction models and formulating new knowledge about it.

Keywords Computational intelligence · Softcomputing · Covid-19 · Acquisition of knowledge

1 Introduction

The United Nations Sustainable Development Goals (SDGs) from a multidimensional perspective seek the development of mankind through ending poverty, protecting the planet, and guaranteeing that all people enjoy peace and prosperity [1–5]. For that, 17 sustainable development objectives and 169 goals are established.

R. Bello · M. M. García
Universidad Central "Marta Abreu" de Las Villas, Santa Clara, Cuba
e-mail: rbellop@uclv.edu.cu

M. M. García
e-mail: mmgarcia@uclv.edu.cu

Y. Caballero · Y. Rodríguez
Universidad de Camagüey, Camaguey, Cuba
e-mail: yaile.caballero@reduc.edu.cu

Y. Rodríguez
e-mail: yanela.rodriguez@reduc.edu.cu

A. Rosete (✉)
Universidad Tecnológica de La Habana "José Antonio Echeverría", Cujae, Havana, Cuba
e-mail: rosete@ceis.cujae.edu.cu

J. L. Verdegay et al. (eds.), *Computational Intelligence Methodologies Applied to Sustainable Development Goals*, Studies in Computational Intelligence 1036,
https://doi.org/10.1007/978-3-030-97344-5_5

Member states have agreed to try to reach them by 2030. Among the considered aspects in this development perspective, health occupies a fundamental place in the 3 SDG "Guaranteeing a healthy life and promoting the well-being for all in all ages", articulated around 13 goals that cover a wide range of the work carried out by the World Health Organization.

Several SDGs keep a close relationship with health and the accomplishment of these goals will have a very positive impact on the health and well-being of the people. However, Objective 3 is directly focused on responding to several of the most important global sanitary issues. Among them for 2030, to put an end to Acquired Immunodeficiency Syndrome (AIDS), tuberculosis (TB), malaria epidemics, and neglected tropical diseases and fight hepatitis, waterborne diseases, and other communicable diseases, guaranteeing access to essential health services and safe, effective, accessible, and top-quality medicines and vaccines for all.

To achieve these purposes several executing actions have been established:

- Support the research and development activities of vaccines and medicines for communicable and non-communicable diseases that mainly affect developing countries and facilitate access to essential medicines and vaccines that are accessible following the Doha Declaration. This declaration states the right of developing countries to make full use of the provisions of the Agreement on Trade-Related Aspects of Intellectual Property Rights regarding flexibility to protect public health and, particularly provide access to medicines for all.
- Increase substantially the funding of health and contracting, development, training, and holding of healthcare workers in developing countries, especially in less developed countries and small insular developing states.
- Reinforce the capacity of all the countries, particularly developing countries in terms of early warning, risk reduction, and risk management for national and worldwide health.

The need for a development approach like this one has become more evident with the appearance of the last pandemic that has affected humankind: Covid19. It has revealed the weakness of the human being and their environment to be able to face the challenges that they should confront. Since new threats that disturb human life can appear without distinguishing between rich and poor, or between developed and non-developed countries, the entire world's technological arsenal must be put to the test in a short period.

Covid-19 is a new disease whose emergence took place in China at the end of 2019, but it extended very quickly throughout the world, with millions of sick and dead people. Despite all the actions taken by the governments, health and scientific organizations, and other institutions, it has not been possible to overcome it. Because of its novelty, knowledge about it was relatively scarce, and it was necessary to develop studies about its possible behavior, diagnosis means, and treatments. Techniques based on Artificial Intelligence are among the ones used, specifically those from computational intelligence [6, 7].

In this chapter, a summary of some of the conducted studies about Covid19 using techniques of computational intelligence is made. In the second section, a group of developed works oriented to the study of different aspects of the disease and its development is presented. Also, conducted studies for the diagnosis and prognosis are described below.

2 Using Computational Intelligence in Covid-19 Studies

In the context of health and medical systems, the management of uncertainty and the complexity of the decisions are very important challenges that can be addressed by using Artificial Intelligence and Soft Computing techniques [8, 9]. Some studies have demonstrated the interest in the medical applications of intelligent techniques in four directions: machine learning, intelligent robots, image recognition technology, and expert systems [10]. In addition, intelligent techniques have been identified as important components for the management of healthcare-associated infections to support surveillance, to understand risk factors, to improve patient risk stratification, and to identify transmission pathways timely [11].

Indeed, in the current extreme situation caused by the COVID-19 outbreak several examples of these techniques have shown their potential for the battle against the pandemic [12, 13] and they are considered as key components in the post-COVID-19 era [14, 15]. Particularly, Artificial Intelligence and Machine Learning are included among the key component in the Covid-19 context, along with supply chains, communications, surveillance, Internet of Things, robotics, drone-based systems, and mobile applications [15]. This interest has been demonstrated in the publication of reviews, both with a general perspective [16–18] and in particular topics such as medical images analysis [19, 20] or drugs design [21]. Some potential uses of the intelligent techniques in the context of COVID-19 include medical diagnosis, prediction of patient evolution, design of drugs and medical procedures, prediction of outbreaks, and pandemic management [14].

An interesting application is the early classification of patients in terms of the severity of the disease to optimize the use of the medical staff [19, 20]. In the same line, another singular approach is based on the use of Fuzzy Logic to decide the correct procedure for patients that are not in a severe situation [22]. Also, soft computing techniques for uncertainty management have been successfully used to model causal relationships between protocols and relaxation strategies [23].

In general, some of the most used techniques include Deep Learning [7, 20], Support Vector Machines and Random Forest [16], combinations of Fuzzy Logic and Metaheuristics [24], etc. In some contexts, the combination of techniques has demonstrated a remarkable advantage [25]. For example, combinations of fuzzy sets and metaheuristics have been successfully used in cancer detection [26]. Another intelligent technique that has shown a high potential in medical decision systems is the Rough Set Theory [27].

An important aspect that is to be taken into account is to allow diagnosis strategies that minimize the need for laboratory tests [28]. Besides, the importance of considering in the diagnosis aspects such as age, gender, and other conditions (diabetes, hypertension, liver and kidney preconditions) has also been stated as a necessity [18].

3 Creating Scenarios for the Development of Covid-19

The application of mathematical and computational methods to predict the behavior of epidemics allows having an approximation of the number of people who will suffer from the disease, which is the base for decision-making, including logistic aspects to guarantee assistance to the population. The methods based on Statistics and mathematical-physics equations have been some of the most used for this purpose [6, 29–33] are some of the methods based on Artificial Intelligence.

Scenario analysis techniques have been used in the study of the behavior of epidemics, showing that it is another useful approach [34]. In [35] it is stated that since the parameters change continuously in different periods, making predictions difficult, it is useful to analyze different scenarios. In [36] different graphics are shown for the behavior of the epidemic according to different scenarios of reduction in city mobility.

A scenario is a narrative of a situation, frequently accompanied by graphics and limited quantification [37–39]. The scenario analysis helps to explore a range of believable and possible futures. A group of two to five believable narratives that describe possible evolutions in key areas is created. The elaboration of scenarios is not conditioned by the availability of evidence that backs neither the relation of causality of the variables nor the value of each of them.

The scenario analysis in the study of the behavior of epidemics offers a complementary approach to the use of prediction of behavior models. Generally, these models build a function to predict the behavior of the epidemic, its duration, the total accumulated, etc. When using the scenario analysis, other possible behaviors of the development of the epidemic can be visualized according to key factors. Even though we can start from a prediction model to build sceneries, these can also be developed through expected values of the key factors established according to the criteria of experts or decision-makers, and visualize how the behavior of the pandemic would be under those suppositions [40].

Next, a method developed to establish possible scenarios for the development of Covid19 based on computational intelligence techniques is described. The purpose is to build different possible curves for diagnosing positives that satisfy the given values for the key factors for that scenario. Three key factors are taken into account:

- The maximum expected number of positives diagnosed during the epidemic, that is, the accumulated of infected persons at the end of the epidemic (MaxAccum), whose value is defined as an interval.
- The maximum number of days in which positive diagnoses are expected. (Days)

- The maximum number of diagnosed cases in one day (MNC).

The method that is proposed for the analysis of scenarios builds a function ***x*** that shows a possible development of the quantity of diagnosed cases, the arguments of this function are the three key factors. For a given scenario, different forms can exist for the function ***f*** which satisfies the established values for key factors. That forms a search space from which the function or functions that adjust more to the established scenario should be selected. To build the function ***f***, the fuzzy set theory is used. A fuzzy set is defined through a membership function (x) that assigns a value in the interval [0,1] to the elements of the universe U where the set is defined. (*f*: U → [0,1]). In this problem, the universe U is the duration of the pandemic and the function ***f*** should reach the maximum (value 1) when the maximum number of diagnosed cases in a day (MNC) is known and the sum of positive cases at the end of the pandemic (defined in days) should be in the interval of values given by MaxAccum. The problem is finding the mathematical expression for function ***f***, for which a search in the space of possible functions should be made. For that, the use of a metaheuristic (such as Genetic Algorithms or Particle Swarm Optimization) is proposed.

To reduce the search three types of ways for the function *f* are used: triangle function, trapezoid, or bell; since the studies conducted about the behavior of epidemics have evidenced that these usually develop with that form [29, 41, 42]. A cleaning mechanism is included to improve the exploration of the method with which a certain difference is kept among the functions that are generated. In other words, the function ***f*** is built as a fuzzy set, and metaheuristics allows looking for the curve that generates an accumulated closer to the value of MaxAccum, taking into consideration the given values by the other factors Days and MCD. Each slate of three (MaxAccum, Days, and MCD) defines a scenario. Since it is not recommended to present many alternatives, three possible cases for the function ***f*** are presented in the research, which represent possible behaviors of the pandemic.

The series of real values of positive diagnoses reported daily since the beginning of the pandemic until the last date is another information obtained when the scenario analysis is done within the period of development of the pandemic and this information can also be used in the construction of the function ***f***, seeking that the curve approximates the most to that series. To combine the adjustment of the function to the value of *MaxAccum* given and the series of real values of daily reports collected up to that moment a multi-objective function for the heuristic evaluation with two components is formulated: a measure of the proximity of the accumulated generated by the function ***f*** to the value of the key factor *MaxAccum* and the adjustment of the values of daily diagnoses generated by ***f*** to the series of diagnoses that are collected to that date (measure as the sum of the daily differences between the real value of positive cases in a day and the value generated by the function ***f***).

In the article published in [43], several scenarios were proposed to analyze the possible behaviors of Covid-19 in Cuba in the initial evolution of the pandemics. Among the defined scenarios the one defined by the following key factors was considered: Days = 120; MNC = 60; Accumulated = [1850, 2150]. Those predict

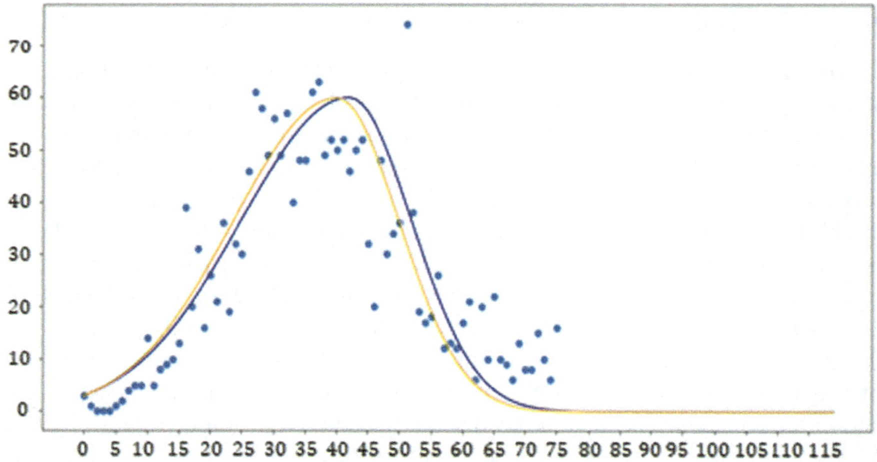

Fig. 1 Constructed functions for a scenario given by key factors (x-axis corresponds to days, y-axis indicates the number of new cases reported in the day)

the development of the epidemic during a period of 120 days, with a daily maximum of positive diagnoses of 60 and a total number of Covid-19 patients at the end of the period between 1850 and 2150 cases.

Using the series of reported positive cases in Cuba until April 24th, 2020 and a function in the form of a bell, the scenario that is shown in Fig. 1 was constructed. There are two possible curves for function f. The series of positive cases until May 26th, 2020 was plotted, which allows us to compare the real data with the curves until then.

Figure 2 presents the predicted accumulated and the real values in that scenario, which evidences an approximation between the constructed functions and the real values. The constructed scenarios using the proposed methods were accurately adjusted to the reality of the behavior of the epidemic of Covid19 during the first stage of its development in Cuba.

From the point of view of a decision maker this soft-computing based tool allows to understand the range where the real pandemic evolution may occur.

4 Acquisition of Knowledge About the Fatality of Covid19

As aforementioned, one of the researched aspects of Covid19 is its possible treatment. Given the ignorance that existed about the disease when it spreads in early 2020, it was necessary to apply different techniques from clinical, mathematical, and computational perspectives for its study. Concerning its diagnosis, one of the aspects of interest was the risk factors for the disease, that is, which ones could have any bearing in its aggressive and risky evolution. Covid19 is characterized by having

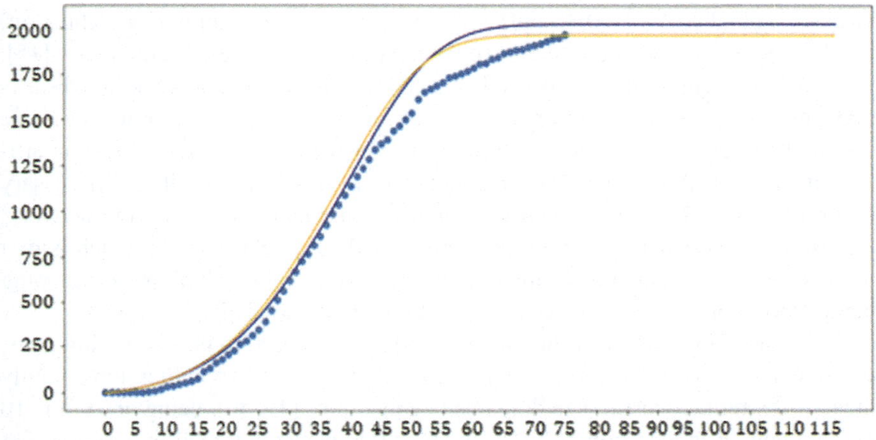

Fig. 2 Accumulated values by the constructed functions and the real accumulated up to May 26 2020 in Cuba (*x*-axis corresponds to days, *y*-axis indicates the number of accumulated cases for each day)

a very changeable behavior, from being completely asymptomatic to causing death very rapidly. Therefore, predicting a possible lethal evolution of the disease is of utmost importance.

The study conducted in [44] follows this purpose. A model for classifying the severity and predicting the tendency of fatal development of the disease is developed in it, in which several hundreds of cases were studied, described initially by more than 300 attributes and those techniques were applied, being identified three relevant attributes, as well as rules for predicting a negative ending for a patient. Intelligent systems can be built with the knowledge generated, and it can be used to increase the existing knowledge on the application domain. This last aspect is especially useful in domains where knowledge is still limited. Such is the case of the Covid19 epidemic.

Another example of knowledge acquisition is presented in [45] through the work done with the data about the development of Covid-19 in Mexico until April 20, 2020. This database was composed of 8775 Covid-19 positive cases with two kinds: 8060 patients who do not die and 712 patients who do. Each case contains 16 attributes or predicting features and the decision or fatality feature. With that database, the process of discovering knowledge was carried out, in which three main tasks were conducted: analysis of the consistency of the data, selection of important features to predict the unfavorable evolution of the patient, and the knowledge acquisition for the prediction.

For the analysis of consistency, the measure of the quality of the classification of Rough Sets Theory (RST) [46] was used, which measures the degree of separation between the objects of the classes, in this problem the classes {Lethal = Yes, Lethal = No}. Using the tool of data analysis based on approximated sets [47], when applying the measures to the studied base of cases, the following results are obtained: Quality of approximation of the class "Lethal = No" = 0.76, quality of the approximation of the class "Lethal = Yes" = 0.37 and Consistency of the data (quality of classification) = 0.73. The value of the consistency of the data can be considered satisfactory [45].

Knowledge representation forms that allow specialists to understand the discovered knowledge were used to carry out the learning process. The acquired knowledge is formalized as decision rules and prototypes for the Fatality class = yes; an artificial neural net (ANN) was implemented also. In this last case, it was necessary to combine the ANN model with a method for rule extraction from the network, because the neural net fails in interpretability, called a "black box"; which is part of the so-called Explainable Artificial Intelligence (XAI) [48]. It allows developing methods to improve the interpretability of discovered knowledge.

In the case of the rule learning, the method J48 was used to build decision trees, which represents one of the most referenced classifying systems in literature, mainly due to its extreme robustness in domains like medicine [49, 50], being one of the 10 most influential in data mining [51]. The resulting decision tree after applying J48 (with a confidence factor 0.85 and 10 as a minimum number of objects) allowed to generate six rules with more certainty CNF (values between 0.7 and 0.76, from a possible maximum of 1.0) for the case of "Lethal = Yes", as shown in Table 1.

The process of knowledge acquisition also included the construction of a multilayer perceptron network, with an input layer with the number of neurons that matches the predicting features of the problem, a hidden layer of 10 neurons, and an output layer with two neurons representing the outputs Lethal = yes and Lethal = no, trained using backward propagation of errors with an adaptation of it for unbalanced classes. Such is the case of the minority class associated with the main interest, i.e., Lethal = yes. This trained network represents a model of knowledge domain that allows the prediction for a given patient to determine if they have or do not have a tendency

Table 1 Obtained rules from the decision trees [45]

No.	Rule	CNF
R1	If COPD = YES AND AGE > 63 AND OBESITY = YES AND DIABETES = YES then Lethal = YES	0.71
R2	if COPD = YES AND AGE > 63 OR HYPERTENSION = YES then Lethal = Yes	0.72
R3	if COPD = NO AND CHRONIC_KIDNEY = YES AND HYPERTENSION = YES OR OBESITY = YES then Lethal = Yes	0.7
R4	if COPD = NO AND CHRONIC_KIDNEY = NO AND HYPERTENSION = YES AND AGE > 58 then Lethal = Yes	0.75
R5	if COPD = NO AND CHRONIC_KIDNEY = NO AND SMOKING = YES AND AGE < 60 then Lethal = Yes	0.74
R6	If COPD = NO AND CHRONIC_KIDNEY = NO AND AGE > 64 AND OBESITY = YES then Lethal = Yes	0.76

to get over the disease. Since Covid19 is a relatively new disease, it is necessary to provide specialists with the biggest possible knowledge of it. Being the built artificial neuronal network a black box, it was complemented with a knowledge extraction method. For that purpose, the proposed method in [52] for the rule extraction from a neural network was used. This method determines the features that are significant for the prediction, eliminating the other ones. For that, it calculates how much the effectiveness of the built network is affected if an entry neuron representing the feature is removed from it.

After this process, a decision rule is built. In this case:

if (AGE > 57 AND DIABETES = YES AND HYPERTENSION = YES AND OBESITY = YES) **then** Lethal = Yes.

Another interpretable way for specialists is the formalization of knowledge through the construction of prototypes [53]. A method that is based on making a granulation of the universe, that is, determining the case sets that are similar, and then building a prototype for each granule or set of similar cases. To build the granulation the proposed method in [54] is used to determine the most adequate relation of similarity with the problem.

The classical algorithms of construction of prototypes treat the positive and negative class symmetrically, hence they are not prepared to deal with the class imbalance. Taking this into consideration, the method for the construction of prototypes according to IMBNPBASIR SEL-CLASS for the classification of imbalanced datasets was used [55]. The best-built prototypes for the class Lethal = Yes are presented in [45].

When doing the classification based on prototypes it is necessary to take into consideration that these are representative of a group of cases so that the calculation of the precision of the discovered knowledge, that is, the set of prototypes is carried out looking for which prototype is the most similar to each case. Selected prototypes by the IMBNPBASIR SEL-CLASS method allow to carry out a classification with a general accuracy of 79.14. Taking into account that the relevant class in this study is the one that denotes a high lethality possibility of the disease, only the best-built prototypes for the class LETHAL = Yes are presented in Table 2.

To evaluate the quality of prototypes (CNF), the set of similar cases is determined, and how many of them have the decision value yes. The quality or certainty (CNF) of the selected prototype for each granule is calculated in function of the decision of it and regarding all the cases of the granule. When all the cases are from the same class of the selected prototype, CNF = 1; otherwise, and depending on the number of cases that are from different classes, the value of CNF will be 0 < CNF < 1. The bigger the CNF is, the quality will be bigger or certainty of knowledge in the form of a prototype, $\mathrm{CNF}(p_i) = \frac{|V_{ii}|}{|V_i|}$, where V_{ii} is the set of instances located in the granule of p_i with the same class as p_i. V_i represents all the instances located in the granule of p_i.

The group of representative prototypes of the different granules or groups is used later in the classification or prediction of a new case. Then, the general accuracy of

Table 2 Obtained prototypes [45]

No.	Prototype	CNF
P1	AGE = 60, DIABETES = yes, COPD = no, ASTHMA = yes, IMMUSUPPR = no, HYPERTENSION = yes, OTHER_COM = no, CARDIOVASCULAR = yes, OBESITY = yes, CHRONIC_ KIDNEY = no, SMOKING = no, LETHAL = yes	0.63
P2	AGE = 76, DIABETES = yes, COPD = no, ASTHMA = no, IMMUSUPPR = yes, HYPERTENSION = yes, OTHER_COM = no, CARDIOVASCULAR = no, OBESITY = yes, CHRONIC_KIDNEY = yes, SMOKING = no, LETHAL = yes	1.00
P3	AGE = 65, DIABETES = yes, COPD = no, ASTHMA = no, IMMUSUPPR = yes, HYPERTENSION = yes, OTHER_COM = no, CARDIOVASCULAR = no, OBESITY = no, CHRONIC_ KIDNEY = no, SMOKING = no, LETHAL = yes	0.61
P4	AGE = 65, DIABETES = yes, COPD = no, ASTHMA = no, IMMUSUPPR = no, HYPERTENSION = no, OTHER_COM = no, CARDIOVASCULAR = no, OBESITY = no, CHRONIC_KIDNEY = yes, SMOKING = no, LETHAL = yes	0.57
P5	AGE = 60, DIABETES = yes, COPD = no, ASTHMA = no, IMMUSUPPR = no, HYPERTENSION = yes, OTHER_COM = no, CARDIOVASCULAR = no, OBESITY = no, CHRONIC_KIDNEY = no, SMOKING = no, LETHAL = yes	0.55
P6	AGE = 72, DIABETES = yes, COPD = no, ASTHMA = no, IMMUSUPPR = yes, HYPERTENSION = no, OTHER_COM = no, CARDIOVASCULAR = no, OBESITY = no, CHRONIC_KIDNEY = yes, SMOKING = no, LETHAL = yes	0.55

the classifier is calculated as the average of the obtained precisions with all the test subsets.

In the case of the Age feature, the value represents an average of the age values for the cases engaged in the prototype. Therefore, it should be seen as an approximate value for that feature. For instance, prototype 2 can be described in general as "patients whose average age is 76 years and suffer from diabetes, immunosuppression, hypertension, obesity, and chronic kidney failure, has a risk (0.63) of a lethal decision". It is wort noting that the average age in all prototypes does not correspond to young people, as it seems to be a singular characteristic of COVID-19.

Furthermore, as a result of the construction of the relation of similarity the importance of the features of the domain of application is also obtained, showing complete coincidence with the selection methods and ordering of the features that were previously used. Table 3 shows the most important features and the weight of each to the feature [45].

The global pandemic of Covid-19 has affected the entire world. Cuba has not escaped this lash despite affectations have been outstandingly fewer concerning other countries due to the high priority given to this issue by the country. Similarly, it is convenient to evaluate what was done in the study with the data about the development of Covid-19 in Mexico [45], how personal pathological backgrounds impact fatality in the case of Cuba, as well as in patients who develop serious and critical conditions of the disease.

To develop this analysis, data from patients in serious and critical conditions as well as dead were taken, from 11/1/2020 to 3/11/2021, reported in the daily reports

Table 3 The most important features and the weight of each to the feature for the case study of Mexico

Weight	Features
0.191323	COPD
0.168401	CARDIOVASCULAR
0.140204	CHRONIC_KIDNEY
0.111942	DIABETES
0.085462	IMMUSUPPR
0.081247	PREGNANCY
0.055963	AGE
0.044351	OBESITY
0.043986	ORIGIN
0.027925	SMOKING
0.021732	HYPERTENSION
0.010395	ASTHMA
0.008574	OTHER CASE
0.008495	OTHER_COM

given by the Public Ministry of Health. A database composed of 4321 cases described by 30 features that are described in Table 4 was obtained.

It was necessary to carry out an information filtering in the case of the patients who died and were previously reported as seriously ill and critical to eliminate the base of cases, the instances associated with this patient, and just leave the one regarding Dead. After filtering the information, 1155 instances were obtained, 230 dead, 291 critically 630 seriously ill.

It was decided to discretize the value of age feature, as it is observed in Table 5, according to the age groups that the Ministry of Public Health (MINSAP) considers in the daily reports that it announces to the population about the behavior of Covid19 in the country.

Obtained prototypes that describe dead patients or those whose condition could cause them to die (seriously ill or critical patients) are shown in Table 6. They have an accuracy of classification of 78.79. It is worth mentioning that in Table 6 are not only included the prototypes related to the patients who died (CLASS = dead, prototypes P3 and P5) but also the persons corresponding to other classes (CLASS = seriously ill or CLASS = critical).

Furthermore, as a result of the construction of the relation of similarity the importance of the features of the domain of application is also obtained. Table 7 shows the most important features and the weight of each to the feature for the case study of Cuba.

The most important features for the case study of Cuba coincided with the most important features for the case study of Mexico, as it is shown in Table 3.

Among other personal pathological backgrounds in Cuban patients, the most relevant risk factors are hypertension, cerebrovascular diseases, cancer, dementia, hepatic diseases, hyperthyroidism, colitis, Parkinson's disease, and chronic liver disease.

Table 4 Description of predicting features for the case study of Cuba from 11/1/2020 to 3/11/2021

No.	Name of the feature	Description	Format or source
1	UPDATE_DATE	Date of the last report at 12 am	MM/DD/YY
2	NATIONALITY	It identifies the patient's nationality	Text
3	SEX	It identifies the sex of the patient	F, M
4	AGE	It identifies the age of the patient	Numerical
5	PROVINCE_RES	It identifies the province of residence of the patient	Text
6	MUNICIPALITY_RES	It identifies the municipality of residence of the patient	Text
7	PREGNANCY	It identifies if the patient is pregnant	YES, NO
8	DIABETES	It identifies if the patient is diagnosed with diabetes	YES, NO
9	COPD	It identifies if the patient is diagnosed with COPD	YES, NO
10	ASTHMA	It identifies if the patient is diagnosed with asthma	YES, NO
11	IMMUSUPPR	It identifies if the patient is diagnosed with immunosuppression	YES, NO
12	HYPERTENSION	It identifies if the patient is diagnosed with hypertension	YES, NO
13	CARDIOVASCULAR	It identifies if the patient is diagnosed with cardiovascular diseases	YES, NO
14	OBESITY	It identifies if the patient is diagnosed with obesity	YES, NO
15	CHRONIC_KIDNEY	It identifies if the patient is diagnosed with chronic kidney failure	YES, NO
16	SMOKING	It identifies if the patient smokes	YES, NO
17	OTHER_APP	It identifies if the patient is diagnosed with other diseases	YES, NO
18	HOSPITAL_STAY	It identifies the number of days the patient has been hospitalized	Numeric

(continued)

Table 4 (continued)

No.	Name of the feature	Description	Format or source
19	ICU	It identifies if the patient has been hospitalized in an Intensive Care Unit	YES, NO
20	FEVER	It identifies if the patient has had a fever	YES, NO
21	ARRHYTHMIA	It identifies if the patient has had arrhythmia	YES, NO
22	INTUBATED	It identifies if the patient has been intubated	YES, NO
23	RESP_DISTRESS	It identifies if the patient has had respiratory distress	NO, SLIGHT, MODERATE, SEVERE
24	HEMODYNAMICS	It identifies the hemodynamic values of the patient	Stable, Unstable
25	DIURETIC_RHYTHM	It identifies the diuretic rhythm the patient has had	Anuria, Preserved, Normal, oliguric, Oligoanuric, Polyuric
26	BLOOD GASES	It identifies the values of blood gases of the patient	Text
27	PNEUMONIA	It identifies if the patient has had pneumonia	YES, NO
28	EVOL_ PCR	It identifies if the patient has got an evolutionary PCR	NO, Positive, Negative
29	OTHER_COMP	It identifies if the patient has presented other complications	Text
30	STATE	It identifies the state of the patient at the end of the report	SERIOUSLY ILL, CRITICAL, DEAD

Table 5 Range of values of age categories according to how it is established by MINSAP

Age group	Age categories (years)
G1	0–19
G2	20–39
G3	40–59
G4	60–69
G5	≥ 70

Based on this initial evidence, it seems that the lethality prediction is also influence by national conditions or the medical procedures used. The inferior lethality of the pandemic in Cuba may provoke a different proportion of conditions in each class.

This type of comparison between the obtained prototypes obtained by using computational intelligence techniques may be useful to improve the understanding of this complex disease.

Table 6 Obtained prototypes for the case study of Cuba

No	Prototype	CNF
P1	AGE = between 20–39, DIABETES = no, COPD = no, ASTHMA = yes, IMMUSUPPR = no, HYPERTENSION = yes, OTHER_COM = yes, CARDIOVASCULAR = no, OBESITY = no, CHRONIC_KIDNEY = no, SMOKING = no, CLASS = seriously ill	0.75
P2	AGE = between 60–69, DIABETES = yes, COPD = no, ASTHMA = no, IMMUSUPPR = no, HYPERTENSION = yes, OTHER_COM = no, CARDIOVASCULAR = yes, OBESITY = no, CHRONIC_KIDNEY = no, SMOKING = no, CLASS = seriously ill	0.84
P3	AGE = $\geq$ 70, DIABETES = no, COPD = no, ASTHMA = yes, IMMUSUPPR = no, HYPERTENSION = yes, OTHER_COM = no, CARDIOVASCULAR = yes, OBESITY = no, CHRONIC_KIDNEY = no, SMOKING = no, CLASS = dead	0.63
P4	AGE = $\geq$ 70, DIABETES = no, COPD = no, ASTHMA = no, IMMUSUPPR = no, HYPERTENSION = yes, OTHER_COM = no, CARDIOVASCULAR = yes, OBESITY = no, CHRONIC_KIDNEY = no, SMOKING = no, CLASS = seriously ill	0.82
P5	AGE = between 40–59, DIABETES = yes, COPD = no, ASTHMA = yes, IMMUSUPPR = no, HYPERTENSION = yes, OTHER_COM = no, CARDIOVASCULAR = yes, OBESITY = yes, CHRONIC_KIDNEY = no, SMOKING = no, CLASS = dead	0.67
P6	AGE = $\geq$ 70, DIABETES = yes, COPD = yes, ASTHMA = no, IMMUSUPR = no, HYPERTENSION = yes, OTHER_COM = yes, CARDIOVASCULAR = yes, OBESIDAD = yes, CHRONIC_KIDNEY = no, SMOKING = no, CLASS = critical	1

Table 7 The most important features and the weight of each to the feature for the case study of Cuba

Weight	Features
0.1820705	AGE
0.1726605	HYPERTENSIN
0.1639233	OTHER_COM
0.1433021	CARDIOVASCULAR
0.0945617	DIABETES
0.0826513	COPD
0.0718882	CHRONIC_ KIDNEY
0.0387562	OBESITY
0.0351651	ASTHMA
0.0075744	IMMUSUPPR
0.0074468	SMOKING

5 Conclusions

Taking into consideration the novelty of Covid19, it has been necessary to urgently develop a high volume of research to gain knowledge about it, including the creation of prediction models of behavior and diagnosis of the disease. Furthermore, it has been essential to discover knowledge about it, highlighting the study of risk factors that increase its fatality. In this context, computational intelligence methods have been used in several pieces of research with positive results. In this chapter, some developed studies have been presented, including results based on fuzzy sets, metaheuristics, and different techniques for knowledge discovery.

References

1. Alfaro-Amieiro, M., Arias-Careaga, S., Gamba-Romero, A.: Agenda 2030: Claves Para la Transformación Sostenible: Investigación y Debate 256, Jun 2019
2. Fundación Seminario de Investigación para la Paz. El mundo que queremos: La Agenda 2030 (Estudios para la paz) – 1 oct 2016, (SIP) (2016)
3. Lorenzo, J.A.S.: El reto de cambiar el mundo: La Agenda 2030 de desarrollo sostenible – 7 dic 2015
4. Naciones Unidas: Report of the inter-agency and expert group on sustainable development goals indicators. Consejo Económico y Social, E/CN.3/2016/2/Rev.1, 19 de febrero (2016)
5. ONU: "Resolución A/RES/70/1 Transformar nuestro mundo: la Agenda 2030 para el Desarrollo Sostenible", 25 de noviembre de 2015. [edición electrónica] http://www.un.org/es/comun/docs/?symbol=A/RES/70/1.2015
6. Hu, Z., Ge, O., Li, S. Jin, L., Xiong, M.: Artificial Intelligence Forecasting of Covid-19 in China (2020). arXiv preprint arXiv:2002.07112
7. Yan, T., Wong, P.K., Ren, H., Wang, H., Wang, J., Li, Y.: Automatic distinction between COVID-19 and common pneumonia using multi-scale convolutional neural network on chest CT scans. Chaos Solitons Fract. **140**, 110153 (2020). https://doi.org/10.1016/j.chaos.2020.110153
8. Cruz-Corona, C.: Soft computing techniques and sustainability science, an introduction. In: Cruz Corona, C. (ed.) Soft Computing for Sustainability Science. Cham: Springer International Publishing, pp. 1–10 (2018)
9. Di Vaio, A., Palladino, R., Hassan, R., Escobar, O.: Artificial intelligence and business models in the sustainable development goals perspective: a systematic literature review. J. Bus. Res. **121**, 283–314 (2020)
10. Liu, R., Rong, Y., Peng, Z.: A review of medical artificial intelligence. Glob. Health J. **4**(2), 42–45 (2020)
11. Scardoni, A., Balzarini, F., Signorelli, C., Cabitza, F., Odone, A.: Artificial intelligence-based tools to control healthcare associated infections: a systematic review of the literature. J. Infect. Publ. Health **13**, 1061–1077 (2020)
12. Vaishya, R., Javaid, M., Khan, I.H., Haleem, A.: Artificial intelligence (AI) applications for COVID-19 pandemic. Diabetes Metab. Syndr. **14**, 337–339 (2020). https://doi.org/10.1016/j.dsx.2020.04.012
13. Vinod, D.N., Prabaharan, S.R.S.: Data science and the role of Artificial Intelligence in achieving the fast diagnosis of COVID-1. Chaos Solitons Fract. **140**, 110182 (2020). https://doi.org/10.1016/j.chaos.2020.110182
14. Bullock, J., Luccioni, A., Hoffmann-Pham, K., Nga-Lam, C.S., Luengo-Oroz, M.: Mapping the landscape of artificial intelligence applications against Covid-19. United Nations Global

Pulse, Durham University, Universite de Montreal, NYU Stern School of Business, World Health Organization, arXiv:2003.11336v1cs.CY. (2020)

15. Elavarasan, R.M., Pugazhendhi, R.: Restructured society and environment: a review on potential technological strategies to control the COVID-19 pandemic. Sci. Total Environ. **725**, 138858 (2020). https://doi.org/10.1016/j.scitotenv.2020.138858
16. Lalmuanawma, S., Hussain, J., Chhakchhuak, L.: Applications of machine learning and artificial intelligence for COVID-19 (SARS-CoV-2) pandemic: a review. Chaos Solitons Fractals **139**, 110059 (2020). https://doi.org/10.1016/j.chaos.2020.110059
17. Sipior, J.C.: Considerations for development and use of AI in response to COVID-19. Int. J. Inf. Manage. (2020). https://doi.org/10.1016/j.ijinfomgt.2020.102170
18. Swapnarekha, H., Behera, H.S., Nayak, J., Naik B.: Role of intelligent computing in COVID-19 prognosis: a state-of-the-art review. Chaos, Solitons Fract. **138**, 109947 (2020). https://doi.org/10.1016/j.chaos.2020.109947
19. Shaikh, F., Brun-Andersen, M., Sohail, M.R., Mulero, F., Awan, O., Dupont-Roettger, D., Kubassova, O., Dehmeshki, J., Bisdas, S.: Current landscape of imaging and the potential role for artificial intelligence in the management of COVID-19. Curr. Probl. Diagn. Radiol. **000**(1–6), 2020 (2020). https://doi.org/10.1067/j.cpradiol.2020.06.009
20. Suri, J.S., Puvvula, A., Biswas, M., Majhail, M., Saba, L., Faa, G., Singh, I.M., Oberleitner, R., Turk, M., Chadha, P.S., Johri, A.M., Sanches, J.M., Khanna, N.N., Viskovic, K., Mavrogeni, S., Laird, J.R., Pareek, G., Miner, M., Sobel, D.W., Balestrieri, A., Sfikakis, P.P., Tsoulfas, G., Protogerou, A., Misra, D.P., Agarwal, V., Kitas, G.D., Ahluwalia, P., Kolluri, R., Teji, J., Maini, M.A., Agbakoba, A, Dhanjil, S.K., Sockalingam, M., Saxena, A., Nicolaides, A., Sharma, A., Rathore, V., Ajuluchukwu,, J.N.A., Fatemi, M., Alizad, A., Viswanathan, V., Krishnan, P.K., Naidu, S.: COVID-19 pathways for brain and heart injury in comorbidity patients: a role of medical imaging and artificial intelligence-based COVID-19 severity classification: a review. Comput. Biol. Med. (2020) https://doi.org/10.1016/j.compbiomed.2020.103960
21. Kaushik, A.C., Raj, U.: AI-driven drug discovery: a boon against COVID-19? AI Open **1**, 1–4 (2020). https://doi.org/10.1016/j.aiopen.2020.07.001
22. Ren, Z., Liao, H., Liu Y.: Generalized Z-numbers with hesitant fuzzy linguistic information and its application to medicine selection for the patients with mild symptoms of the COVID-19. Comput. Indust. Eng. **145**, 106517 (2020). https://doi.org/10.1016/j.cie.2020.106517
23. Ocampo, L., Yamagishi, K.: Modeling the lockdown relaxation protocols of the Philippine government in response to the COVID-19 pandemic: An intuitionistic fuzzy DEMATEL analysis. Socioecon. Plann. Sci. (2020). https://doi.org/10.1016/j.seps.2020.100911
24. Togacar, M., Ergen, B., Comert, Z.: COVID-19 detection using deep learning models to exploit Social Mimic Optimization and structured chest X-ray images using fuzzy color and stacking approaches. Comput. Biol. Med. **121**, 103805 (2020). https://doi.org/10.1016/j.compbiomed.2020.103805
25. Gomes, R.R., Mariani, M.H.D.M., dos Santos, L.: Forecasting Brazilian and American COVID-19 cases based on artificial intelligence coupled with climatic exogenous variables. Chaos, Solitons and Fractals **139**, 110027 (2020). https://doi.org/10.1016/j.chaos.2020.110027
26. Salmeron, J.L., Arevalo, I.A.: Privacy-preserving, distributed and cooperative FCM-based learning approach for cancer research. In: Bello R et al (eds) Proceedings of IJCRS 2020, Lecture Notes in Artificial Intelligence LNAI 12179, pp. 477–487 (2020). https://doi.org/10.1007/978-3-030-52705-1_35
27. Yao, J., Azam, N.: Web-based medical decision support systems for three-way medical decision making with game-theoretic rough sets. IEEE Trans. Fuzzy Syst. **23**(1), 3–15 (2014)
28. Oyelade, O.N., Ezugwu, A.E.: A case-based reasoning framework for early detection and diagnosis of novel coronavirus. Inform. Med. Unlocked 20100395 (2020). https://doi.org/10.1016/j.imu.2020.100395
29. Chowell, G., Luo, R., Sun, K., Roosa, K., Tariq, A., Viboud C.: Real-time forecasting of epidemic trajectories using computational dynamic ensembles. Epidemics **30**, 100379 (2016). https://doi.org/10.1016/j.epidem.2019.100379

30. Guo, Q., Li, M., Wang, C., Wang, P., Fang, Z., tan, J., Wu, S., Xiao, Y., Zhu, H.: Host and infectivity prediction of Wuhan 2019 novel coronavirus using deep learning algorithm. bioRxiv (2020). https://doi.org/10.1101/2020.01.21.914044. https://www.biorxiv.org/content/early/2020/02/02/2020.01.21.914044.full.pdf
31. Mohammed, A.A., Al-qaness, A.A., Ewees, H.F., El Abd El Aziz, M.A.: Optimization method for forecasting confirmed cases of COVID-19 in China. J. Clin. Med. **9**(3), 674 (2020)
32. Mu, T.C., et al.: A mathematical model for simulating the phase-based transmissibility of a novel coronavirus. Infect. Dis. Poverty **9**, 24 (2020). https://doi.org/10.1186/s40249-020-00640-3
33. Viboud, C., Simonsen, L., Chowell, G.: A generalized-growth model to characterize the early ascending phase of infectious disease outbreaks. Epidemics **2016**(15), 27–37 (2016)
34. Magdon-Ismail, M.: Machine Learning the Phenomenology of COVID-19 From Early Infection Dynamics (2020). arXiv: 2003.07602v2
35. Guoping, Z., et al.: Forecasting and analysis of time variation of parameters of COVID-19 infection in China using an improved SEIR model. In: Public Meteorological Service Center of China Meteorological Administration (2020)
36. Wu, J.T., Leung, K., Leung, G.M.: Now casting and forecasting the potential domestic and international spread of the 2019-nCoV outbreak originating in Wuhan, China: a modelling study. The Lancet. (2020). https://doi.org/10.1016/S0140-6736(20)30260-9
37. Goodwin, P., Wright, G.: Enhancing strategy evaluation in scenario planning: a role for decision analysis. J. Manage. Stud. **38**, 1 (2001)
38. Sorousha, M., Bahri, P.A.: Hybrid intelligent scenario generator for business strategic planning by using ANFIS. Expert Syst. Appl. **36**, 7729–7737 (2009)
39. Sutcliffe, A.G., Gregoriades, A.: Automating scenario analysis of human and system reliability. IEEE Trans. Syst. Man Cybern. Part A: Syst. Hum. **37**(2), 249–261 (2007)
40. Kosow, H., Gaßner, R.: Methods of future and scenario analysis. In: Overview, Assessment, and Selection Criteria. DIE Research Project "Development Policy: Questions for the Future", p. 133 (2008). Studies/DeutschesInstitutfürEntwicklungspolitik. ISSN 1860-0468
41. Chowell, G.: Fitting dynamic models to epidemic outbreaks with quantified uncertainty: a primer for parameter uncertainty, identifiability, and forecasts. Infect. Dis. Model. (2017). https://doi.org/10.1016/j.idm.2017.08.001
42. Yang, Z.F., et al.: Modified SEIR and AI prediction of the epidemics trend of COVID-19 in China under public health interventions. J. Thoracic Disease **12**(3), 165–174 (2020). https://doi.org/10.21037/jtd.2020.02.64
43. Hernandez, A.R., Bello García, B., Bello García, M., García Lorenzo, M., Bello Pérez, R.: Análisis de escenario utilizando técnicas de Inteligencia Artificial para estudiar el posible comportamiento de la Covid-19. Revista Anales de la Academia de Ciencia de Cuba **10**(2), 831–846 (2020). ISSN: 2304-0106: http://www.revistaccuba.cu/index.php/revacc/article/view/831/846
44. Li, Y. et al.: Prediction of criticality in patients with severe Covid-19 infection using three clinical features: a machine learning-based prognostic model with clinical data in Wuhan (2020). preprint doi: https://doi.org/10.1101/2020.02.27.20028027
45. García-Lorenzo, M., Rodriguez-Alvarez, Y., Ramon-Hernandez, A., Bello García, B., Filiberto, Y., Rosete, A., Caballero, Y., Bello, R.: Adquisición de conocimiento sobre la letalidad de la Covid-19 usando técnicas de Inteligencia Artificial. Revista Anales de la Academia de Ciencia de Cuba. **10**(3), 891–912 (2020). ISSN: 2304-0106. Disponible: http://www.revistaccuba.cu/index.php/revacc/article/view/891/912
46. Caballero, Y., Bello, R., et al.: Knowledge discovery using rough set theory. In: Koronacki, J., Ras, Z.W., Wierzchon, S.T., Kacprzyk, J. (eds.) Advances in Machine Learning I Dedicated to the memory of Professor Ryszard S. Michalski in Series: Studies in Computational Intelligence. ISBN 978-3-642-05176-0. **262**, 367–383 (2010)
47. Bello-García, B. et al.: Implementación de métodos para el pre-procesamiento de datos usando Teoría de los conjuntos aproximados (RST) en Python. Memoria de la Conferencia Internacional de Procesamiento de la Información (CIPI2019), Cuba, ISBN 978-959-312-372-3 (2019)

48. Montavon, G., Samek, W., Müller, K.R.: Methods for interpreting and understanding deep neural networks. Digit. Signal Proces. **73**, 1–15 (2018)
49. Hamsagayathri, P., Sampath, P.: Priority based decision tree classifier for breast cancer detection. In: Advanced Computing and Communication Systems (ICACCS). 2017 4th International Conference. IEEE, pp. 1–6 (2017). https://doi.org/10.1109/ICACCS.2017.8014598
50. Jhajharia, S., Verma, S., Kumar, R.: A cross-platform evaluation of various decision tree algorithms for prognostic analysis of breast cancer data. In: En Inventive Computation Technologies (ICICT), International Conference, vol. 3, pp. 1–7 (2016). IEEE. https://doi.org/10.1109/INVENTIVE.2016.7830107
51. Bashir, U., Chachoo, M.A.: Performance evaluation of J48 and Bayes algorithms for intrusion detection system. Int. J. Netw. Secur. Appl. **9**, 01–11 (2017)
52. Augasta, M.G., Kathirvalavakumar, T.: Reverse engineering the neural networks for rule extraction in classification problems. Neural Process. Lett. **35**(2), 131–150 (2012)
53. Bello-García, M., García-Lorenzo, M.M., Bello, R.: A method for building prototypes in the nearest prototype approach based on similarity relations for problems of function approximation. Lectures Notes on Computer Sciences 7629 (2012)
54. Filiberto, Y., et al.: An analysis about the measure quality of similarity and its applications in machine learning. In: Fourth International Workshop on Knowledge Discovery, Knowledge Management and Decision Support. Atlantis Press (2013)
55. Rodríguez, Y., et al.: Similar prototype methods for class imbalanced data classification. In: Uncertainty Management with Fuzzy and Rough Sets. Springer, pp. 193–209 (2019)

Quantification of the Risk of ASF Appearance Using OWA Operators

E. H. Cables, F. Moreno, M. T. Lamata, and F. Gómez

Abstract African Swine Fever (ASF) is an infectious and devastating viral disease that affects pigs. Measures to control this disease may have critical socioeconomic consequences, especially for low and mediumincome countries. Therefore, the implementation of actions to control ASF may positively impact several Sustainable Development Goals, such as ending poverty, zero hunger, and responsible consumption and production. Countries that are yet to be affected by the disease suffer from a lack of historical data with which to devise policies to prevent the appearance of ASF. This work introduces a novel methodology to quantify risk factors related to ASF appearing using expert knowledge. The methodology is based on operator weight average, in which weights are determined by using linguistic labels and a consensus mechanism. The proposed approach does not require particular quantitative knowledge from the experts about how weights are determined. The methodology was used to quantify the relevance of 21 different factors and 6 categories (groups of factors) linked to the appearance of ASF in Colombia, using the opinions of 15 domain experts. The results suggest that this approach may help quantify factors and categories without relying on any historical data of occurrence, resulting in consistent quantifications similar to those reported in the literature.

Keywords Decision making · African Swine Fever · Sustainable development goals · Operator weight average · Preference quantifiers

---

E. H. Cables
Universidad Antonio Nariño, Bogotá, Bogota, Colombia
e-mail: ehcables@uan.edu.co

F. Moreno · F. Gómez (✉)
Universidad Nacional de Colombia, Bogota, Colombia
e-mail: fagomezj@unal.edu.co

F. Moreno
e-mail: fcmorenov@unal.edu.co

M. T. Lamata
Universidad de Granada, 18071 Granada, Spain
e-mail: mtl@decsai.ugr.es

J. L. Verdegay et al. (eds.), *Computational Intelligence Methodologies Applied to Sustainable Development Goals*, Studies in Computational Intelligence 1036,
https://doi.org/10.1007/978-3-030-97344-5_6

1 Introduction

African Swine Fever (ASF) is a highly contagious viral disease that affects pigs and generates a substantial economic impact. The disease is characterized by its high dissemination capacity and high mortality rate. Therefore, the presence of ASF presence in a territory requires severe sanitary control measures to stop the virus spreading, including the mass slaughter of animals, limiting production, and the prohibition of the commercialization of products derived from animals [20]. These control measures may have catastrophic socioeconomic consequences for small and mediumscale pig producers. In particular, this may occur in areas with inadequate biosecurity measures, which commonly cover the pork demand in domestic markets of developing countries.

To day, ASF has affected some countries. However, in many low and mediumincome countries, it is still rare [11]. For instance, in the Andean Community, there are few reports of its presence. In countries in these regions, where the pork economic sector is mainly composed of small producers, the appearance of the disease would generate negative scenarios beyond animal health. For example, the presence of ASF may increase risks related to food safety and may significantly affect production systems [13].

In this context, the design and implementation of actions to prevent ASF appearance may positively impact several Sustainable Development Goals (SDG) [1]. In particular, these actions may contribute to avoid poverty (SDG 1: End Poverty), reducing the cases of sick animals, helping to eradicate hunger by contributing to the supply of affordable protein, improving animal health and welfare in humans, reducing inequalities, and fostering economic growth. Additionally, these measures may help ensure the proper functioning of this value chain, contributing positively to animal health, impacting the meat production and productivity levels, contributing to food safety (SDG 2: Zero Hunger). Finally, due to its direct impact on production and the possible ways in which the presence of the disease damages the productive apparatus, preventing the disease improves food safety and reduces adverse effects (SDG 12: Responsible Consumption and Production).

The design of objective policies for preventing and controlling animal diseases relies on objectively quantifying the disease risks. The risk in animal health refers to something likely to be harmful or contribute to the animal's illness. The risks can be categorized depending on the disease degree of advance, for instance, appearance, contagion, and dissemination risks. For their quantification, these risks are commonly associated with different factors [12]. For example, the risk of ASF appearance links to quantifiable factors, such as proximity to airports and low biosecurity measures in animal productions. These factors are commonly grouped into categories, such as environmental, economic, and social, resulting in a hierarchical organization suitable to perform the quantification tasks [21].

The relevance of particular categories and factors depends on individual local characteristics that may vary across different territories. The most common strategy to quantify these risks is to use historical records on disease occurrences and weigh the relevance of factors and categories based on odd ratios. However, there are no historical records to perform such a quantification in cases where the disease has not appeared yet. This lack of knowledge about the relevance of factors and categories limits the possibility of achieving an objective quantification of the risk of ASF appearance [7].

This work studies how expert knowledge on the current territory situation and the disease's features may help quantify the risk of ASF appearance in the absence of reports of the disease presence. In particular, we aimed to formulate a new methodology for quantifying the relevance of risk factors and categories based on the knowledge of experts in the area. This methodology is applied to quantify the risk of ASF appearing, or indeed other animal and human diseases. The quantification is constructed by combining weighted values of a set of categories and individual risk factors associated with the disease appearance. These weights are obtained through linguistic labels defined by a group of experts in the area. This strategy defines levels of importance for different factors linked to the disease. The proposed methodology was applied to determine the risk of the appearance of ASF in Colombia, using data from a group of domain experts, and it is currently being used as a reference for decision making in the pork sector in Colombia.

2 Materials and Methods

The proposed strategy aims to quantify the risk of ASF appearing, and the relevance of different factors and categories related to the disease. For this, we proposed a method aimed at determining a set of weights or relevancies for a group of predefined categories (factors) related to the appearance of ASF. For each expert, weights that reflected their individual opinions about each category and factor significance were first determined. These weights were combined into a consensus vector used to weight quantitative values reflecting the risk in each category (factor). Values on each factor were hierarchically combined into a single and general risk quantitative value used by the expert to drive decision making in controlling the disease.

2.1 Operator Weight Average

Yager [26] proposes a family of aggregation operators called OWA, based on the ordered weights average principle. The operator (OWA) considers a vector of weights to perform the aggregation process, which fulfills the following properties:

$$w_i \in [0, 1] \tag{1}$$

$$\sum_{i=1}^{n} w_i = 1 \tag{2}$$

To determine the weights vector associated with the OWA operator, different methods and procedures have been developed, such as:

- Mathematical programming models: which take expressions (1) and (2) as a reference, in addition to some properties of the OWA operator, such as entropy and Orness, to determine the weights vector [19, 22].
- Oriented by quantifiers: it bases its working principle using fuzzy sets, and distinguishes two classes of linguistic quantifiers, absolute and relative [28].
- Analytical functions: Use analytical functions to determine the weights vector [2, 3, 9, 16, 23].
- Algorithmic procedures: Through a sequence of steps the weights vector is obtained [10, 24, 25].
- Operator Based on Ratio Scale: This method is based on the specification of a vector that represents the proportionality ratio between weights [4].
- The equidifferent OWA operator: This method is characterized by obtaining a vector of weights such that the adjacent weights have a common difference [18].
- The priority of the criteria: Called a prioritized average operator, where there is an order of priority among the criteria, thus, without loss of generality, it can be assumed that there is a linear order [27, 28].

These weights were used to model the relevance of each category (factor) related with the risk of ASF appearing.

2.2 OWA Weights Based on Preference Quantifiers

Several strategies have been proposed to determine the OWA weights. Most of them require a solid quantitative background to be used by experts. The aim here was to quantify expert knowledge about the ASF disease. However, these experts were not familiar with the quantitative strategies for weight determination. Therefore, the preference quantifier with linguistic labels was used to indicate how important one category (factor) was compared to another. In particular, the *Algorithm 1* proposed by Lamata and Cables [15] was used to obtain a weight vector for n categories (factors).

Algorithm 1. Computation of individual weights in OWA operators

1. Computation of the initial weights using the Borda-Kendall expression:

$$w_i = \frac{2(n+1-i)}{n(n+1)} \tag{3}$$

2. Computation of the vector of distances between adjacent weights (4) and the vector of maximum weights (5):

$$D = (d_1, d_2, \ldots, d_{n-1}, 0), \; where\, d_i = \begin{cases} w_i - w_{i+1} & if\ i \neq n \\ 0 & if\ i = n \end{cases} \tag{4}$$

$$W_1 = (w_1, w_1, \ldots, w_1) \tag{5}$$

3. Computation of a quantity vector from the preference quantified relationship between the Criteria.

Definition 1 Let $L = (l_0, l_1, \ldots, l_h)$ be a set of linguistic labels which define a relationship of quantified preference. Let $l_p \in L$ be a linguistic label, where $p = 0, 1, \ldots, h$ then the function of quantity δ is:

$$\delta : L \to [0, 1] \tag{6}$$

with $\delta(l_p) = \frac{1}{h}.p$, such that p is the subindex of the linguistic label l_p in the set L.

After defining the set of linguistic labels and a set of ordered criteria it is possible to define a quantified relationship between consecutive criteria using the following expression:

$$\psi : C^n \to L^{n-1}$$

$$\psi\,(C_1, \ldots, C_n) = (R_1, \ldots, R_{n-1}) \tag{7}$$

where $R_i \in L$, with $i = 1, \ldots, n-1$ represents the preference relationship quantified between two criteria C_j and C_{j+1}, and is represented by $C_j R_i C_{j+1}$.

Finally, the quantity vector is obtained using:

$$\lambda : L^{n-1} \to [0, 1]^{n-1}$$

$$\lambda\,(R_1, \ldots, R_{n-1}) = (\delta(R_1), \ldots, \delta(R_{n-1})) = (v_1, \ldots, v_{n-1}) \tag{8}$$

4. Compute the redistribution vector as:

$$V_{CD} = (f_d, f_d, \ldots, f_d)$$

5. Compute the vector of adjustment differences.

$$D^* = \left(M^*_{\Delta inf}.D^T\right)^T + V_{CD} \tag{9}$$

6. Compute the final weight vector.

$$W^* = W_1 - D^* \tag{10}$$

2.3 Consensus of Experts' Individual Weight Vectors

The following strategy was used in order to obtain a consensus weight vector out of the individuals vectors provided by each expert:

1. For each expert steps 1 and 2 in *Algorithm 1* were computed.
2. The quantity vector for each expert was determined based on the linguistic labels.
3. The consensus vector is computed as follows:

$$\overline{V} = \frac{1}{m}\left(\sum_{j=1}^{m} v_1^j, \sum_{j=1}^{m} v_2^j, \ldots, \sum_{j=1}^{m} v_{n-1}^j\right) = (\overline{v}_1, \overline{v}_2, \ldots, \overline{v}_{n-1}) \qquad (11)$$

4. Steps 4–6 were computed (*Algorithm 1*) using the consensus vector $\overline{V}$.

2.4 Quantifying Risk Using a Hierarchical Representation

A hierarchical scheme for integrating factors and experts' weights provided a quantitative representation for the risk of disease appearance. As previously discussed, ASF appearance risk depends on several factors that can be independently quantified for particular territories. The quantification of risk per category resulted from a linear weighting of factors in the same category using the consensus of experts' factor relevancies. Similarly, the appearance risk resulted from a linear weighting of all categories quantification also using the consensus weights of categories constructed using the experts' knowledge. Figure 1 illustrates the hierarchical representation using for the aggregation of information of particular factors related to ASF appearance.

It is worth noting that two kinds of weights were computed using the expert input weights for categories and weights for factors. Importantly, both types of weights verify the $\sum_{i=1}^{n} w_i = 1$ property.

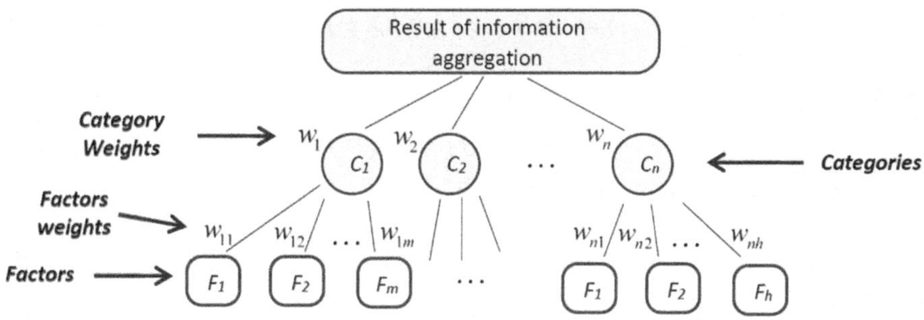

Fig. 1 Hierarchical structure for categories and factors used to aggregate individual factor information

Table 1 Table for the preference orders proposed by the experts

	Order of categories/Factors				
Categories/factors risks	1	2	…	n	Preference index
C1 o F1			…		
C2 o F2			…		
…	…	…	…	…	…
Cn o Fn			…		

In order to determine the weights for categories and factors, the following process was considered:

1. Categories (Factors) order determination, a set of 6 categories and 21 factors (6 factors for category 1; 2 factors for category 2; 9 factors for category 3; 0 factors for category 4; 2 factors for category 5; and 2 factors for category 6) were characterized from the literature used. Factors and categories were chosen with the help of expert animal epidemiologists to characterize the local territory conditions. Secondly, each expert defined a preference of order of among the categories, i.e., each expert defined which was the most relevant category for appearance of ASF, the second most important, etc. A similar ordering process was performed for the factors inside each category. No two categories (factors) were allowed to have the same relevance. Therefore, for each expert a set of orders $(1, 2, \ldots, n)$ was obtained per category (factor).
2. Categories (Factors) tabulation. The orders proposed by the experts for the different categories and factors were represented into tables containing the number of experts that preferred a certain order for each category (factor), as ahown Table 1.
   Each cell in Table 1 stored the number of times each category (factor) was selected by the set of experts. The last column contains a preference index defined as:

$$Ipt_i = \sum_{j=1}^{n} (n + 1 - j)\, x_{ij} \tag{12}$$

where:

- n, is the number of preference or order.
- x_{ij}, is the number of times the category C_i or risk factor was selected in the order j.

This preference index summarizes the preferences given by the set of experts for each category (factor). The preference indices computed for each category (factor) were arranged in descending order. Importantly, these tables were computed

for each category, containing the corresponding factors, and for the appearance risk factor, containing the complete set of categories.
3. Categories (factors) weights computation, the method described in Sect. 2.2 was used to assign weights to the categories (factors).

The resulting weights were linearly combined with the factors individual quantification computed for each territory, resulting in a quantification for the different categories and a single quantification for the ASF appearance.

3 Results and Discussion

Our aim was to quantify the risk of ASF appearance in Colombia. In this section, results related to this quantification are organized as follows. First, the preference indices obtained for the different categories and factors are reported. Later, the weights computed for the factors and categories are reported and described.

3.1 OWA Weights Based on Preference Quantifiers

A total of 15 experts were considered for quantifying categories and factors relevance. These experts were selected based on thier experience in disease management in Colombia and their knowledge of diseases and included animal epidemiologists, zootechnicians, veterinarians, among others, from the academy, the pork industry, and the sanitary authority in Colombia. A total of 6 categories and 21 factors were considered.

For each expert, a total of 6 selections of orders were captured, 5 for each category, and 1 selection of orders for the 6 categories. There was a category without any factor that was not presented to the experts. The orders preferred by experts were captured using a GoogleForms survey, which allowed each expert to inform their preferences about the order of each category. The existence of records for all experts and preferences was manually verified. Because of the structure of the survey used, there were no anomalous or absent registers.

As observed (see Table 2), the proposed method allowed to quantify the relevance of each category and factor considered herein. Importantly, these indices reflect the consensus of a set of experts about the significance of each factor. Commonly this relevancy is determined using historical data of disease occurrences, which is missing here. Notably, the order herein obtained resembles previous orders computed for the same disease. In particular, different studies have reported biosecurity issues as being one of the most relevant factors for the appearance of ASF [5, 6, 8, 14, 17].

Subsequently, the Categories and Risk Factors were ordered, and weights were assigned, based on the application of the algorithm described in Sect. 2.3, so that:

Table 2 shows the categories and factors herein considered together with the preference indices resulting from the order determination process

Category	Preference index	Factors	Preference index
Biosecurity	85	Feeding type	83
		Infrastructure and use of facilities	74
		Indirect contact associated with people	68
		Dead animal management	63
		Management of breeding animals	55
		Keeping other animals on the pig farm	53
Animal mobilization	84	Animal mobilization	27
		Mobilization of products and by products of the pork meat chain	26
Farm location	64	Proximity to borders	122
		Proximity to ports, airports, and others	118
		Proximity to and density of pig farms	114
		Proximity to roads	111
		Proximity to dumps and landfills	109
		Proximity to population centers	103
		Proximity to trade fairs and exhibitions, auctions, and auctions houses	102
		Proximity to abattoirs/slaughter slabs	93
		Proximity to packing and processing plants of the pork meat chain	92
Habitat	59		
Pig management systems	58	Pig production type	23
		Type of swine farm and intensification of swine production	21
Productive and socioeconomic environment	57	Entry of swine, genetic material, meat, meat by products, pig products and nonedible meat products	29
		International movement of people	25

Table 3 The vectors obtained in steps 1 and 2 of the algorithms described in Sect. 2.2

Types of vectors	Vectors
Linear weights vector (Borda-Kendall)	$W = (0.286, 0.238, 0.19, 0.143, 0.095, 0.048)$
Distance vector between adjacent weights	$D = (0.048, 0.048, 0.048, 0.048, 0.048, 0)$
Maximum weights vector	$W_1 = (0.286, 0.286, 0.286, 0.286, 0.286, 0.286)$

1. For each expert, steps 1 and 2 of the algorithms described in Sect. 2.2 were carried out, and the weights vector was obtained through the Borda-Kendall expression, the distance vector between adjacent weights and the maximum weights vector, for all categories and factors. In this case, only the three vectors are shown for the Categories, which total six (see Table 3).

Table 4 Linguistic labels and their meaning

Linguistic labels	Meaning of the preference relationship $C_i l_p C_{i+1}$
$l_0 = \{\cong\}$	C_i and C_{i+1} have the same preference
$l_1 = \{>^1\}$	C_i is more preferred than C_{i+1}
$l_2 = \{>^2\}$	C_i has a greater preference C_{i+1}
$l_3 = \{>^3\}$	C_i is much more preferred than C_{i+1}
$l_4 = \{>^4\}$	C_i has the highest preference over C_{i+1}

Table 5 Quantified preference relationship issued by 15 experts and their respective quantity vector, for the 6 categories

Experts	Quantified preference relationship between the categories	Quantity vectors
Expert-1	$C_1 \cong C_2 >^2 C_3 >^2 C_4 >^2 C_5 >^4 C_6$	(0, 0.5, 0.5, 0.5, 1)
Expert-2	$C_1 >^3 C_2 >^2 C_3 >^4 C_4 \cong C_5 >^2 C_6$	(0.75, 0.5, 1, 0, 0.5)
Expert-3	$C_1 \cong C_2 >^4 C_3 >^2 C_4 >^1 C_5 >^2 C_6$	(0, 1, 0.5, 0.25, 0)
Expert-4	$C_1 >^4 C_2 >^3 C_3 >^2 C_4 >^1 C_5 \cong C_6$	(1, 0.75, 0.5, 0.25, 0)
Expert-5	$C_1 >^3 C_2 >^2 C_3 \cong C_4 >^1 C_5 >^2 C_6$	(0.75, 0.5, 0, 0.25, 0.5)
Expert-6	$C_1 >^3 C_2 >^3 C_3 >^1 C_4 \cong C_5 \cong C_6$	(0.75, 0.75, 0.25, 0, 0)
Expert-7	$C_1 \cong C_2 >^3 C_3 >^4 C_4 \cong C_5 >^1 C_6$	(0, 0.75, 1, 0, 0.25)
Expert-8	$C_1 \cong C_2 >^3 C_3 >^4 C_4 \cong C_5 >^1 C_6$	(0, 0.75, 1, 0, 0.25)
Expert-9	$C_1 >^1 C_2 >^2 C_3 \cong C_4 >^1 C_5 >^3 C_6$	(0.25, 0.5, 0, 0.25, 0.75)
Expert-10	$C_1 \cong C_2 >^3 C_3 \cong C_4 >^1 C_5 >^2 C_6$	(0, 0.75, 0, 0.25, 0.5)
Expert-11	$C_1 \cong C_2 >^3 C_3 \cong C_4 >^1 C_5 >^2 C_6$	(0, 0.75, 0, 0.25, 0.5)
Expert-12	$C_1 >^2 C_2 >^2 C_3 \cong C_4 >^1 C_5 \cong C_6$	(0.5, 0.5, 0, 0.25, 0)
Expert-13	$C_1 \cong C_2 >^1 C_3 >^3 C_4 >^3 C_5 >^1 C_6$	(0, 0.25, 0.75, 0.75, 0.25)
Expert-14	$C_1 >^3 C_2 >^3 C_3 >^1 C_4 \cong C_5 \cong C_6$	(0.75, 0.75, 0.25, 0, 0)
Expert-15	$C_1 >^4 C_2 \cong C_3 >^2 C_4 >^2 C_5 \cong C_6$	(1, 0, 0.5, 0.5, 0)

2. Each expert indicated how important each category (Factor) is with respect to another, using the linguistic labels defined in Table 4.
   To capture the corresponding data, a GoogleForms survey was used for the same experts. The captured data was verified manually, and the existence of all records was verified.
   This data collected allowed to obtain a vector of quantity for each expert (see Table 5), which was constructed for values belonging to the unit interval.
3. The consensus of the quantity vectors of the experts was calculated using expression (11).

$$\overline{V} = (0.38, 0.6, 0.42, 0.23, 0.33)$$

Table 6 Weights of the categories and risk factors calculated with the algorithm in Sect. 2.3, where 15 experts participate

No.	Category	Weights	No.	Factors	Weights
1	Biosecurity	0.21720	1	Feeding type	0.21005
			2	Infrastructure and use of facilities	0.18624
			3	Indirect contact associated with people	0.17751
			4	Dead animal management	0.16323
			5	Management of breeding animals	0.14181
			6	Keeping other animals on the pig farm	0.12116
2	Animal mobilization	0.19894	1	Animal mobilization	0.55278
			2	Mobilization of products and by products of the pork meat chain	0.44722
3	Farm location	0.17037	1	Proximity to borders	0.14070
			2	Proximity to ports, airports, and others	0.13403
			3	Proximity to and density of pig farms	0.12292
			4	Proximity to roads	0.11514
			5	Proximity to dumps and landfills	0.10996
			6	Proximity to population centers	0.10255
			7	Proximity to trade fairs and exhibitions, auctions, and auctions houses	0.09774
			8	Proximity to abattoirs/slaughter slabs	0.09070
			9	Proximity to packing and processing plants of the pork meat chain	0.08626
4	Habitat	0.15053			
5	Pig management systems	0.13942	1	Pig production type	0.53889
			2	Type of swine farm and intensification of swine production	0.46111
6	Productive and socioeconomic environment	0.12354	1	Entry of swine, genetic material, meat, meat by products, pig products and nonedible meat products	0.58611
			2	International movement of people	0.41389

4. Steps 4–6 of the algorithm described in session 2.2 are performed using the quantity vector $\overline{V}$, which enabled the weights vector for the Categories to be obtained.

This algorithm was also applied for each of the Risk Factors, which enabled the weights for the Categories and Risk Factors to be obtained (see Table 6).

As observed Biosecurity was revealed as the most relevant category for the appearance of ASF in Colombia, followed by animal mobilization and biophysical space. The weights linked to these categories probably reflect the importance of the ASF transmission mechanisms and the importance of having good biosecurity measurements in the appearance of the disease.

Importantly, the proposed method also provides individual information on the weights inside each category. For instance, in the Biosecurity category, the most critical factor affecting ASF contagion risk was feeding type. This factor is related to the use of inadequate practices for feeding animals, such as feed scraps or human food waste to feed pigs. ASF virus may survive these wastes, which animals can consume, thereby increasing the risk of transmission. In contrast, the second most relevant factor is related to the bio protection measurements of the productions. Notably, the proposed method disentangles both factors based on expert knowledge to obtain quantitative relevancies for both aspects.

3.2 Another Variant of Solution

The example solved above performs the calculation of the weights associated with the Risk Criteria and Factors following the algorithmic flow shown in Fig. 2, where the process of unifying the information of the experts is carried out at the end of step 3, with obtaining the consensus of the Quantity vectors and from there, proceeding with the remaining steps of the algorithm. However, it is also possible to calculate the weights vector by applying the algorithmic flow described in Fig. 3.

When the algorithmic flow described in Fig. 3 is used, the algorithm shown in Sect. 2.2 is applied for each expert and the consensus of all the weight vectors is obtained through expression (13).

$$\overline{W} = \frac{1}{m}\left(\sum_{j=1}^{m} w_1^j, \sum_{j=1}^{m} w_2^j, \ldots, \sum_{j=1}^{m} w_n^j\right) = (\overline{w}_1, \overline{w}_2, \ldots, \overline{w}_n) \tag{13}$$

By applying the algorithmic flow represented in Fig. 3, then the weights vectors of the categories according to the judgments of each expert are obtained (see Table 7).

Finally, by applying expression (13) to the set of weights vectors shown in Table 7, the resulting weights vector $\overline{W}$ through the consensus of the weights vectors considered by each expert is obtained.

$$\overline{W} = (0.21719, 0.19894, 0.17037, 0.15053, 0.13942, 0.12355)$$

When comparing the weights vectors obtained through the two algorithms (see Table 8), it can be observed that the greatest difference between the elements of the weights vectors was very small (<0.00001), therefore, it can be concluded that the weights vectors are almost equal. Given this result, it can be stated that it is feasible

Fig. 2 Algorithms *A* to determine the consensus weights vector

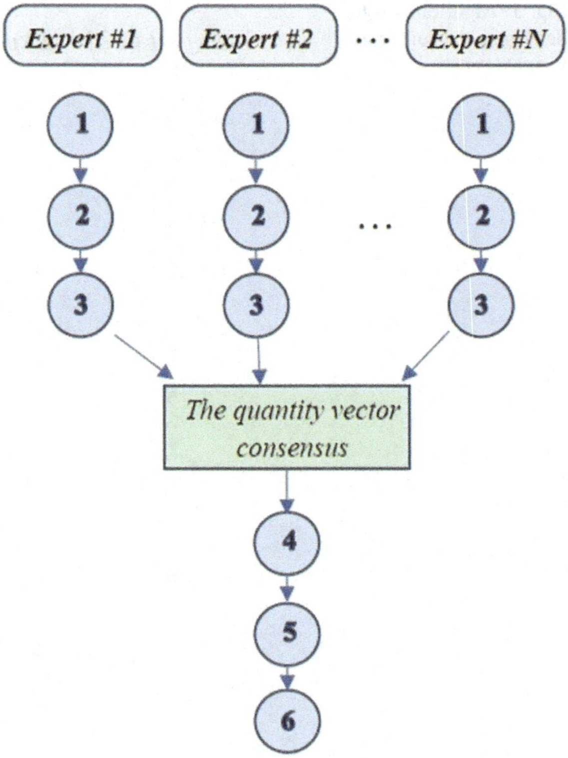

to apply the algorithm of Fig. 2a, since it performs a smaller number of steps to determine the weights vector based on the judgments of the experts.

4 Conclusions

We presented a novel method to quantify the risk factors of the appearance of ASF. The method is based on the construction of OWA operators using linguistic labeling provided by experts to reflect their preferences on relevancies of categories and factors related to the appearance of the disease. Two variants of the consensus algorithm were studied, resulting in almost the same result. Importantly, the proposed methodology is extendible to the quantification of other diseases.

The proposed method quantifies categories and risk factors for ASF appearing without relying on historical data of disease occurrences and without requiring from the particular expert any quantitative knowledge about the weights assignation. In places where there is no historical data for performing a statistical analysis of risk factors, this approach may serve as an alternative to design preventive control mea-

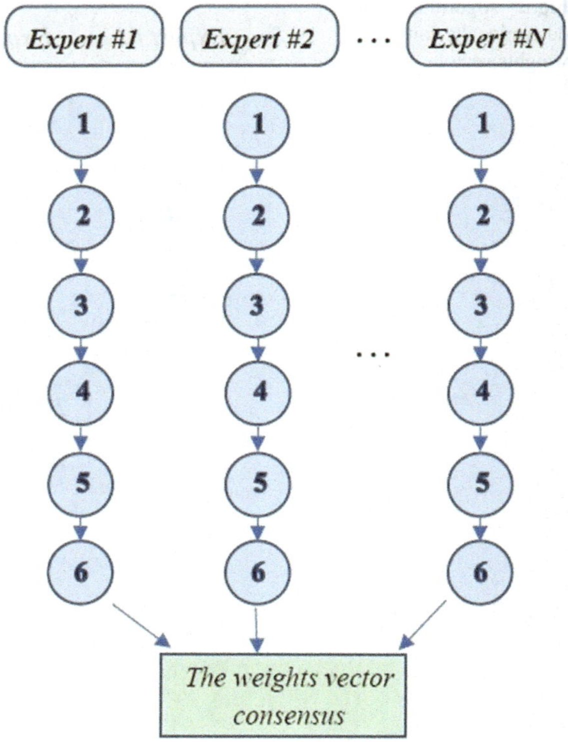

Fig. 3 Algorithms *B* to determine the consensus weights vector

Table 7 The weights vectors of the categories from the quantity vector obtained for each expert

Experts	Quantity vectors	Weights vector
Expert-1	(0, 0.5, 0.5, 0.5, 1)	(0.21032, 0.21032, 0.18651, 0.1627, 0.13888, 0.09127)
Expert-2	(0.75, 0.5, 1, 0, 0.5)	(0.24008, 0.20436, 0.18055, 0.13294, 0.13294, 0.10912)
Expert-3	(0, 1, 0.5, 0.25, 0)	(0.21825, 0.21825, 0.17063, 0.14683, 0.13493, 0.11111)
Expert-4	(1, 0.75, 0.5, 0.25, 0)	(0.24603, 0.19841, 0.16271, 0.13889, 0.12698, 0.12698)
Expert-5	(0.75, 0.5, 0, 0.25, 0.5)	(0.22024, 0.18453, 0.16071, 0.16071, 0.14881, 0.125)
Expert-6	(0.75, 0.75, 0.25, 0, 0)	(0.22619, 0.19047, 0.15476, 0.14286, 0.14286, 0.14286)
Expert-7	(0, 0.75, 1, 0, 0.25)	(0.21627, 0.21627, 0.18055, 0.13294, 0.13294, 0.12103)
Expert-8	(0, 0.75, 1, 0, 0.25)	(0.21627, 0.21627, 0.18055, 0.13294, 0.13294, 0.12103)
Expert-9	(0.25, 0.5, 0, 0.25, 0.75)	(0.20238, 0.19047, 0.16667, 0.16667, 0.15476, 0.11905)
Expert-10	(0, 0.75, 0, 0.25, 0.5)	(0.19841, 0.19841, 0.1627, 0.1627, 0.15079, 0.12699)
Expert-11	(0, 0.75, 0, 0.25, 0.5)	(0.19841, 0.19841, 0.1627, 0.1627, 0.15079, 0.12699)
Expert-12	(0.5, 0.5, 0, 0.25, 0)	(0.20634, 0.18254, 0.15873, 0.15873, 0.14683, 0.14683)
Expert-13	(0, 0.25, 0.75, 0.75, 0.25)	(0.20635, 0.20635, 0.19444, 0.15873, 0.12302, 0.11111)
Expert-14	(0.75, 0.75, 0.25, 0, 0)	(0.22619, 0.19048, 0.15476, 0.14285, 0.14286, 0.14286)
Expert-15	(1, 0, 0.5, 0.5, 0)	(0.22619, 0.17857, 0.17857, 0.15477, 0.13095, 0.13095)

Table 8 Weights vectors obtained through the algorithms shown in Figs. 2 and 3

Algorithms	Weights vectors obtained
Figure 2	(0.21720, 0.19894, 0.17037, 0.15053, 0.13942, 0.12354)
Figure 3	(0.21719, 0.19894, 0.17037, 0.15053, 0.13942, 0.12355)

surements. These quantifications may address decision making, and positively impact several SDG.

Future work may include increasing the number of experts involved in the study, studying the stability of results, and integrating with quantitative values computed for each factor.

Acknowledgements The support offered by PorkColombia-FNP, the Data Analytics Laboratory of the National University of Colombia (DataLab) and the Antonio Nariño University of Colombia, as well as the support of the project PID2020-112754GB-I00 from the Spanish Govern financed with FEDER funds are appreciated.

References

1. Adamowicz, M.: Agricultural development processes in the context of globalization challenges and new approaches to the concept of sustainable development (2021). Zagadnienia Ekonomiki Rolnej/Problems Agric. Econ. **1** (2021)
2. Ahn, B.S.: On the properties of OWA operator weights functions with constant level of orness. IEEE Trans. Fuzzy Syst. **14**, 511–515 (2006)
3. Ahn, B.S., Park, C.: Least-squared ordered weighted averaging operator weights. Int. J. Intell. Syst. **23**, 33–49 (2008)
4. Cables, E., Lamata, M.T., Verdegay, J.L.: Relation between AHP and operators based on different scales. In: Calvo Sánchez, T., Torrens Sastre, J. (eds.) Fuzzy Logic and Information Fusion. Studies in Fuzziness and Soft Computing, vol. 339. Springer, Cham (2016)
5. Costard, S., Jones, B.A., Martínez-López, B., Mur, L., de la Torre, A., Martínez, M., Wie-land, B.: Introduction of African Swine Fever into the European Union through Illegal Importation of Pork and Pork Products. PLoS One **8**(4) (2013)
6. Dione, M.M., Akol, J., Roesel, K., Kungu, J., Ouma, E.A., Wieland, B., Pezo, D.: Risk factors for African swine fever in smallholder pig production systems in Uganda. Transbound Emerg. Dis. **64**, 872–882 (2017)
7. Dufour, B., Plee, L., Moutou, F., Boisseleau, D., Chartier, C., Lancelot, R., Toma, B.: A qualitative risk assessment methodology for scientific expert panels. Revue scientifique et Technique - Office International des Epizooties **30**(3), 673–681 (2011)
8. Fasina, F.O., Agbaje, M., Ajani, F.L., Talabi, O.A., Lazarus, D.D., Gallardo, C., Bastos, A.D.S.: Risk factors for farm-level African swine fever infection in major pig-producing areas in Nigeria, 1997–2011. Prev. Vet. Med. **107**(1), 65–75 (2012)
9. Filev, D.P., Yager, R.R.: On the issue of obtaining OWA operator weights. Fuzzy Set Syst. **94**, 157–169 (1998)
10. Fullér, R., Majlender, P.: An analytic approach for obtaining maximal entropy OWA operator weights. Fuzzy Sets Syst. **124**, 53–57 (2001)
11. Giraldo-Ramirez, S., Rendon-Marin, S., Ruiz-Saenz, J.: A concise review on certain important veterinary viruses in the Americas. Revista MVZ Córdoba **26**, e1965 (2021)

12. Hertz-Picciotto, I.: Epidemiology and quantitative risk assessment: a bridge from sci-ence to policy. Am. J. Public Health **85**(4), 484–491 (1995)
13. Jongeneel, R., Gonzalez Martinez, A., Hoste, R.: An uncertain fate for the EU pig sector: potential consequences of the 2019 African Swine Fever Outbreak in East Asia. EuroChoices **20**, 22–29 (2021)
14. Kabuuka, T., Kasaija, P.D., Mulindwa, H., Shittu, A., Bastos, A.D.S., Fasina, F.O.: Drivers and risk factors for circulating African swine fever virus in Uganda, 2012–2013. Res. Vet. Sci. **97**(2), 218–225 (2014)
15. Lamata, M.T., Cables, E.: Obtaining OWA Operators starting from a linear order and preference quantifiers. Int. J. Intell. Syst. **27**, 242–258 (2012)
16. Lamata, M.T., Cables, E.: OWA weights determination by means of linear functions. Mathware Soft Comput. **16**, 107–122 (2009)
17. Lichoti, J.K., Davies, J., Maru, Y., Kitala, P.M., Githigia, S.M., Okoth, E., Bishop, R.P.: Pig traders' networks on the Kenya-Uganda border highlight potential for mitigation of African swine fever virus transmission and improved ASF disease risk management. Prev. Vet. Med. **140**, 87–96 (2017)
18. Liu, X.: On the properties of equidifferent OWA operator. Int. J. Approx. Reason. **43**, 90–107 (2006)
19. O'Hagan, M.: Aggregating template rule antecedents in real-time expert systems with fuzzy set logic. In: Twenty-Second Asilomar Conference on Signals, Systems and Computers, vol. 2, pp. 681–689 (1988)
20. Penrith, M.L., Lopes Pereira, C., Lopes da Silva, M.M R., Quembo, C., Nhamusso, A., Banze, J.: African swine fever in Mozambique: review, risk factors and considerations for control. Onderstepoort J. Veterin. Res. **74**(2), 149–160 (2007)
21. Taylor, N., Rushton, J.: A value chain approach to animal diseases risk management: technical foundations and practical framework for field application. FAO **4** (2011)
22. Xu, Z.S., Da, Q.L.: Approaches to obtaining the weights of the ordered weighted aggregation operators. J. Southeast Univ. **33**, 94–96 (2003)
23. Xu, Z.: An overview of methods for determining OWA weights. Int. J. Intell. Syst. **20**, 843–865 (2005)
24. Yager, R.R., Filev, D.F.: Parameterized andlike and orlike OWA operators. Int. J. Gen. Syst. **22**, 297–316 (1994)
25. Yager, R.R., Filev, D.P.: Induced ordered weighted averaging operators. IEEE Trans. Syst., Man, Cybern. Part B: Cybern. **29**, 141–150 (1999)
26. Yager, R.R.: On ordered weighted averaging aggregation operators in multi-criteria decision making. IEEE Trans. Syst. Man Cybern. **18**(1), 183–190 (1988)
27. Yager, R.R.: Prioritized aggregation operators. Int. J. Approx. Reason. **48**, 263–274 (2008)
28. Yager, R.R.: Prioritized OWA aggregation. Fuzzy Optim. Decis. Making **8**, 245–262 (2009)
29. Zadeh, L.A.: A computational approach to fuzzy quantifiers in natural languages. Comput. Math. Appl. **9**(1), 149–184 (1983)

An Optimization Model for Location-Allocation of Health Services Under Uncertainty

Juan Carlos Figueroa–García, Carlos Franco, and Roman Neruda

Abstract This work presents a uncertainty-based optimization model for allocation of healthcare facilities to serve patients with different needs. Fuzzy uncertainty is considered in the location-allocation costs, utility and the available budget which are commonly defined by experts and are subject to adjustments and negotiation over time. A fuzzy optimization method based on the cumulative membership function of a fuzzy set is applied to solve the problem where an equilibrium between a fuzzy utility goal and fuzzy-budgets, covering and service constraints is reached.

Keywords Fuzzy optimization · Healthcare location-allocation · Health services

1 Introduction and Motivation

Healthcare costs are constantly increasing, so health system managers are highly concerned with properly (or even optimal) delivering service to patients. Patient demand is complex due to different uncertainties, so most of healthcare models use discrete event simulation to incorporate probabilistic uncertainty in order to realistically describe the whole system.

Uncertain-based decision-making models are alternative tools to model problems with noise, uncertainties and risks generated by different sources which produce aversion in the stakeholders. For this reason, a well suited optimization approach should provide solutions to problems with uncertain "black-box" goals and constraints. To

J. C. Figueroa–García (✉)
Universidad Distrital Francisco José de Caldas, Bogotá, Colombia
e-mail: jcfigueroag@udistrital.edu.co

C. Franco
Universidad del Rosario, Bogotá, Colombia
e-mail: carlosa.franco@urosario.edu.co

R. Neruda
Informatics Department, Charles University, Prague, Czech Republic
e-mail: neruda@cs.caz.cz

J. L. Verdegay et al. (eds.), *Computational Intelligence Methodologies Applied to Sustainable Development Goals*, Studies in Computational Intelligence 1036,
https://doi.org/10.1007/978-3-030-97344-5_7

do so, a robust optimal staffing model for healthcare facilities allocation is proposed to minimize the risk to patients while considering system constraints and uncertain objective functions.

The main problem concerns allocating healthcare facilities across geographically distributed areas with uncertain allocation costs. This problem highlights the trade-offs between allocating costs and patient access to required services which is modeled using fuzzy sets and iterative mathematical programming models. While most of the available fuzzy optimization methods look for a set of possible solutions, the proposed approach looks for an overall satisfaction degree between fuzzy costs (goal), budget constraints and accessibility of patients to required services by allocating healthcare facilities in different areas.

A key feature of the proposal is to find solutions for allocating healthcare facilities under uncertainty in order to practically improve real-world decision making. With the proposed method, healthcare managers are able to generate optimal strategies for staffing and resource allocation with a better understanding of trade-offs in resource allocation that enable risk-averse decision makers to better serve patients.

The work is divided into six sections; Sect. 1 introduces the location–allocation of healthcare facilities problem; Sect. 2 presents some basics on fuzzy numbers; Sect. 3 introduces the fuzzy location–allocation problem and a method to solve it; Sect. 4 presents an application example and Sect. 5 presents some concluding remarks of the study.

2 Location–Allocation of Healthcare Facilities

The analysis and optimization of health systems is highly dependent on its context. For instance, Syam and Côté [18] focalized their efforts on solving the location–allocation of specialized services problems including real–world issues using a multi–objective optimization model which was solved with a simulated-annealing algorithm. Similarly, Karakaya and Meral [14] applied location–allocation concepts to a maternal/neonatal care problem by proposing a bi–objective mathematical optimization model with three kinds of hierarchical analyses whose solution was given as a combination of top–down heuristics and Lagrange relaxation.

Other approaches have addressed the location-allocation problem applied to healthcare services by assuming all parameters are either known or can be properly estimated (an interesting review about the topic was provided by Ahmadi-Javid, Seyedi and Syam [2], Bazo et al. [3] and Dell'Ovo, Oppio and Capolongo [7, 8] who published a contributed book with different works on location-allocation for healthcare centers). For instance, an application of telemedicine services for earthquake response is developed by Gulzari and Tarakci [13] who propose a framework to optimize the process of locating temporary facilities to allocate healthcare workers while minimizing unsatisfied demands. Similarly, Ahmadi-Javid and Ramshe [1] propose a network model for primary healthcare cross-training personnel services formulated as a mix between queuing networks and non-linear programming.

Another approach considering risk aversion was addressed by Taymaz et al. [19] who solved walk–in clinic services whose objective is to offer continuous monitoring via mobile network to patients where risk aversion is considered as the possibility of not providing required services. Shariff, Moin and Omar [17] addressed the location of new healthcare facilities/services problem in Malaysia formulated as a maximal covering location problem with capacity constraints which was solved with genetic algorithms. Ye and Kim [20] solved a covering location model for maximizing the access to healthcare services given a limited number of facilities.

Most of these works assume totally known/complete information to perform decision-making, but many problems lack of well estimated data and parameters, so we consider some of them as uncertain. To do so, we involve uncertainty in both opening locations costs and available budget since these values are not fully certain in the healthcare sector (budgets are often affected by public policies and adjustments while opening a location costs are usually subject to unexpected issues and unforeseen conditions).

2.1 Mathematical Model

The mathematical location-allocation model considers a set of P patients who require different types of S services. Those patients should be allocated in a location to be treated, where L possible locations are considered. Each patient requires different medical services represented by a parameter a_{sp} where every patient has an associated allocation cost c_{pl}. Each location $l \in L$ offers specific services expressed as b_{sl} and each one can be opened at an associated cost oc_l; every location has a limited capacity of patients to be attended, a limited number of T locations to be opened and also a limited budgets B and C for location-allocation respectively. All index sets, variables and parameters are described next.

Index sets:
L is the set of health facilities.
P is the set of patients.
S is the set of required services.

Decision variables:
x_{pl} : binary variable for the assignment of patients to a health facility.
y_l : binary variable for the opening of a health facility.

Parameters:
a_{sp} : type of required services by a patient.
u_l : utility function of a patient after allocation.
c_{pl} : cost of assignment of a patient to a health facility.
oc_l : opening cost of a facility.
b_{sl} : services available at each facility.

T : maximum number of health facilities to be opened.
c_l : capacity of each health facility.
B : available budget for locating health facilities.
C : available budget for allocating patients.

And the mathematical model is as follows.

$$\text{Max } z = \sum_{l \in L} u_l \cdot y_l \quad (1)$$

$$s.t.$$

$$\sum_{p \in P} x_{pl} \leqslant M \cdot y_l \;\forall\, l \in L \quad (2)$$

$$\sum_{l \in L} y_l \leqslant T \quad (3)$$

$$a_{sp} \cdot x_{pl} \leqslant b_{sl} \cdot y_l \;\forall\, l \in L, p \in P, s \in S \quad (4)$$

$$\sum_{p \in P} x_{pl} \leqslant c_l y_l \;\forall\, l \in L \quad (5)$$

$$\sum_{p \in P} x_{pl} = 1 \;\forall\, p \in P \quad (6)$$

$$\sum_{l \in L} oc_l \cdot y_l \leqslant B \quad (7)$$

$$\sum_{l \in L} \sum_{p \in P} c_{pl} \cdot y_l \leqslant C \quad (8)$$

where $x_{pl} \in \{0, 1\}$, $y_l \in \{0, 1\}$.

This model is composed by two different binary variables x_{pl} and y_l where x_{pl} defines the allocation of patients to health care facilities and y_l denotes the decision of opening of a health care facility. The objective function (see Eq. (1)) aims to maximize the utility function of allocating patients. Constraint (2) ensures no assignment of patients to unopened facilities. Constraint (3) stablishes T as the maximum facilities to be opened. Constraint (4) assigns patients to health care facilities that offers the required service. Constraint (5) establishes the capacity of every health care facility (the amount of patients assigned to every facility cannot exceed $c_l y_l$). Each patient must be assigned to a health care facility which is included in constraint (6), and finally the opening of healthcare facilities and assignment cannot exceed the available budgets as defined in constraints (7) and (8). Figure 1 shows an example of the healthcare allocation problem.

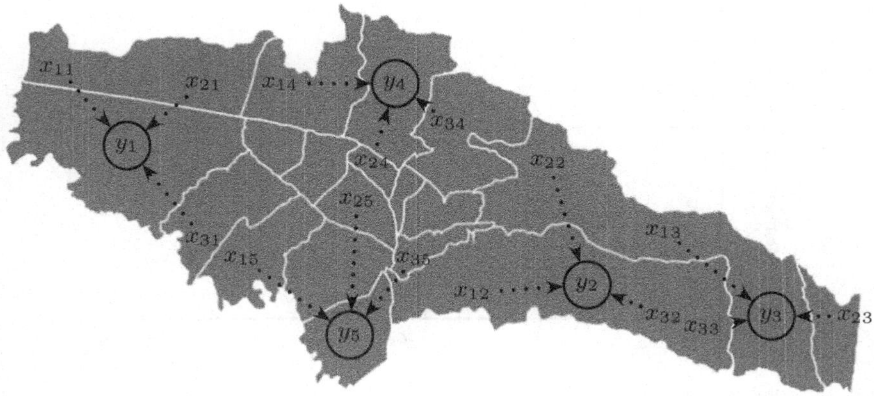

Fig. 1 Example of assignment of patients x_{pl} to locations y_l

3 Basics on Fuzzy Numbers

A fuzzy set $A = \{(x, \mu_A(x)) \mid x \in X\}$ is defined by a membership function $\mu_A : X \to [0, 1]$ which measures the membership of a value $x \in X$ regarding the concept/word/label A. $\mathcal{P}(X)$ is the class of all crisp sets, $\mathcal{F}(X)$ is the class of all fuzzy sets, $\mathcal{F}(\mathbb{R})$ is the class of all real-valued fuzzy sets, and $\mathcal{F}_1(\mathbb{R})$ is the class of all fuzzy numbers. A fuzzy number $A \in \mathcal{F}_1(\mathbb{R})$ is then defined as follows.

Definition 1 Let $A : \mathbb{R} \to [0, 1]$ be a fuzzy subset of the reals. Then $A \in \mathcal{F}_1(\mathbb{R})$ is a Fuzzy Number (FN) iff there exists a closed interval $[x_l, x_r] \neq \emptyset$ with a membership function $\mu_A(x)$ such that:

$$\mu_A(x) = \begin{cases} c(x) \;\; for \; x \in [c_l, c_r], \\ l(x) \;\; for \; x \in [-\infty, x_l], \\ r(x) \;\; for \; x \in [x_r, \infty], \end{cases} \tag{9}$$

where $c(x) = 1$ for $x \in [c_l, c_r]$, $l : (-\infty, x_l) \to [0, 1]$ is monotonic non-decreasing, continuous from the right, i.e. $l(x) = 0$ for $x < x_l$; $l : (x_r, \infty) \to [0, 1]$ is monotonic non-increasing, continuous from the left, i.e. $r(x) = 0$ for $x > x_r$.

The α-*cut* of a fuzzy number $A \in \mathcal{F}_1(\mathbb{R})$ namely ${}^{\alpha}A \triangleq \{x \mid \mu_A(x) \geqslant \alpha\}$ $\forall$ $x \in X$ is defined as follows:

$${}^{\alpha}A = \left[\inf_x {}^{\alpha}\mu_A(x), \sup_x {}^{\alpha}\mu_A(x)\right] = \left[\check{a}_{\alpha}, \hat{a}_{\alpha}\right]. \tag{10}$$

The cumulative function $F(x)$ of a probability distribution $f(x)$ is normalized i.e. $F(x) \in [0, 1]$ and is monotonically increasing:

$$F(x) \triangleq \int_{-\infty}^{x} f(t)\, dt, \;\; x \in \mathbb{R} \tag{11}$$

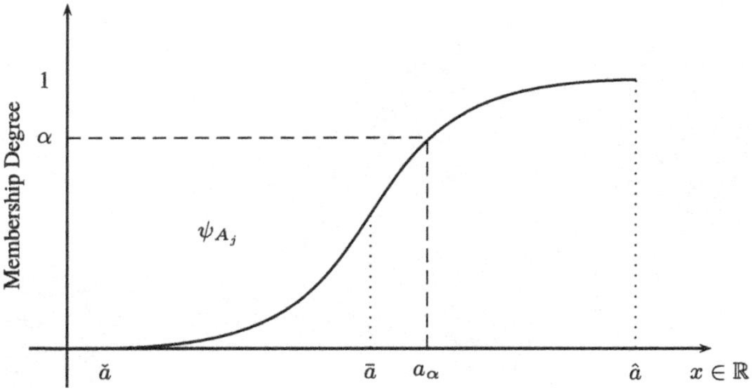

Fig. 2 Cumulative membership function for a fuzzy number A

whose fuzzy counterpart was defined by Figueroa-García & López-Bello [9, 12] and Figueroa-García [10]:

Definition 2 (*Cumulative Membership Function*) Let $A \in \mathcal{F}(X)$ be a fuzzy set. The Cumulative Membership Function (CMF) of A, $\psi_A(x)$ is:

$$\psi_A(x) \triangleq \mu_A(X \leqslant x) \tag{12}$$

and its complement $\neg\psi_A(x)$ is defined as follows:

$$\neg\psi_A(x) \triangleq 1 - \psi_A(x). \tag{13}$$

Equation (12) is the cumulative possibility of all $X \leqslant x$ i.e. $Ps(X \leqslant x)$. Then $\psi_A(x)$ is normalized by the cardinality (or total area) of A namely $|A|$, as follows:

$$\psi_A(x) \triangleq \frac{1}{|A|} \int_{-\infty}^{x} \mu_A(t)\, dt = \frac{\displaystyle\int_{-\infty}^{x} \mu_A(t)\, dt}{\displaystyle\int_{-\infty}^{\infty} \mu_A(t)\, dt} \tag{14}$$

The CMF $\psi_A(x)$ provides a single-valued α-level function which is very convenient in decision making because $\alpha \in [0, 1]$ can be incorporated as an auxiliary decision variable in order to find values $\psi^{-1}(\alpha) \rightarrow x$. Figure 2 presents the CMF ψ_A of a fuzzy number.

Instead of using $\mu_A(x)$ which has interval-valued α-cuts, the CMF $\psi_A(x)$ can reduce computations and helps to find solutions to optimization problems (see Kreinovich & Figueroa-García [15], Figueroa-García [9] and [10]).

4 A Fuzzy Location-Allocation Model

The model introduced in (1) can be extended to a model that considers fuzzy opening costs namely $\tilde{oc}_l \in \mathcal{F}_1(\mathbb{R})$ and budget $\tilde{B} \in \mathcal{F}_1(\mathbb{R})$ since they often show uncertainty coming from experts who define their values, so an appropriate way to represent this kind of uncertainty is by fuzzy sets. The extended fuzzy location-allocation model is as follows.

$$\text{Max } \tilde{z} = \sum_{l \in L} \tilde{u}_l \cdot y_l \tag{15}$$

$$s.t.$$

$$g(x_{pl}, y_l) \leqslant, = b$$

$$\sum_{l \in L} \tilde{oc}_l \cdot y_l \precsim \tilde{B} \tag{16}$$

$$\sum_{l \in L} \tilde{c}_{pl} \cdot y_l \precsim \tilde{C} \tag{17}$$

where $y_l, x_{pl} \in \{0, 1\}$, $\tilde{u}_l \in \mathcal{F}_1(\mathbb{R})$, $\tilde{c}_{pl} \in \mathcal{F}_1(\mathbb{R})$, $\tilde{oc}_l \in \mathcal{F}_1(\mathbb{R})$, $\tilde{B} \in \mathcal{F}_1(\mathbb{R})$, $g(x_{pl}, y_l) \leqslant, = b$ denotes Eqs. (2), (3), (4), (5), (6), and the binary relation $\precsim$ for fuzzy sets as proposed by Ramík and Řimánek [16] is as follows:

Definition 3 Let $\tilde{A}, \tilde{B} \in \mathcal{F}_1(\mathbb{R})$ be two fuzzy numbers. Then $\tilde{A} \precsim \tilde{B}$ if and only if $\sup {}^\alpha\tilde{A} \leqslant \sup {}^\alpha\tilde{B}$ and $\inf {}^\alpha\tilde{A} \leqslant \inf {}^\alpha\tilde{B}$ for each $\alpha \in [0, 1]$, where ${}^\alpha\tilde{A} \triangleq [\inf {}^\alpha\tilde{A}, \sup {}^\alpha\tilde{A}]$ and ${}^\alpha\tilde{B} \triangleq [\inf {}^\alpha\tilde{B}, \sup {}^\alpha\tilde{B}]$ are its α–cuts. This binary relation is called the fuzzy max order.

The main goal is to maximize $\sum_{l \in L} \tilde{u}_l y_l$ using the Zimmermann's method (see Zimmermann [21], Verdegay [6], Campos & Verdegay [4] and Cardenas & Verdegay [5]) to solve the problem with the iterative method proposed by Figueroa-García & Tenjo [11], Figueroa-García [10] and Figueroa-García & López-Bello [9, 12] which is summarized as follows:

Method 1

1- Setup:
- Set $\alpha \in [0, 1]$ and compute $\psi_{\tilde{u}_l}, \psi_{\tilde{c}_{pl}}, \psi_{\tilde{oc}_l}, \mu_{\tilde{B}}, \mu_{\tilde{C}}$
- Set $u_l(\alpha) = \psi_{\tilde{u}_l}^{-1}(\alpha)$, $c_{pl}(\alpha) = \psi_{\tilde{c}_{pl}}^{-1}(1-\alpha)$, $oc_l(\alpha) = \psi_{\tilde{oc}_l}^{-1}(1-\alpha)\ \forall\, l$, $B(\alpha) = \mu_{\tilde{B}}^{-1}(\alpha)$ and $C(\alpha) = \mu_{\tilde{C}}^{-1}(\alpha)$

2- Soft constraints method:
- Compute

$$\check{z}(\alpha) = \text{Max}\{z(\alpha) = \sum_{l\in L} u_l(\alpha)y_l : g(x_{pl}, y_l) \leqslant,$$
$$= b; \sum_{l\in L} oc_l(\alpha)y_l \leqslant \check{B}, \sum_{l\in L} c_{pl}(\alpha)y_l \leqslant \check{C}\}$$

– Compute

$$\hat{z}(\alpha) = \text{Max}\{z(\alpha) = \sum_{l\in L} u_l(\alpha)y_l : g(x_{pl}, y_l) \leqslant,$$
$$= b; \sum_{l\in L} oc_l(\alpha)y_l \leqslant \hat{B}, \sum_{l\in L} c_{pl}(\alpha)y_l \leqslant \hat{C}\}$$

– Define the conditional set $\tilde{z}(\alpha)$ with the following membership function:

$$\mu_{\tilde{z}}(z, \check{z}, \hat{z}|\alpha) = \begin{cases} 1, & z(\alpha) \geqslant \hat{z}(\alpha) \\ \dfrac{z(\alpha) - \check{z}(\alpha)}{\hat{z}(\alpha) - \check{z}(\alpha)}, & \check{z}(\alpha) \leqslant z(\alpha) \leqslant \hat{z}(\alpha) \\ 0, & z(\alpha) \leqslant \check{z}(\alpha) \end{cases} \tag{18}$$

where $z(\alpha) = \sum_{l\in L} u_l(\alpha)y_l$

– Thus, solve the following LP model:

$$\text{Max}_{\{\lambda\,|\,\alpha\}}\ \lambda$$
$$s.t.$$
$$\sum_{l\in L} u_l(\alpha)y_l - \lambda(\hat{z}(\alpha) - \check{z}(\alpha)) = \check{z}(\alpha), \tag{19}$$
$$g(x_{ij}) \leqslant, = b, \tag{20}$$
$$\sum_{l\in L} oc_l(\alpha)y_l + \lambda(\hat{B} - \check{B}) \leqslant \hat{B}, \tag{21}$$
$$\sum_{l\in L} c_{pl}(\alpha)y_l + \lambda(\hat{C} - \check{C}) \leqslant \hat{C}, \tag{22}$$
$$x_{pl}, y_l \in \{0, 1\};\ \lambda \in [0, 1]$$

3- Convergence:
– If $\lambda^* = \alpha$ then stop and return λ^* as the overall satisfaction degree of $\tilde{z}$, $\tilde{u}_l$, $\tilde{c}_{pl}$, $\tilde{oc}_l$, $\tilde{B}$, and $\tilde{C}$; if $\lambda^* \neq \alpha$ then go to Step 1 and update $\alpha = \lambda^*$.

The set $\tilde{z}(\alpha)$ is defined as $z(\alpha) = \sum_l u_l(\alpha)y_l, \alpha \in [0, 1]$ i.e. *maximum allocation utility* whose highest membership degree is given by $\check{z}(\alpha)$ and its lowest membership degree is given by $\hat{z}(\alpha)$. On the other hand, the budgets $\tilde{B}, \tilde{C}$ get their highest membership degree at $\check{B}, \check{C}$ and its lowest membership degree at $\hat{B}, \hat{C}$ so the method maximizes the overall satisfaction degree between $\tilde{z}(\alpha)$ (via $\tilde{u}_l(\alpha) = \psi_{\tilde{u}_l}^{-1}(1-\alpha)$) $\tilde{B}$ and $\tilde{C}$ using the auxiliary variable λ.

5 Application Example

An instance of 30 patients with six different required services (taken as geographically identified groups of patients sharing the same healthcare needs) to be allocated in 15 possible healthcare facilities (in the city of Bogotá - Colombia) whose allocation costs interacts with the contribution of each patient group and affects the sustainability of the whole system. All fuzzy parameters $\tilde{u}_l, \tilde{c}_{pl}, \tilde{oc}_l$ are defined as Gaussian since it represents the nonlinear nature of experts perception about the costs/utility of location–allocation of patients to a healthcare facilities while $\tilde{B}, \tilde{C}$ are linearly shaped since they are easy to be defined using two values: pessimistic/optimistic budgets $\check{B}, \hat{B}$ and $\check{C}, \hat{C}$.

The method has been tested using three different initial $\alpha_0 = \{0.2, 0.5, 0.8\}$ in order to evaluate the stability of the results which are presented in Fig. 3 where the overall optimal satisfaction degree λ^* is obtained in 4 iterations. It is important to remark that λ looks for the optimal costs that satisfy budgets for a given α which defines the utility structure, so the bigger $oc_l(\alpha), c_{pl}(\alpha)$ the bigger budgets are needed at a global utility $z(\lambda^*)$. Figure 4 shows the set $\tilde{z}$ of optimal location-allocation for a global optimal degree $\lambda^* = 0.6996$.

Also note that the obtained optimal $\lambda^* = 0.6996$ is the same no matter where $\alpha \in [0, 1]$ starts from. Figure 3 shows λ_k^* for three different starting values $\alpha_0 = \{0.2, 0.5, 0.8\}$, all of them converge to $\lambda^* = 0.6996$ whose optimal values are $\check{z} = 6'545.884.66$, $\hat{z} = 25'248.411.23$ and $z(\lambda^*) = 19'630.623.27$. It is important to note that any global value $\lambda \leqslant \lambda^*$ is feasible as well while any $\lambda > \lambda^*$ is not which in practice means that $u_l(\lambda), c_{pl}(\lambda), oc_l(\lambda), B(\lambda)$ and $C(\lambda)$ leads to feasible solutions

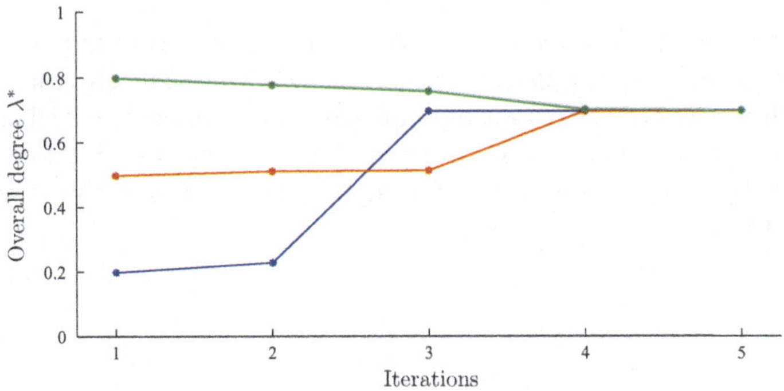

Fig. 3 Convergence of the optimal utility

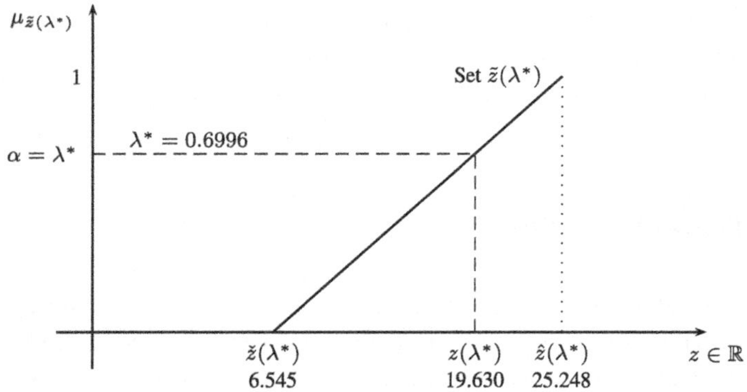

Fig. 4 Optimal solution of the problem

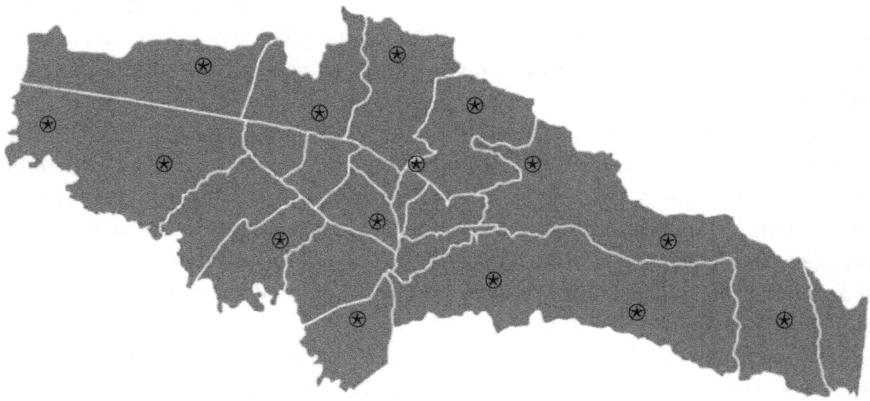

Fig. 5 Optimal location of the y_l healthcare facilities in Bogotá—Colombia

if $\lambda \leqslant \lambda^*$ i.e. the whole system is sustainable (not optimum but feasible) for $\lambda \leqslant 0.6996$. The obtained optimal location of the 15 healthcare centers is displayed in Fig. 5.

Figure 5 shows the location of all facilities on its real–world approximated coordinates where all groups of patients are geographically close to a healthcare facility in which their needs can be reasonably supplied. It is important to note that all zones (and all groups of patients which are geographically disperse over different zones) are covered by a facility whose sustainability is guaranteed by a utility (cost/benefit equilibrium) function.

6 Concluding Remarks

Location-allocation problems under uncertainty can be solved using the method proposed by Figueroa-García [10] and Figueroa-García & López-Bello [9, 12] with satisfactory results. The optimal global satisfaction degree λ^* works as a defuzzifier for all fuzzy parameters $\tilde{z}$, $\tilde{u}_l$, $\tilde{c}_{pl}$, $\tilde{oc}_l$, $\tilde{B}$ and $\tilde{C}$ to optimize a utility function subject to location-allocation budgets losing feasibility. The provided example illustrates how the iterative method works and the obtained results for selected instance also show that the algorithm converges to an optimal λ^* and subsequently to optimal values for $\tilde{z}$, $\tilde{u}_l$, $\tilde{c}_{pl}$, $\tilde{oc}_l$, $\tilde{B}$ and $\tilde{C}$.

The proposed method is able to deal with nonlinear fuzzy costs and linear fuzzy budgets using an iterative method to get a global optimal λ^* which is the maximum satisfaction degree between the overall utility and budgets i.e. $u_l(\alpha) \leqslant u_l(\lambda^*)$ and/or $B(\lambda) \leqslant B(\lambda^*)$, $C(\lambda) \leqslant C(\lambda^*)$ are feasible for λ^*. This is useful in practical applications since $u_l(\lambda^*)$ are the maximum utilities for minimum budgets $B(\lambda^*)$, $C(\lambda^*)$, so the analyst can know how to allocate patients to healthcare centers at a maximum utility while satisfying budgets.

The optimal solution provides the location-allocation of patients to healthcare facilities and crisp values for $\tilde{z}$, $\tilde{u}_l$, $\tilde{c}_{pl}$, $\tilde{oc}_l$, $\tilde{B}$ and $\tilde{C}$. The optimal λ^* leads to an equilibrium between sustainability and location-allocation budgets which is useful in decision making since managers can handle uncertain information and allocate patients to healthcare centers at a maximum allowable utility without breaking budgets or incurring in costs overrunning.

References

1. Ahmadi-Javid, A., Ramshe, N.: A stochastic location model for designing primary healthcare networks integrated with workforce cross-training. Oper. Res. Health Care **24**(1), 100226 (2020)
2. Ahmadi-Javid, A., Seyedi, P., Syam, S.S.: A survey of healthcare facility location. Comput. Oper. Res. **79**(1), 223–263 (2017)
3. Bazo, R., da Costa, C.A., Adams-Seewald, L., da Silveira-Jr., L.G., Stoffel-Antunes, R., da Rosa-Righi, R., Facco-Rodrigues, V.: A survey about real-time location systems in healthcare environments. J. Med. Syst. **45**(1), 35 (2021)
4. Campos, L., Verdegay, J.: Linear programming problems and ranking of fuzzy numbers. Fuzzy Sets Syst. **32**(1), 1–11 (1989)
5. Cardenas, J., Verdegay, J.: Towards a new strategy for solving fuzzy optimization problems. Fuzzy Optim. Decis. Mak. **8**, 231–244 (2009)
6. Delgado, M., Verdegay, J.L., Vila, M.: A general model for fuzzy linear programming. Fuzzy Sets Syst. **29**(1), 21–29 (1989)
7. Dell'Ovo, M., Oppio, A., Capolongo, S.: Decision support system for the location of healthcare facilities. Springer, Cham (2020)
8. Dell'Ovo, M., Oppio, A., Capolongo, S.: The location problem. Addressing decisions about healthcare facilities. Springer, Cham (2020)

9. Figueroa, J.C., López, C.A.: Linear programming with fuzzy joint parameters: a cumulative membership function approach. In: 2008 Annual Meeting of the IEEE North American Fuzzy Information Processing Society (NAFIPS) (2008)
10. Figueroa-García, J.C.: Mixed production planning under fuzzy uncertainty: a cumulative membership function approach. In: IEEE Workshop on Engineering Applications (WEA). vol. 1, pp. 1–6. IEEE (2012). https://doi.org/10.1109/WEA.2012.6220081
11. Figueroa-García, J.C.: Solving linear programming problems with interval type–2 fuzzy constraints using interval optimization. In: Proceedings of NAFIPS 2013. pp. 1–6. IEEE (2013)
12. Figueroa-García, J.C., López, C.A.: Pseudo-optimal solutions of FLP problems by using the cumulative membership function. In: Annual Meeting of the North American Fuzzy Information Processing Society (NAFIPS). vol. 28, pp. 1–6. IEEE (2009). https://doi.org/10.1109/NAFIPS.2009.5156464
13. Gulzari, A., Tarakci, H.: A healthcare location-allocation model with an application of telemedicine for an earthquake response phase. Int. J. Disas. Risk Reduct. **55**(1), 102100 (2021)
14. Karakaya, Ş., Meral, S.: A biobjective hierarchical location-allocation approach for the regionalization of maternal-neonatal care. Socio-Economic Planning Sciences p. In Press (2021)
15. Kreinovich, V., Figueroa-Garcia, J.C.: Optimization under fuzzy constraints: from a heuristic algorithm to an algorithm that always converges. In: Proceedings of WEA 2018, pp. 3–16. Springer (2018)
16. Ramík, Ř.: Inequality relation between fuzzy numbers and its use in fuzzy optimization. Fuzzy Sets Syst. **16**, 123–138 (1985). https://doi.org/10.1016/S0165-0114(85)80013-0
17. Shariff, S.R., Moin, N.H., Omar, M.: Location allocation modeling for healthcare facility planning in Malaysia. Comput. Ind. Eng. **62**(4), 1000–1010 (2012)
18. Syam, S.S., Côté, M.J.: A comprehensive location-allocation method for specialized healthcare services. Oper. Res. Health Care **1**(4), 73–83 (2012)
19. Taymaz, S., Iyigun, C., Bayindir, Z., Dellaert, N.: A healthcare facility location problem for a multi-disease, multi-service environment under risk aversion. Socio-Econ. Plan. Sci. **71**(1), 100755 (2020)
20. Ye, H., Kim, H.: Locating healthcare facilities using a network-based covering location problem. GeoJournal **81**(1), 875–890 (2016)
21. Zimmermann, H.J.: Fuzzy programming and Linear Programming with several objective functions. Fuzzy Sets Syst. **1**(1), 45–55 (1978)

Computational Intelligence and Tikhonov Regularization with Reduced Dimension Model: Applications in Health, Renewable Energy and Climate Heat Transfer Inverse Problems

Lucas Correia da Silva Jardim, Diego Campos Knupp, Taciano Magela de Souza Monteiro de Barros, Luiz Alberto da Silva Abreu, Carlos Cruz Corona, and Antônio J. Silva Neto

Abstract In this chapter we present a method to predict the optimal value of the Tikhonov's regularization parameter by solving simplified versions of the inverse problems considered. This can be of great benefit since methods such as the L-curve and the Fixed Point Iteration require the inverse problem to be solved several times in order to determine the optimal value for the regularization parameter. The main idea that supports the proposed approach is to solve the problem of interest using a low set of dimensions to represent the function to be estimated and, then, this solution is used to obtain an estimate for the regularization parameter of the complete model, based on the Fixed Point Iteration method. Tests are performed on three inverse heat transfer problems: estimation of the variable thermal conductivity of a biological tissue, estimation of the inlet temperature in parallel plates channel, and the estimation of the variable scattering albedo of a radiative transfer participating medium. The results obtained demonstrate the feasibility of the technique in three

L. C. S. Jardim (✉) · D. C. Knupp · T. M. S. Monteiro de Barros · L. A. S. Abreu · A. J. Silva Neto
LEMA—Patrícia Oliva Soares Laboratory of Experimentation and Numerical Simulation in Heat and Mass Transfer, Polytechnic Institute, Rio de Janeiro State University, Nova Friburgo, RJ, Brazil
e-mail: ljardim@iprj.uerj.br

D. C. Knupp
e-mail: diegoknupp@iprj.uerj.br

T. M. S. Monteiro de Barros
e-mail: taciano@iprj.uerj.br

L. A. S. Abreu
e-mail: luiz.abreu@iprj.uerj.br

A. J. Silva Neto
e-mail: ajsneto@iprj.uerj.br

C. C. Corona
Department of Computer Science and Artificial Intelligence, Granada University, Granada, Spain
e-mail: carloscruz@decsai.ugr.es

J. L. Verdegay et al. (eds.), *Computational Intelligence Methodologies Applied to Sustainable Development Goals*, Studies in Computational Intelligence 1036,
https://doi.org/10.1007/978-3-030-97344-5_8

problems with potential practical applications in bioengineering, renewable energy and climate in alignment with the Sustainable Development Goals (SDG) 3, 4, 7, 9, 13, and 17 of the United Nations 2030 Agenda established in 2015.

Keywords Tikhonov regularization · Inverse heat transfer problem · Function estimation · Fixed point iteration · Reduced dimension model

1 Introduction

Inverse problems can be formulated as minimization problems, where an objective function is defined as the sum of the squared differences between information calculated with a model and experimental data, which corresponds to the maximum likelihood approach [29]. The class of inverse problems that deals with function estimation is often ill-posed due to the presence of noise on the data used as input [28, 29, 35, 46]. Such experimental noise is amplified and generates undesired oscillations on the obtained solution, making it very difficult to retrieve any useful information and, to overcome this problem, the classical technique known as Tikhonov Regularization [45] can be used. This technique works by adding a term to the objective function which penalizes oscillatory solutions. This term comes multiplied by a positive non-zero scalar, namely the Tikhonov regularization parameter λ. The fine tuning of λ is crucial to the outcome of the method: if λ is too large, only overregularized solutions will prevail, on the other hand, if it is too small, not enough regularization will happen, and the solution will still be too oscillatory.

Therefore, the solution of the inverse problem may incorporate it in the search of an adequate value for the regularization parameter. To do so, several methods have been proposed along the years, such as the L-curve method [13, 15], Morozov's Discrepancy Principle [27], Fixed Point Iteration method [2], among others. For a more detailed review of these methods, the reader can see the work of Hansen [14].

In this chapter we focus on the Fixed Point Iteration method (FP), as proposed by Bazán [2]. This method does not require previous information on the experimental error variance and works by finding the parameter λ that balances the values of the objective function terms, i.e., the residue and the regularization term [8]. The FP finds the best λ by reaching the convergence of a series where each new term is calculated by solving the inverse problem with the previous λ. When convergence is reached, the newly found λ is considered the optimal one. In practice, the iterative process stops when a small prescribed tolerance is reached.

Although FP is very efficient, it obviously can be affected by the number of dimensions of the problem, i.e., the number of points used to represent the sought function. In this chapter we present a technique to find an estimate for the best value of λ in a problem with a large number of variables by solving an alternative problem with reduced order, i.e., smaller number of dimensions. This low order solution is extrapolated in order to find an estimate for the complete problem regularization parameter λ and, then, this estimate is used as an initial value to restart the series.

Furthermore, when divergence is detected on the series, no value of λ can be retrieved. Bazán [2], the author of the method, suggested different approaches to overcome this problem, but in this chapter we present a step-by-step algorithm based on the recent work of Aucejo and De Smet [1], which introduced an idea to overcome this problem based on the objective function terms values.

The methodology developed in this work is demonstrated in three inverse heat transfer problems with potential practical applications in bioengineering, renewable energy and climate, as follows:

- **Identification of spatially variable thermal conductivity of biological tissues** The bioheat transfer model proposed by Pennes [36] is used in a steady-state condition to predict the thermal conductivity of tissues. Thermal techniques and their analysis are common in the treatment of cancer [21, 25], laser ablation procedures [38] and even brain related diseases therapy [12], to name a few. Furthermore, using this model with spatially variable thermal conductivity of biological tissue has attracted the attention of the scientific community [11, 20, 31, 33].
- **Estimation of spatially variable inlet temperature profiles in parallel plates channel** The model used takes into account the velocity field of a fluid inside a parallel plates channel. The heat transfer process by convection considered has many applications in the design of optimal heat exchangers, energy transformation, among many others. Spatially variable inlet temperature profiles may be of special interest in micromixers.
- **Spatially dependent single scattering albedo** The process of radiative transfer is investigated in steady state for one dimensional participating media. Such problem, related model and analysis is of great importance in many areas of interest [23, 24, 32, 42, 43], including those related to climate research, such as atmospheric and oceanic models [48], oceanography [9] and atmosphere studies [16].

Such problems and corresponding applications are in alignment with the Sustainable Development Goals (SDG) 3, 4, 7, 9, 13 and 17 of the United Nations 2030 Agenda established in 2015.

2 Fixed Point Iteration Method with Reduced Model

Consider an inverse problem formulated with the maximum likelihood approach. This formulation leads to a minimization problem where the squared norm of the difference between the vectors of observed data and values obtained from a model must be minimized. This objective function can be written as

$$Q_{TR}(\mathbf{u}) = \left|\left|\mathbf{T}^{\exp} - \mathbf{F}(\mathbf{u})\right|\right|^2 \tag{1}$$

where $\mathbf{u}$ is the vector of unknown parameters, $\mathbf{T}^{\exp}$ is the vector of experimental data, $\mathbf{F}(\mathbf{u})$ is the vector of values obtained with the mathematical model, and the operator $||.||$ is the l_2 Euclidean norm.

When a function intrinsic to the model **F** is unknown and must be estimated, one possible approach is to represent this function as linearly interpolated discrete points, where such points compose the vector **u** in Eq. (1). This formulation of the inverse problem is often ill-posed and the solutions obtained for **u** may contain heavy oscillations [3, 4, 43].

The Tikhonov Regularization (TR) is one of the many techniques proposed to overcome this problem [6, 29, 46]. It consists on adding a term to the objective function Q that penalizes oscillating solutions of **u**. The modified Eq. (1) with the regularization term is written as

$$Q(\mathbf{u}) = \left|\left|\mathbf{T}^{\text{exp}} - \mathbf{F}(\mathbf{u})\right|\right|^2 + \lambda^2 \left|\left|\mathbf{L}\mathbf{u}\right|\right|^2 \tag{2}$$

where λ is the regularization parameter, and **L** is the derivative operator matrix. Particularly, we call the term $||\mathbf{T}^{\text{exp}} - \mathbf{F}(\mathbf{u})||$ as the residue norm and $||\mathbf{L}\mathbf{u}||$ as the regularization norm.

The derivative operator **L** can be of zero, first or second order: this choice depends on desired aspects of the sought solution [2, 8]. The regularization parameter λ controls the balance between how much regularized the solution must be [6], in other words, if λ is large, the solution will tend to be constant (over regularized) and, on the other hand, if λ tends to zero, no regularization is performed.

The following sections explain the Fixed Point Iteration method (FP) used to calculate the optimal value of λ, and how it is possible to perform the associated iterative procedure using a simpler version of the inverse problem, i.e., using a lower dimension vector of unknowns **u**.

2.1 The Fixed Point Iteration Method (FP)

The main idea of FP is to select the value of λ that minimizes the product between the residue norm and the regularization norm as a function of λ [8]. This minimization corresponds, in other words, to the value of λ that balances the two terms growth. The iterative process is built with the expression

$$\lambda_{i+1} = \sqrt{\phi}\, \left.\frac{||\mathbf{T}^{\text{exp}} - \mathbf{T}(\mathbf{u})||}{||\mathbf{L}\mathbf{u}||}\right|_{\lambda_i}, \quad i = 0, 1, 2, \ldots \tag{3}$$

where $\phi > 0$ is a parameter that plays a role on the convergence of the series, and the residue and regularization norms are obtained from the solution of the inverse problem with the previous value for the regularization parameter, i.e. λ_i. Therefore, it is necessary to perform the optimization cycle in order to obtain the next term of the sequence.

A few aspects of this technique must be pointed out. First of all, to start the series, the author of the method suggests in Ref. [2] that the initial regularization parameter should be close to zero, i.e., $\lambda_0 \to 0^+$. Furthermore, it is possible that the series fails to converge. Initially, it is suggested to use $\phi = 1$ as an initial choice and, if divergence is detected, ϕ must be updated to overcome this problem. Bazán [2] proposed different ways for this parameter tuning, but in this chapter we use the approach recently proposed by Aucejo and De Smet [1], which takes into account the balance of the terms of the objective function, and does not require any information of the physical problem. We implemented this approach with some modifications on the stopping criterion, but the core of the actualization of ϕ holds, as shown in Algorithm 1 next:

Algorithm 1: Fixed Point Iteration method with adapted stopping criterion

1: Set $\phi = 1$.
2: Set $i = 0$, $\lambda_0 \to 0^+$, and $R_0 \to 0^+$.
3: Compute λ_{i+1} with Eq. (3).
4: Compute the ratio $R_{i+1} = \lambda_i^2 \left. \frac{||\mathbf{Lu}||^2}{||\mathbf{T}^{\exp} - \mathbf{T}(\mathbf{u})||^2} \right|_{\lambda_i}$.
5: Compute $\Delta = R_{i+1}/R_i$.
6: If $\Delta < 1$, the series is considered divergent, then make $\sqrt{\phi} = \sqrt{R_i}$, and restart the process from Step 2, otherwise, go to Step 7.
7: If $\Delta - 1 \leq \epsilon$, the series is considered convergent, λ_{i+1} is the optimal regularization parameter, and the iterative process is stopped. Otherwise, make $i = i + 1$, and go to Step 3.

where ϵ is a relatively small prescribed tolerance. Here it is important to make a few remarks. The first term R_1 will most likely be very small because $\lambda_0 \to 0^+$. Furthermore, the sequence R_i increases until divergence is detected on Step 6. This indicates that the regularization parameter is increasing and making the regularization norm decrease to a point that it is not balanced with the residue norm.

Algorithmically, we introduce a stopping criterion on the convergence of the ratio R on Step 7, which corresponds to the convergence of λ, whereas Aucejo and De Smet [1] suggested a stopping criterion on the convergence of ϕ, which is a robust option, despite being more computationally expensive.

2.2 *The Reduced Dimension Model Approach*

The objective of this approach is to find an estimate for the initial value of λ to start the algorithm presented in Sect. 2.1, by solving the inverse problem with a coarser discretization of the sought function, i.e., a lower dimension vector $\mathbf{u}$, identified as $\mathbf{u}_{\text{low}}$. Once the corresponding optimal regularization parameter λ_{low} and the regularized solution $\mathbf{u}_{\text{low}}$ are determined, the initial regularization parameter λ_{high} for a problem with higher dimension can be estimated by equalizing the regularization norms with their respective regularization parameters, yielding

$$\lambda_{\text{low}}^2 \, ||\mathbf{Lu}||_{\text{low}}^2 = \lambda_{\text{high}}^2 \, ||\mathbf{Lu}||_{\text{high}}^2 \tag{4}$$

where "low" and "high" are the indexes indicating the lower and the higher dimension inverse problem, respectively. Then:

$$\lambda_{\text{high}} = \lambda_{\text{low}} \frac{||\mathbf{Lu}||_{\text{low}}}{||\mathbf{Lu}||_{\text{high}}} \tag{5}$$

where λ_{low} and $||\mathbf{Lu}||_{\text{low}}$ are known from the solution of the lower dimension problem. The matrix $\mathbf{L}_{\text{high}}$ is simply the finite difference operator with the corresponding dimension. Yet to be determined is the term $\mathbf{u}_{\text{high}}$. Here some assumptions must be introduced, since we do not have any solution $\mathbf{u}$ obtained with the the high dimension model. The suggestion is to perform a linear interpolation of $\mathbf{u}_{\text{low}}$ in order to generate $\mathbf{u}_{\text{high}}$.

Now it is possible to evaluate the whole expression given by Eq. (5) to obtain an initial value for λ_{high}, since all the terms on the right-hand side are determined. In terms of computational cost, this can be of great advantage if the estimated value of λ_{high} is reasonably good, meaning that the higher dimension problem will be solved fewer times.

Another important aspect of this approach is the choice of the dimension size for $\mathbf{u}_{\text{low}}$. A small dimension size will lead to an easier problem to optimize, but relevant information of the real solution can be lost, which can lead to a poor estimate for λ_{high}. On the other hand, if this dimension is increased, the problem will become more cumbersome to solve, since more variables are included, leading to higher computational time and, therefore, more severe regularization issues.

3 Direct Problems

In this section, the three Direct Problems of interest in this work are presented. As already mentioned, the first one is concerned with the bioheat transfer in steady state regime on a biological tissue with spatially variable thermal conductivity in order to model different layers. The second problem deals with the inlet temperature profile of a fluid inside a parallel plate channel aiming at applications in microchannels. The third problem addresses the radiative transfer problem in participating media, in which the spatially variable single scattering albedo must be reconstructed.

3.1 Direct Problem 1: Spatially Varying Thermal Conductivity of a Biological Tissue

The bioheat transfer model given by Pennes' equation [36] is a well established approach in determining the temperature field of biological tissues [11]. In this

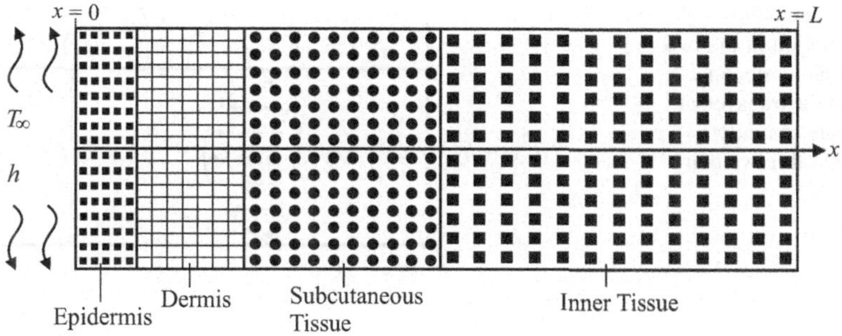

Fig. 1 Schematic representation of the heterogeneous biological tissue. Adapted from Ref. [11]

chapter, we consider the steady-state model with one spatial dimension. Also, the thermal conductivity $k(x)$ is modeled as a spatially dependent parameter, i.e., it varies across the different types of tissues. As presented in Fig. 1, there is heat exchange with the ambient at $x = 0$, and no external heating. The Pennes' model, in such conditions, can be written as

$$\frac{\mathrm{d}}{\mathrm{d}x}\left[k(x)\frac{\mathrm{d}T(x)}{\mathrm{d}x}\right] + \omega_\mathrm{b}\rho_\mathrm{b}c_\mathrm{b}\left[T_\mathrm{b} - T(x)\right] + q_\mathrm{m} = 0 \tag{6a}$$

$$-k(x)\left.\frac{\mathrm{d}T(x)}{\mathrm{d}x}\right|_{x=0} = h\left[T_\infty - T(x)|_{x=0}\right] \tag{6b}$$

$$\left.\frac{\mathrm{d}T(x)}{\mathrm{d}x}\right|_{x=L} = 0 \tag{6c}$$

where ω_b is the blood perfusion parameter, ρ_b and c_b are the specific mass and specific heat of the blood, respectively, T_b is the blood temperature, q_m is the internal heat generation due to metabolic activity, L is the domain length, T_∞ is the ambient temperature, and h is the heat transfer coefficient of the convection phenomenon at $x = 0$.

The problem given by Eq. (6) is solved using the built-in "NDSolve" routine of the Wolfram Mathematica software with the Implicit Runge Kutta method option [47], and fixed step size. The spatially varying thermal conductivity is considered constant for each tissue, and, therefore, this can be implemented as a step function, i.e. constant by parts. In this chapter we use an exponential approximation to write this step function as a continuous function, in order to avoid stiffness of the numerical solution.

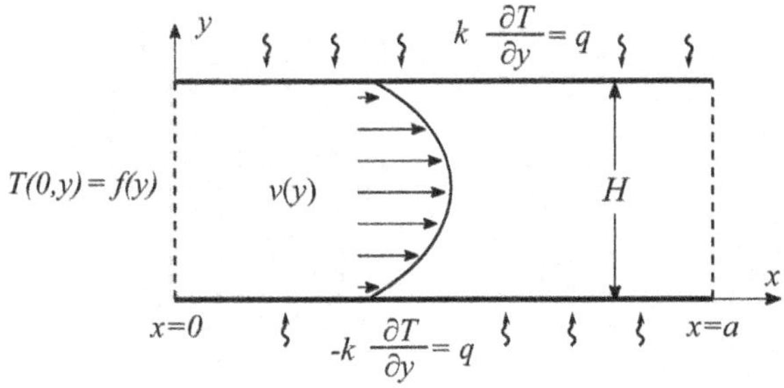

Fig. 2 Schematic representation of the two parallel plates channel showing the transversal coordinate spatially dependent fluid velocity

3.2 Direct Problem 2: Inlet Temperature Profile in a Parallel Plates Channel

Consider the problem of determining the steady state temperature field of a hydrodynamically developed and thermally developing fluid inside a two parallel plates channel subjected to a constant heat flux q at both boundary plate surfaces, as schematically presented in Fig. 2. Neglecting viscous dissipation, free convection and the axial conduction effects, the mathematical formulation that describes the temperature field $T(x, y)$ can be written as [5, 7, 39].

$$k\frac{\partial^2 T(x, y)}{\partial y^2} = v(y)\rho c_{\mathrm{p}}\frac{\partial T(x, y)}{\partial x} \tag{7a}$$

$$k\left.\frac{\partial T(x, y)}{\partial y}\right|_{y=H} = q \tag{7b}$$

$$-k\left.\frac{\partial T(x, y)}{\partial y}\right|_{y=0} = q \tag{7c}$$

$$v(y) = 6v_{\mathrm{m}}\frac{y}{H}\left(1 - \frac{y}{H}\right) \tag{7d}$$

$$T(0, y) = f(y) \tag{7e}$$

where the thermophysical parameters k, ρ and c_{p} are the thermal conductivity, and specific mass and specific heat, respectively. The parameter v_{m} is the average velocity of the fluid inside the channel.

The problem described by Eq. (7) is also solved using the built-in "NDSolve" routine of the Mathematica software [47]. Here, the Finite Element Method option is used, and the mesh generated to discretize the domain is rectangular with all the elements having the same dimensions, i.e., the mesh is uniform. In this approach, the velocity field $v(y)$ and the inlet temperature $f(y)$ are implemented as continuous functions, and, therefore, the discretization needed to solve the differential equation is performed automatically by the routine.

3.3 Direct Problem 3: Radiative Transfer with Variable Space Dependent Scattering Albedo

Consider a steady-state regime for an one-dimensional medium that absorbs, emits and scatters radiation isotropically. The spatial coordinate (optical variable) and the optical thickness are represented by $d\tau = \beta dx$ and $\tau_0 = \beta L$, respectively, where β is the extinction coefficient, and x and L are the physical spatial coordinate and total length of the medium, respectively. External incident radiation with isotropic constant intensities A_1 at $\tau = 0$ and A_2 at $\tau = \tau_0$ are considered as shown in Fig. 3.

The mathematical formulation of the steady-state radiative transfer problem, for a given wavelength, considering the case of azimuthal symmetry, isotropic scattering, and neglecting the radiation emission by the medium, is given by the linear version of the Boltzmann equation [34],

$$\mu \frac{\partial I(\tau,\mu)}{\partial \tau} + I(\tau,\mu) = \frac{\omega(\tau)}{2} \int_{-1}^{1} I(\tau,\mu')\mathrm{d}\mu' \text{ in } 0 < \tau < \tau_0, \ -1 \le \mu \le 1 \quad \text{(8a)}$$

$$I(0,\mu) = A_1, \ \text{for } \mu > 0 \quad \text{(8b)}$$

$$I(\tau_0,\mu) = A_2, \ \text{for } \mu < 0 \quad \text{(8c)}$$

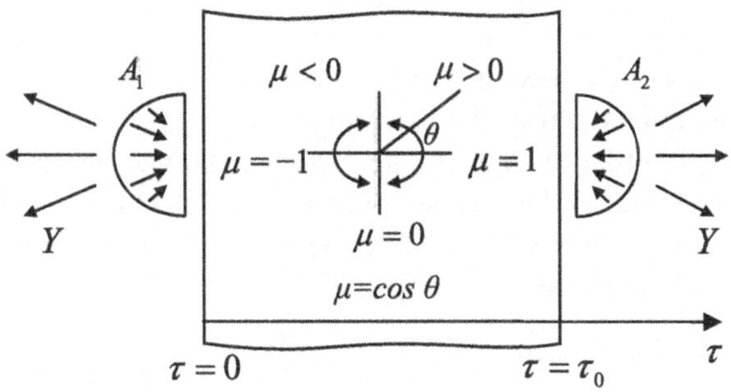

Fig. 3 Schematic representation of a radiative transfer participating medium. Here Y represents the exit radiation intensity at both boundaries of the medium. Adapted from Ref. [41]

where μ is the cosine of the polar angle θ, i.e., the angle of the radiation beam with the positive τ axis, and $\omega(\tau)$ is the spatially varying single scattering albedo.

The solution of the radiative transfer problem given by Eq. (8) is obtained using Chandrasekhar's discrete ordinates method [10], in which the angular domain is discretized and the integrals replaced by Gauss-Legendre quadratures, as performed in [17, 22–24, 29, 42, 43]. The built-in routine "NDSolve" [47] of the Wolfram Mathematica software [47] is then used to solve the system of differential equations generated by this discretization, with the method set as automatic, and a precision goal of 10^{-4} in order to control the computational cost in obtaining the radiative transfer problem solution.

4 Inverse Problems

The Inverse Problem 1 (IP-1) consists of estimating the thermal conductivity $k(x)$ in Eq. (6). The Inverse Problem 2 (IP-2) is formulated similarly to IP-1, but here we are interested in estimating the inlet temperature profile $f(y)$ in Eq. (7e). Lastly, in the Inverse Problem 3 (IP-3) the varying scattering albedo $\omega(\tau)$ in Eq. (8a) is considered unknown, and must be determined.

For the three cases, the corresponding vector of unknowns, here called $\mathbf{u}_1$, $\mathbf{u}_2$ and $\mathbf{u}_3$, contains values of the sought function at discrete points of the computational domain, which can be represented as

$$\mathbf{u}_1 = \{k_1, k_2, \ldots, k_n\}^{\mathrm{T}}, \; k_1 = k(x_1), k_2 = k(x_2), \ldots, k_n = k(x_n) \tag{9a}$$

$$\mathbf{u}_2 = \{T_1, T_2, \ldots, T_n\}^{\mathrm{T}}, \; T_1 = f(y_1), T_2 = f(y_2), \ldots, T_n = f(y_n) \tag{9b}$$

$$\mathbf{u}_3 = \{\omega_1, \omega_2, \ldots, \omega_n\}^{\mathrm{T}}, \; \omega_1 = \omega(\tau_1), \omega_2 = \omega(\tau_2), \ldots, \omega_n = \omega(\tau_n) \tag{9c}$$

where n is the total number of points used in the discretization of the sought function within the physical medium domain.

In order to simulate the experimental measurements required for the solution of each one of the inverse problems considered, noise is added to the solution of the respective direct problem using the exact values for the unknown functions that will be determined:

$$T_i^{1,\mathrm{exp}} = T(x_i^{\mathrm{exp}}, k^*) + \upsilon_1, \; \upsilon_1 \sim N(0, \sigma_1^2), \; i = 1, 2, \ldots, N_1 \tag{10a}$$

$$T_i^{2,\exp} = T(x^{\exp}, y_i^{\exp}, f^*) + \upsilon_2,\ \upsilon_2 \sim N(0, \sigma_2^2),\ i = 1, 2, \ldots, N_2 \tag{10b}$$

$$I_i^{\exp} = I(\tau_0, \mu_i^{\exp}, \omega^*) + \upsilon_3,\ \upsilon_3 \sim N(0, \sigma_3^2),\ i = 1, 2, \ldots, N_3 \tag{10c}$$

where for IP-1, IP-2 and IP-3, the subscript 1, 2 and 3 holds, respectively, for the: total number of experimental data N, and the random numbers υ drawn from a normal distribution centered at zero with standard deviation σ. $T^{1,\exp}$, $T^{2,\exp}$ and $I^{\exp}$ represent the simulated (synthetic) experimental data to be used for the corresponding inverse problems, i.e. IP-1, IP-2 and IP-3, respectively. The terms k^*, f^* and ω^* represent the exact thermal conductivity, inlet temperature and single scattering albedo functions, respectively. Then, the objective functions are formulated similarly as presented in Eq. (2), where, for each inverse problem, the respective direct problem model is used.

5 Results and Discussion

To obtain the solutions of the inverse problems, the optimization is performed initially with a combination of both the Differential Evolution (DE) [44] and the Nelder-Mead (NM) [30] methods. For every solution obtained, DE starts with a random population of five times the dimension of $\mathbf{u}$, i.e. $5 \times n$. The crossover probability CR and the scaling factor W, i.e. control parameters for DE, are set as $\mathrm{CR} = 0.25$ and $W = 0.85$. The DE method stops when a number of internal loops, which is specific for each problem, are completed, and, then, all the points from DE are used to start the NM simplex method. The NM reflection, expansion, contraction and shrink coefficients are set as 1.0, 2.0, 0.5 and 0.5, respectively. The iterative procedure of NM is stopped when

$$100 \times \frac{\max(\mathbf{F}) - \min(\mathbf{F})}{\bar{F}} \leq \gamma \tag{11}$$

where $\mathbf{F}$ is the vector of objective function values for each NM point, $\bar{F}$ is the calculated mean of such values, and γ a relative small tolerance defined by the user. This dimensionless stopping criterion represents the percentage of how the values of $\mathbf{F}$ are spread relative to its average values, making it more reasonable to use across the three different inverse problems.

5.1 Results for the Inverse Problem 1 (IP-1)

The values of the parameters for the bioheat model are obtained from the literature [11, 18]. Such values are displayed in Table 1. Experimental data are simulated in the range $0 \leq x \leq L$ with steps of $\Delta x = L/200$, which in a real application

Table 1 Exact values of the parameters for the Direct Problem 1 (DP-1)

Parameter	Value	Dimension	Description
L	0.042	[m]	Sample thickness
c_b	3770.0	[J/kg K]	Blood specific heat
ρ_b	1060.0	[kg/m^3]	Blood specific mass
T_b	37.0	[°C]	Blood temperature
ω_b	0.00125	[m^3/s m^3]	Blood perfusion coefficient
q_m	420.0	[W/m^3]	Metabolic heat generation
T_∞	20.0	[°C]	Ambient temperature

can be obtained experimentally with the Photoacoustic Thermography [19, 40] or near-field microwave radiometry techniques [26], for instance. The exact calculated temperatures are corrupted with a noise level of $\sigma_1 = 0.05\,°\text{C}$ in Eq. (10a). In order to damp these errors, and make the retrieval of the sought function more robust, we applied the Total Variation filter (TV) on the simulated experimental data [6]. This filtering method requires the definition of its own regularization parameter, which was obtained in such way to satisfy the following criterion

$$\text{SD}\left(\text{TV}(\mathbf{T}^{1,\text{exp}}) - \mathbf{T}^{1,\text{exp}}\right) = \sigma_1 \tag{12}$$

where SD and TV are the standard deviation and Total Variation functions, respectively, leading to a TV regularization parameter of 1.25.

The epidermis layer was neglected due to its small thickness, thus not influencing the quality of the results exposed in this section [11]. Therefore, the exact thermal conductivity function considered in this work has three constant step values: $k_1 = 0.45$ W/m K in $0 \le x < 0.002$ m for the dermis, $k_2 = 0.19$ W/m K in $0.002 \le x < 0.012$ m for the subcutaneous tissue and $k_3 = 0.50$ W/m K in $0.012 \le x \le L$ for the inner tissue. All the searches are performed considering the stopping criterion tolerance $\gamma = 0.001$, see Eq. (11), the search region bounds of $0.1 \le \mathbf{u_1} \le 0.8$ W/m K and 20 internal loops in DE.

Using an initial parameter of $\sqrt{\phi} = 1.0$, see Eq. (3), and Algorithm 1, divergence was detected at the 4th iteration, as can be seen in the behavior of the R_i sequence in Fig. 4 (solid dots), which changes its growth path at the 4th iteration. The process then restarted with $\sqrt{\phi} = 0.7595$—notice in Fig. 4 the converging behavior of this R_i sequence (solid triangles). The series with $\sqrt{\phi} = 0.7595$ is considered converged with $|\Delta - 1| = 0.0092$, which is ≤ 0.01 (see Step 7 in Algorithm 1).

Using this value of $\sqrt{\phi}$ for $n = 7$, the regularization parameter obtained was $\lambda_{\text{low}} = 1.6577$, and the solution obtained for IP-1 is displayed in Fig. 5a. It is possible to observe that the estimation is in general poor, but mainly in $0.012 \le x \le L$. This may be due to two facts: the coarse mesh of points used to discretize the sought

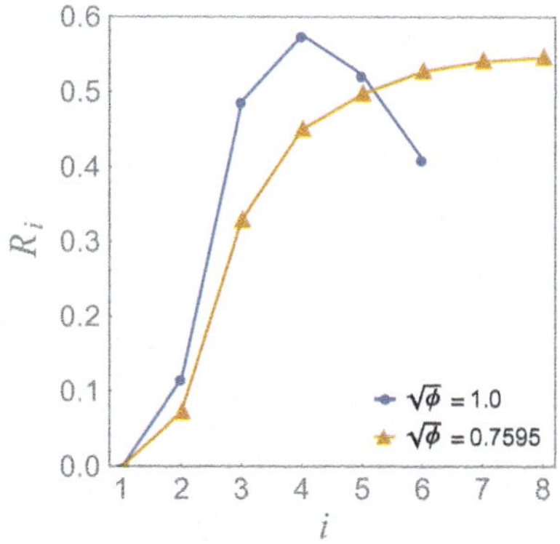

Fig. 4 Evolution of the sequence R_i for the low dimension model with $n = 7$ showing the divergent sequence for $\sqrt{\phi} = 1.0$, and the convergent sequence of the found $\sqrt{\phi} = 0.7595$

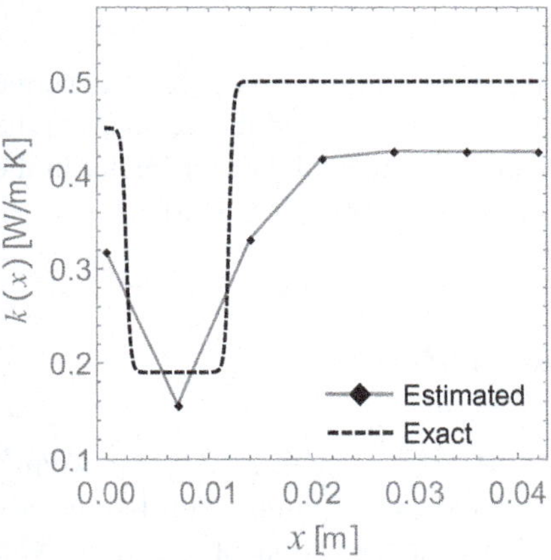

(a) Solution with n = 7 obtained with λ_{low} = 1.6577

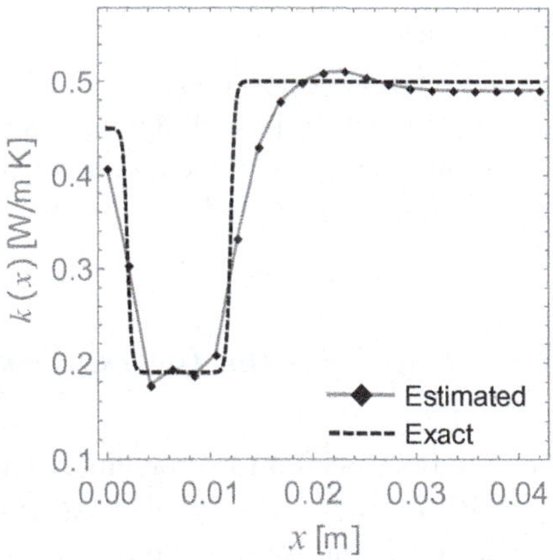

(b) Solution with n = 21 obtained with λ_{high} = 1.3410

Fig. 5 Solutions for IP-1 with the low dimension model (**a**) that generated the estimative for the regularization parameter of the high dimension model (**b**)

function, and a lack of sensitivity, specially in the region of the domain farther from the surface.

Nevertheless, this low order solution generated an initial regularization parameter for the high dimension model with $n = 21$ of $\lambda_{\text{high}} = 3.3650$ calculated according to Eq. (5). Keeping $\sqrt{\phi} = 0.7595$, the process restarts with the new λ and convergence is obtained on the 5th iteration, with $\lambda = 1.3410$. Finally, the recovered thermal conductivity obtained with $\lambda = 1.3410$ is presented in Fig. 5b, where it is possible to see the improvement in the quality of the estimation all over the physical domain.

In Fig. 6 the temperature profile obtained with the solution of the inverse problem with $n = 21$ points is displayed along with the simulated experimental data. The error

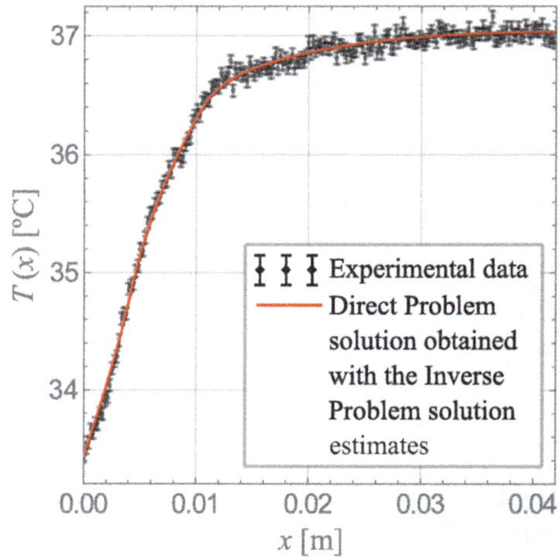

Fig. 6 Simulated experimental data and temperature profile obtained with the solution of IP-1 with $n = 21$

bars of 0.05 °C correspond to the added noise standard deviation used to generate the simulated data. In such figure, it is possible to notice the good agreement of the temperature profile obtained with the simulated experimental data, specially in the regions of the dermis and the subcutaneous tissue, i.e. $0.00 \le x \le 0.012$ m.

5.2 Results for the Inverse Problem 2 (IP-2)

The fluid considered for Problem 2 is air ($k = 0.0265$ W/m K, $\rho = 1.1614$ kg/m^3, $c_p = 1007.0$ J/kg K), and the experimental data are acquired with 101 evenly distributed points in the y direction at the horizontal position of $x = a/5$, with $a = 0.635$ m, representing an acquisition of temperature measurements that can be achieved in real experiments by the use of infrared thermography [37]. The imposed heat flux and average fluid velocity are $q = 500$ W/m^2 and $v_m = 0.025$ m/s. In this chapter, the exact inlet temperature profile $f(y)$ is a step function, which represents a stratified flow of magnitude 26.85 °C in $0 < y \le H/2$ and 46.85 °C in $H/2 < y \le h$, with $H = 0.128$ m. The experimental noise is then added to the exact solution by using a noise level of $\sigma_2 = 0.5$ °C in Eq. (10b). Since the direct solution is less computationally intensive, DE is set to perform 100 internal loops, and NM stopping criterion as $\gamma = 0.005$ in Eq. (11). The search space bounds considered for every point ranges from −73.15 to 326.85 °C.

Using $\sqrt{\phi} = 1$, every test performed for IP-2 converged. For this problem, we present in Table 2 the obtained values for the low dimension model and comparisons with their respective estimates for λ_{high}. Also in Table 2, the values of $\lambda_{21} = 0.4923$ and $\lambda_{31} = 0.5957$ were obtained without the auxiliary reduced model.

As it is possible to observe in Table 2, the estimations for λ_{high} gets closer to the value found by solving the high dimension problem itself. This is expected since the

Table 2 Results obtained for several values for the low dimension model λ, and their respective λ_{high} estimates for IP-2

n	λ_{low}	$\lambda_{21} = 0.4923$	$\lambda_{31} = 0.5957$
6	0.2280	0.4570	0.5597
7	0.3232	0.6263	0.7225
8	0.2807	0.5135	0.5998
9	0.3030	0.5119	0.6048
10	0.3249	0.4996	0.6017
11	0.3402	0.4809	0.5893
12	0.3509	0.4957	0.5991
13	0.3753	0.5052	0.6050

error of the approximation reduces when the dimensions of lower and higher order are closer. Another point to be noticed is that the regularization parameter found for $n = 7$ generated the highest value of λ_{high}. One possible way to explain this is due to the choice of the points in the physical domain, where the error increases depending on how the points are allocated.

Figure 7 shows the solutions obtained for $n = 21$ starting from $\lambda_0 = 0.0001$ (obtained without the auxiliary low dimension problem). It is noticeable how the first solutions are highly oscillatory and, later on, it stabilizes. In fact, these figures demonstrate the behavior of the FP method in converging to the optimal regularization parameter.

The obtained temperature profile with the high dimension model of $n = 21$ points is displayed with the simulated experimental data in Fig. 8. This profile is zoomed at the region of $0.03 \leq y \leq 0.09$ m in order to better visualize the behavior in the middle portion of the channel—the peripheral temperature magnitude is much larger than the ones in such range. The obtained temperature profile displayed shows good agreement with the experimental data.

5.3 *Results for the Inverse Problem 3 (IP-3)*

In the radiative transfer problem, the single scattering albedo considered assumes a parabolic profile along the spatial coordinate, with the exact function given by $\omega^*(\tau) = -0.6\tau^2 + 1.0$. The optical thickness is set as $\tau_0 = 1.0$, and the incident external radiation as $A_1 = 1.0$ and $A_2 = 0.0$, at $\tau = 0$ and $\tau = \tau_0$, respectively, i.e., there is no incident radiation at $\tau = \tau_0$. The experimental data are considered to be evenly acquired in the range $0.1 \leq \mu \leq 0.8$, with steps of $\Delta\mu = 0.05$, and noise level of $\sigma_3 = 0.01$ is used in Eq. (10c). For IP-3, we tested only one low dimension model, with $n = 7$ dimensions. Every solution is obtained with 20 DE internal loops,

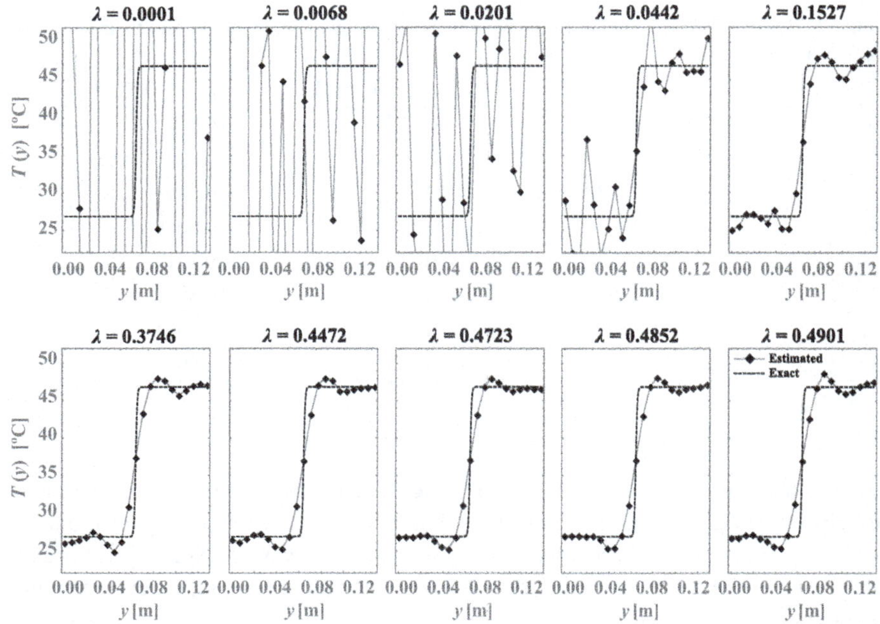

Fig. 7 Evolution of the solution obtained for different values of λ as the FP series converges

Fig. 8 Simulated experimental data and temperature profile obtained with the solution of IP-2 with $n = 21$

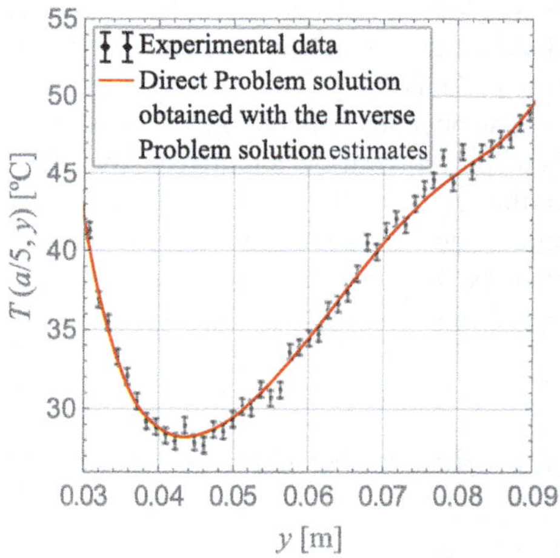

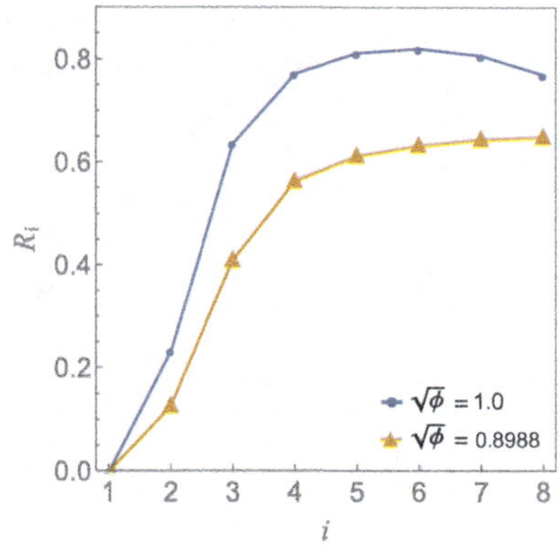

Fig. 9 Evolution of the sequence R_i for the low dimension model with $n = 7$ showing the divergent sequence of $\sqrt{\phi} = 1.0$ (solid dots), and the convergent sequence of the found value $\sqrt{\phi} = 0.8988$ (solid triangles)

and the search is performed in the range $0 \leq \mathbf{u}_3 \leq 1.0$, which are indeed the physical bounds for the single scattering albedo.

Using an initial parameter of $\phi = 1.0$, divergence was detected on the 7th term of the series, as explained in Sect. 2.1. Therefore, using the steps proposed in Sect. 2.1, the new value of $\sqrt{\phi} = 0.8988$ was obtained. With this new ϕ, the series is considered converged with less than 1% of error on the 5th term. In Fig. 9 it is possible to see the evolution of the sequence Ri and how it behaves for each 'phi' used. The solution obtained for this low dimension model is displayed in Fig. 10a. Then, from this low dimension solution, the estimated initial value of the regularization parameter for the high dimension model with $n = 21$ was $\lambda_{\text{high}} = 0.2677$. The problem is solved again with this initial λ_{high}, and the results obtained are displayed in Fig. 10b. For the sake of illustration, the next term of the series calculated by the FP method yielded a regularization parameter of $\lambda_{\text{high}} = 0.2686$, which represents an increase of 0.0009. Therefore, the series can also be considered converged.

It is possible to see in Fig. 10b that the estimate of λ_{high} yielded a good agreement of the estimated values for the single scattering albedo to the exact values. It is important to notice that, due to the lack of sensitivity from $\tau = 0.8$ to $\tau = 1.0$, some information is lost. Although the results displayed in Fig. 10 are equally good in approximating the exact value of the scattering albedo, the increased resolution of $n = 21$ points makes the technique attractive for reducing the computational cost of finding the optimal λ.

Figure 11 presents the radiation intensity profile obtained with the solution of the high dimension model IP-3, i.e. $n = 21$ points. The simulated experimental data is also presented with error bars corresponding to the standard deviation of 0.01. Similarly to IP-1 and IP-2, the solution obtained also shows good approximation with the experimental data.

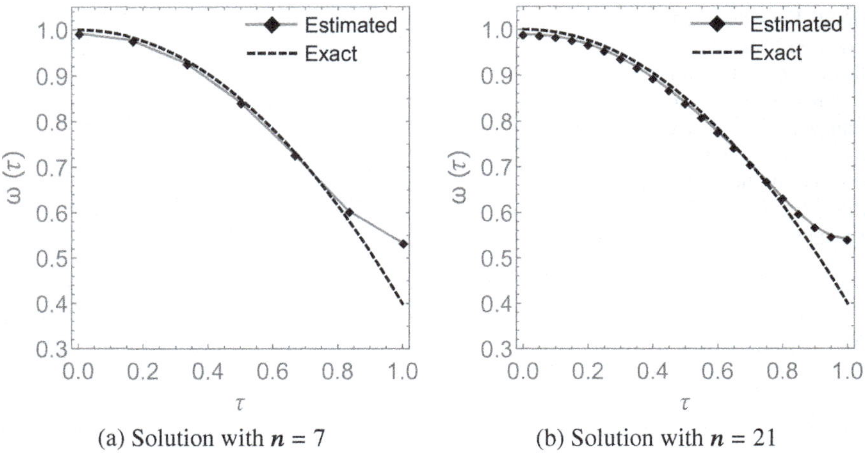

Fig. 10 Solutions for IP-3: **a** low dimension model with $n = 7$, and **b** the high dimension model with $n = 21$, obtained with the estimated value for λ_{high}

Fig. 11 Simulated experimental data and radiation intensity profile obtained with the solution of IP-3 with $n = 21$

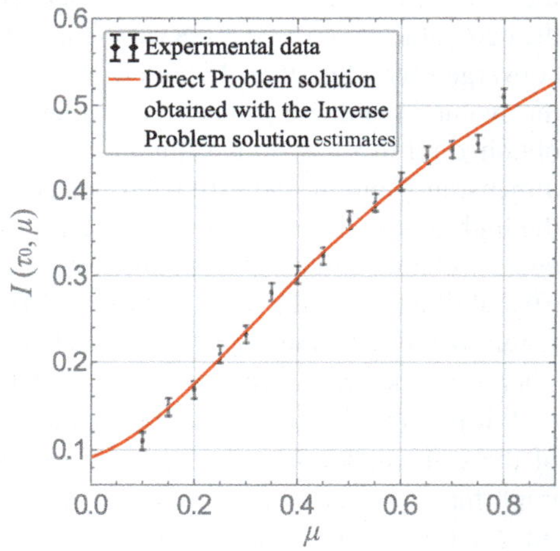

6 Conclusions

In this chapter a technique aimed at obtaining an estimate for the Tikhonov Regularization parameter is presented. This is achieved by finding the optimal parameter of a problem with reduced number of dimensions, and then equalizing the norm of the regularized solutions. The computational experiments were carried out on the scope of inverse heat transfer problems, which are common subjects in the areas of health, renewable energy and climate research. The results have shown that this technique is robust when the low dimension model has sufficient points to represent the sought function, specially when it has several abrupt changes. Investigation must continue on testing this technique on other inverse problems, and also on the development of a more robust and automatic algorithm, especially regarding 2D function estimation problems. Furthermore, the convergence criterion for the Fixed Point Iteration method is discussed in the light of these problems. The presented results reinforce the robustness of the FP method when the proposed series converges to the optimal regularization parameter. However, when the series does not converge, fixing the hyperparameter ϕ of the method is crucial.

Acknowledgements The authors acknowledge the financial support provided by CAPES—Coordination of Superior Level Staff Improvement (Finance Code 001), CNPq—National Council for Scientific and Technological Development, and FAPERJ—Carlos Chagas Filho Foundation for Supporting Research in the State of Rio de Janeiro. The authors acknowledge also the CAPES PrInt program grants (Process Code 88887.469279/2019-00 and 88887.311757/2018-00). This work has been also partially funded by the Spanish Ministry of Economy and Competitiveness with the support of the project TIN2017-86647-P (including funds from the European Regional Development Fund, ERDF).

References

1. Aucejo, M., De Smet, O.: A generalized multiplicative regularization for input estimation. Mech. Syst. Signal Process. **157**, 107637 (2021)
2. Bazán, F.S.V.: Fixed-point iterations in determining the Tikhonov regularization parameter. Inverse Probl. **24**(3), 035001 (2008)
3. Beck, J.V., Arnold, K.J.: Parameter Estimation in Engineering and Science. Wiley, New York (1977)
4. Beck, J.V., Blackwell, B., Clair, C.R.S., Jr.: Inverse Heat Conduction: Ill-Posed Problems. Wiley, New York, NY (1985)
5. Bejan, A.: Convection Heat Transfer, 4th edn. Wiley, Hoboken, NJ (2013)
6. Benning, M., Burger, M.: Modern regularization methods for inverse problems 27, 1–111 (2018)
7. Bokar, J.C., Özisik, M.N.: An inverse analysis for estimating the time-varying inlet temperature in laminar flow inside a parallel plate duct. Int. J. Heat Mass Transf. **38**(1), 39–45 (1995)
8. Bozzoli, F., Cattani, L., Rainieri, S., Bazán, F.S.V., Borges, L.S.: Estimation of the local heat-transfer coefficient in the laminar flow regime in coiled tubes by the Tikhonov regularisation method. Int. J. Heat Mass Transf. **72**, 352–361 (2014)
9. Chalhoub, E.S., Campos Velho, H.F.: Estimation of the optical properties of seawater from measurements of exit radiance. J. Quant. Spectrosc. Radiat. Transf. **72**, 551–565 (2002)
10. Chandrasekhar, S.: Radiative Transfer. Dover Publications, Inc., New York, NY (1960)
11. Cotta, R.M., Cotta, B.P., Naveira-Cotta, C.P., Cotta-Pereira, G.: Hybrid integral transforms analysis of the bioheat equation with variable properties. Int. J. Therm. Sci. **49**(9), 1510–1516 (2010)
12. Elwassif, M.M., Kong, Q., Vazquez, M., Bikson, M.: Bio-heat transfer model of deep brain stimulation-induced temperature changes. J. Neural Eng. **3**(4), 306 (2006)
13. Hansen, P.C.: Analysis of discrete ill-posed problems by means of the L-curve. SIAM Rev. **34**(4), 561–580 (1992)
14. Hansen, P.C.: Rank-Deficient and Discrete Ill-Posed Problems: Numerical Aspects of Linear Inversion. SIAM, Philadelphia (1998)
15. Hansen, P.C., O'Leary, D.P.: The use of the L-curve in the regularization of discrete ill-posed problems. SIAM J. Sci. Comput. **14**(6), 1487–1503 (1993)
16. Horvath, H., Alados Arboledas, L., Olmo, F.J., Jovanovic, O., Gangl, M., Kaller, W., Sánchez, C., Sauerzopf, H., Seidl, S.: Optical characteristics of the aerosol in Spain and Austria and its effect on radiative forcing. J. Geophys. Res. Atmos. **107**, 4386 (2002)

17. Jardim, L.C.S., Knupp, D.C., Sacco, W.F., Silva Neto, A.J.: Solution of a coupled conduction—radiation inverse heat transfer problem with the topographical global optimization method. In: Computational Intelligence in Emerging Technologies for Engineering Applications, pp. 53–71. Springer (2020)
18. Jiang, S.C., Ma, N., Li, H.J., Zhang, X.X.: Effects of thermal properties and geometrical dimensions on skin burn injuries. Burns **28**(8), 713–717 (2002)
19. Ke, H., Tai, S., Wang, L.V.: Photoacoustic thermography of tissue. J. Biomed. Opt. **19**(2), 026003 (2014)
20. Kengne, E., Lakhssassi, A., Vaillancourt, R.: Temperature distributions for regional hypothermia based on nonlinear bioheat equation of Pennes type: dermis and subcutaneous tissues. Appl. Math. **3**(3) (2012)
21. Khanday, M.A., Nazir, K.: Mathematical and numerical analysis of thermal distribution in cancerous tissues under the local heat therapy. Int. J. Biomath. **10**(7), 1750099 (2017)
22. Knupp, D.C., Canato, J.V.M., Silva Neto, A.J., Soeiro, F.J.C.P.: Radiative properties estimation and construction of confidence regions with a combination of the differential evolution algorithm and the likelihood method. Proc. Ser. Braz. Soc. Comput. Appl. Math. **5**(1) (2017)
23. Knupp, D.C., Silva Neto, A.J.: Solution of the inverse radiative transfer problem of simultaneous identification of the optical thickness and space-dependent Albedo using Bayesian inference. Comput. Model. Eng. Sci. **96**(5), 339–360 (2013)
24. Knupp, D.C., Silva Neto, A.J., Sacco, W.F.: Radiative properties estimation with the Luus-Jaakola and the particle collision algorithm. Comput. Model. Eng. Sci. (CMES) **54**(2), 121 (2009)
25. Mellal, I., Oukaira, A., Kengene, E., Lakhssassi, A.: Thermal therapy modalities for cancer treatment: a review and future perspectives. Int. J. Appl. Sci. Res. Rev. **4**(2), 14 (2017)
26. Momenroodaki, P., Haines, W., Fromandi, M., Popovic, Z.: Noninvasive internal body temperature tracking with near-field microwave radiometry. IEEE Trans. Microw. Theory Tech. **66**(5), 2535–2545 (2017)
27. Morozov, V.A.: Regularization Methods for Solving Incorrectly Posed Problems. Springer, New York, NY (1984)
28. Moura Neto, F.D., Silva Neto, A.J.: Two equivalent approaches to obtain the gradient in algorithms for function estimation in heat conduction problems. In: Proceedings of the 34th National Heat Transfer Conference, Pittsburgh, PA (2000)
29. Moura Neto, F.D., Silva Neto, A.J.: An Introduction to Inverse Problems with Applications. Springer-Verlag, Berlin (2013)
30. Nelder, J.A., Mead, R.: A simplex method for function minimization. Comput. J. **7**(4), 308–313 (1965)
31. Ng, E.Y.K., Tan, H.M., Ooi, E.H.: Boundary element method with bioheat equation for skin burn injury. Burns **35**(7), 987–997 (2009)
32. Oliva Soares, P., Silva Neto, A.J., Campos Velho, H.F., Soeiro, F.J.C.P.: A two step inverse problem for vertical temperature profile retrieval in cloudy atmosphere using artificial neural networks. In: Proceedings of the 22nd International Congress of Mechanical Engineering, Ribeirão Preto, Brazil, pp. 4364–4375 (2013)
33. Özen, Ş., Helhel, S., Cerezci, O.: Heat analysis of biological tissue exposed to microwave by using thermal wave model of bio-heat transfer (TWMBT). Burns **34**(1), 45–49 (2008)
34. Özişik, M.N.: Radiative Transfer and Interactions with Conduction and Convection. Wiley, New Jersey (1973)
35. Özişik, M.N., Orlande, H.R.B.: Inverse Heat Transfer: Fundamentals and Applications. Taylor & Francis, New York (2000)
36. Pennes, H.H.: Analysis of tissue and arterial blood temperatures in the resting human forearm. J. Appl. Physiol. **1**, 93–122 (1948)
37. Pradere, C., Joanicot, M., Batsale, J.C., Toutain, J., Gourdon, C.: Processing of temperature field in chemical microreactors with infrared thermography. Quant. InfraRed Thermogr. J. **3**(1), 117–135 (2006)

38. Schena, E., Saccomandi, P., Fong, Y.: Laser ablation for cancer: past, present and future. J. Funct. Biomater. **8**(2), 19 (2017)
39. Shah, R.K.E., London, A.L.: Laminar Flow Forced Convection in Ducts. Advances in Heat Transfer, vol. 1, p. 1. Academic Press, New York (1978)
40. Shah, J., Park, S., Aglyamov, S.R., Larson, T., Ma, L., Sokolov, K.V., Johnston, K., Milner, T., Emelianov, S.Y.: Photoacoustic imaging and temperature measurement for photothermal cancer therapy. J. Biomed. Opt. **13**(3), 034024 (2008)
41. Silva Neto, C.A., Silva Neto, A.J.: Estimation of optical thickness, single scattering albedo and diffuse reflectivities with a minimization algorithm based on an interior points method. In: Proceedings of 17th International Congress of Mechanical Engineering, ABCM, São Paulo, SP, Brazil (2003)
42. Silva Neto, A.J., Özişik, M.N.: An inverse problem of simultaneous estimation of radiation phase function, albedo and optical thickness. J. Quant. Spectrosc. Radiat. Transf. **53**(4), 397–409 (1995)
43. Stephany, S., Becceneri, J.C., Souto, R.P., Campos Velho, H.F., Silva Neto, A.J.: A pre-regularization scheme for the reconstruction of a spatial dependent scattering albedo using a hybrid ant colony optimization implementation. Appl. Math. Model. **34**(3), 561–572 (2010)
44. Storn, R., Price, K.: Differential evolution—a simple and efficient heuristic for global optimization over continuous spaces. J. Glob. Optim. **11**(4), 341–359 (1997)
45. Tikhonov, A.N.: On the solution of ill-posed problems and the method of regularization. Dokl. Akad. Nauk SSSR **151**(3), 501–504 (1963)
46. Wang, J., Silva Neto, A.J., Moura Neto, F.D., Su, J.: Function estimation with Alifanov's iterative regularization method in linear and nonlinear heat conduction problems. Appl. Math. Model. **26**(11), 1093–1111 (2002)
47. Wolfram Documentation Center. https://reference.wolfram.com/. Accessed 15 Mar 2021
48. Zhang, K., Li, W., Eide, H., Stamnes, K.: A bio-optical model suitable for use in forward and inverse coupled atmosphere-ocean radiative transfer models. J. Quant. Spectrosc. Radiat. Transf. **103**, 411–423 (2007)

Emission Control Models

Intelligent Collaborative Freight Distribution to Reduce Greenhouse Gas Emissions: A Review

Christopher Expósito-Izquierdo, Airam Expósito-Márquez, Belén Melián-Batista, José A. Moreno-Pérez, and J. Marcos Moreno-Vega

Abstract Freight distribution suffers from inefficiencies which are responsible for a significant part of greenhouse gas emissions. In addition, they have a negative impact on the performance of companies by reducing profits and increasing costs. Among the measures that can be taken to mitigate these effects are cooperation mechanisms. In this work we review optimization models for horizontal collaborative freight transport that include environmental and economic criteria. Specifically, we consider models for planning delivery routes and for cross-docking.

Keywords Vehicle routing problem · Freight distribution · Horizontal collaborative transport · Cross-docking strategy

1 Introduction

The 2030 Agenda for Sustainable Development, approved by the United Nations in 2015, is articulated in 17 objectives that seek, among others, to "ensure access to affordable, reliable, sustainable and modern energy for all" (objective 7), to "make cities and human settlements inclusive, safe, resilient and sustainable" (goal 11), to "ensure sustainable consumption and production patterns" (goal 12) and to"take

---

C. Expósito-Izquierdo · B. Melián-Batista · J. A. Moreno-Pérez · J. M. Moreno-Vega (✉)
Universidad de La Laguna, San Cristóbal de La Laguna, Spain
e-mail: jmmoreno@ull.edu.es

C. Expósito-Izquierdo
e-mail: cexposit@ull.edu.es

B. Melián-Batista
e-mail: mbmelian@ull.edu.es

J. A. Moreno-Pérez
e-mail: jamoreno@ull.edu.es

A. Expósito-Márquez
Kaizten Analytics, San Cristóbal de La Laguna, Spain
e-mail: airamexpmar@gmail.com

J. L. Verdegay et al. (eds.), *Computational Intelligence Methodologies Applied to Sustainable Development Goals*, Studies in Computational Intelligence 1036,
https://doi.org/10.1007/978-3-030-97344-5_9

urgent action to combat climate change and its impacts" (goal 13). The Agenda is a purposeful commitment to improve the future of the planet, Nature and people by promoting prosperity and the protection of the environment.

Freight last mile logistics is carried out on the transport networks used by the rest of the socio-economic activities related to mobility. The simultaneous presence of freights delivery vehicles, public transport vehicles and private vehicles congests the roads with the consequent increase in travel times and greenhouse gas emissions. It should be noted that, according to data for 2010, road transport was responsible for 35% of total greenhouse gas emissions in France [3]. Furthermore, in large cities, traffic congestion is directly related to the loss of life quality and to the decrease in business competitiveness.

Among the measures that can be implemented to reduce these effects are: (1) those that promote the movement of goods and passengers towards less polluting means of transport, (2) those that try to improve congestion by promoting the shared use of private vehicles, and (3) those that advocate for an intelligent management of the distribution of goods through collaborative transport.

The collaboration seeks to correct inefficiencies in the distribution network, which includes, among others, the design of excessively long routes or the circulation of empty vehicles after delivery. In this regard, it should be noted that a study carried out in the Bangkok area [33] revealed that weekly around 85% of return trips were made with the truck empty. This fact has a negative impact on the efficiency of delivery companies and on the emission of greenhouse gases.

The chapter is structured in 5 sections. In Sect. 2, we describe horizontal, vertical, and lateral collaboration structures. In add ition, we list some of the decisions that can be taken to reduce the environmental impact of logistics activities. Two relevant alternatives are to consider cross-docking mechanisms, and to optimize delivery routes. In Sect. 3, we analyze the scientific literature on vehicle routing problem with cross-docking and in Sect. 4 we review optimization models for planning joint delivery routes. Finally, in Sect. 5 we identify some future research lines.

2 Collaborative Transportation

Collaborative transport models are suitable and efficient alternatives for the freight distribution. These models encourage the shared use of information, processes, tasks, networks, facilities or vehicles by different transport stakeholders.

There are basically three types of collaboration structures [24]. Vertical collaboration occurs between stakeholders belonging to the same supply chain, although they must be located at different levels of the chain (for example, collaboration between suppliers and carriers). In horizontal collaboration, two or more unrelated stakeholders, who usually compete in the same market, decide to share information or capabilities to operate collaboratively (for example, collaboration between carriers from the same geographic region). In lateral collaboration there is a combination of vertical and horizontal collaboration.

Collaboration can take place at the strategic, tactical, and operational levels. Strategic, tactical and operational agreements are related, respectively, to long, medium and short term decisions. As stated in [19], at a strategic level decisions are made regarding the installation of warehouses and distribution centers, at a tactical level joint delivery routes are designed and at an operational level the capacity of delivery vehicles is shared.

The benefits derived from collaborative freight transport include lower costs, improved service quality and reduced environmental impact. In their study, with real data from the city of Bogotá, Montoya-Torres et al. [26] analyzed the collaborative planning and management of urban freight transport. The results obtained show that it is possible both to decrease logistics costs and reduce environmental impact. Another real case is described in [9]. This paper analyzes the supply chain of an agro-food company in the United Kingdom. The adoption of collaboration mechanisms reduces CO_2 emissions by 26%. A list of the main benefits and challenges related to collaborative freight transport can be found in [19].

To reduce the environmental impact of the logistics activities associated with the distribution of freights, a wide variety of decisions can be taken. Some alternatives are listed in [7]: use low-energy vehicles, use inter-modal transport, optimize delivery routes or improve driving. Cross-docking [1] is another interesting alternative to reduce costs and minimize the environmental impact of logistics activities.

In this paper we review optimization models for horizontal collaborative freight transport that include environmental criteria. Specifically, we consider models for planning delivery routes and for cross-docking.

Excellent bibliographic reviews on collaborative transport and on the benefits and challenges of this paradigm can be found in [2, 8, 13, 14, 32]. The reader interested in obtaining a broad perspective on collaborative transport can consult these works.

For the distribution of freights, the models studied so far focus on the design of joint delivery routes to end customers that minimize transportation costs, CO_2 emissions or energy consumed. In general, several variants and extensions of the vehicle routing problem have been considered during modelling. Among the solution techniques used, the heuristic and metaheuristic procedures stand out.

A wide variety of optimization problems related to the cross-docking strategy have been analyzed in the literature in an attempt to reduce greenhouse emissions. Specifically, vehicle routing and scheduling problems with cross-docking have caught the attention of researcher for the last decade. Despite this fact, most papers consider the minimization of operational costs, transportation costs and travel times, instead of environmental criteria. Both exact and metaheuristic algorithms have been proposed for solving these problems.

3 Vehicle Routing Problems with Cross-Docking

Cross-docking is a collaborative strategy of supply chain whereby shipments from several suppliers to a given customer passes through an intermediate transshipment and freight consolidation point, called cross-dock [1, 6, 15, 25, 45]. The incoming products are unloaded, sorted and consolidated according to the final destinations. They are then loaded directly onto outbound vehicles for the shipment to end customers, without cross-dock storage. The time the products stay at the cross-dock does not usually exceed 24 h, although in most cases it is usually less than one hour [4].

Cross-docking is a strategy that allows simultaneously considering the economic, environmental and social requirements to obtain an approach aligned with the sustainable development goals.

In the scientific literature, several cross-docking optimization problems have received the attention of researchers. Among them, problems related to location, truck scheduling, trucks assignment to doors, internal flow and vehicle routing have been widely considered. This book chapter is particularly focused on vehicle routing problems [16, 17, 31, 37] due to their impact on the reduction of the emissions of greenhouse gases. Therefore, the literature on the Vehicle Routing Problem with Cross-Docking (VRPCD) [17, 31] and its variants is analyzed.

The vehicle routing problem with cross-docking consists of designing vehicle routes for carrying out the pickup and delivery of products at minimum cost using a single intermediate transfer point. The vehicles carry the products from the pickup points to the cross-dock, where they are unloaded and loaded before continuing on their delivery routes to the final customers. The literature on cross-docking is quite recent, having been published more than half of the articles that study variants of the vehicle routing problem with cross-docking after year 2005.

Lee et al. [21] are the first authors that consider a variant of this problem, in which all vehicles have to arrive at the cross-dock at the same time in order to guarantee synchronization between input and output operations. The work tackles an integrated model that considers cross-docking and vehicle routing scheduling with the aim of minimizing the transportation cost. The authors develop a tabu search algorithm for solving the problem.

Wen et al. [46] solve the vehicle routing problem with cross-docking considering time windows in the nodes and a time horizon for the performance of all operations, with the purpose of minimizing the total required time. This is the variant that has been referred to as Vehicle Routing Problem with Cross-Docking (VRPCD) in the literature. In this problem, each request associates a supplier with a customer. The vehicles have to perform the pickup operations at the suppliers and reach the cross-dock to carry out the unloading before the delivery operations begin. They propose a mathematical model and an algorithm based on tabu search with adaptive memory. They solve instances with up to 200 requests. Liao et al. [22] also propose a tabu search algorithm to solve the VRPCD in order to minimize the sum of operational and transportation costs, obtaining better results than those reported in [21].

Dondo et al. [11] propose a mathematical model for solving the vehicle routing problem with cross-docking in supply chain management (VRPCD-SCM), in which customers can receive products from different suppliers. The products can be sent either directly to the final customer or through intermediate warehouses or cross-docks. The goals are to minimize the total routing cost and the travel time.

Vahdani et al. [44] propose a hybrid algorithm that combines components of variable neighborhood search, simulated annealing and particle swarm optimization to solve the vehicle routing scheduling with cross-docking introduced by Lee et al. [21]. The hybrid algorithm developed by the authors outperforms the tabu search proposed in [21].

Santos et al. [40] introduce the pickup and delivery problem with cross-docking (PDPCD), in which two kinds of routes are allowed with the purpose of minimizing the total cost. On one hand, the pickup and delivery routes, that pickup and deliver the same products without stopping at the cross-dock. On the other hand, the routes that carry out unloading and loading operations at the cross-dock. The authors propose exact optimization algorithms and corroborate that using different kinds of routes the total costs are lower.

Mousavi et al. [28] develop two mathematical models to solve the integrated problem of locating several cross-docks and vehicle routing scheduling. They consider uncertainty in the input data of the problem, which is dealt by making use of fuzzy logic.

Dondo and Cerdá [10] propose a mathematical model to solve the integrated problem of designing the pickup and delivery routes and assigning vehicles to doors, as well as the scheduling of the vehicles in a single cross-dock. They provide near-optimal solutions for medium-sized problems. Therefore, instances with realistic sizes are not considered.

Morais et al. [27] adapt a constructive heuristic and several local search procedures from the VRP literature to solve the VRPCD. In addition, they propose an iterative local search algorithm. They compare the results with those of the literature, improving some better solutions, and propose a benchmark of larger instances to corroborate the efficiency of the proposed algorithms.

Yu et al. [48] solve the open VRP with cross-dock. They deal with the retail case, where the acquisition of vehicles may not be feasible for retailers who must outsource services to a logistics company. In this situation, the products from different suppliers share vehicles for the subsequent delivery to the final customers. They propose a mathematical model and a simulated annealing algorithm for solving this problem of creating an open flow network of routes.

Maknoon and Laporte [23] propose a mathematical model and an adaptive large neighborhood search to solve the VRPCD. The computational experience carried out demonstrates the efficiency of the proposed methodology. Grangier et al. [16] propose a matheuristic based on large neighborhood search to solve the VRPCD.

Enderer et al. [12] solve the integrated problem of vehicle routing and trucks assignment to doors. Incoming vehicles are assigned to doors of the cross-dock, products are transferred between doors, and routes from exit doors are planned for subsequent delivery to the final customers. They develop mathematical models for their resolution.

Most of the papers that tackle the VRPCD consider that every transportation service connects a single supplier with a single customer and a homogeneous fleet of vehicles. Nikolopoulou et al. [31] introduce a vehicle routing problem with cross-docking, in which a many-to-many relationship between suppliers and customers exist. Moreover, they make use of different fleets to carry out the input and output routes. They propose an adaptive memory method integrated within a tabu search algorithm. The method is competitive compared to the literature for one-to-one VRPCD instances, reaching even some new better solutions. In addition, they make available a new set of instances with various characteristics to solve the many-to-many VRPCD.

Grangier et al. [17] tackle the VRPCD in which the number of docks that can be simultaneously used at the cross-dock is restricted. With the purpose of solving this extended problem, the authors adapt the matheuristic proposed by them in the paper [16]. The work analyzes the impact on the final solutions due to the docks limitation constraint.

From this review, it can be extracted that several variants of vehicle routing and scheduling problems with cross-docking have been considered in the recent literature. Despite the fact that cross-docking has been introduced as a transport strategy with a high impact on reducing greenhouse emissions, environmental criteria have not been properly addressed. On the contrary, most considered criteria attempt to minimize operational costs, transportation costs and travel times.

4 Collaborative Freight Distribution

Different procedures have been proposed to incorporate environmental criteria into freight distribution models. Some procedures estimate the amount of energy required by the vehicle fleet and others measure the carbon footprint. Bektas and Laporte [5] estimate the energy used by a vehicle to traverse an arc as a function of the length of the arc and the weight, load and speed of the vehicle. A standard procedure to measure the carbon footprint can be found at [47].

As mentioned above, in the distribution of freights to end customers, the most used models are variants and extensions of the vehicle routing problem.

In [34] the vehicle routing problem with backhauls and time windows with a homogeneous fleet of vehicles is considered. Three different scenarios are analyzed: total competition (or non-cooperation), total cooperation and mixed system. The energy required is estimated following the procedure proposed by Bektas and Laporte [5]. The authors propose a Scatter Search to solve the previous scenarios. From the results obtained, it is concluded that, in the best cooperative scenario, the solutions obtained reduce greenhouse gas emissions and costs by approximately 30%.

To analyze the benefits of logistics collaboration, a bi-objective vehicle routing problem with time windows is considered in [39]. The proposed model includes economic and environmental criteria (carbon footprint measured according to a standard procedure [47]). The authors consider four different collaboration scenarios that are solved using Scatter Search. The results show that the more resources are shared, the greater the economic and environmental benefits. In the best scenario, greenhouse gas emissions were reduced by 60% and economic savings were 55%.

An inventory routing problem that incorporates environmental and total cost objectives is considered in [41]. The model allows to analyze the benefits of horizontal collaboration between suppliers and customers with respect to emissions, driving time, total cost of route and cost of waste. The model is solved using the commercial CPLEX package. In the practical case under consideration, a reduction of the total cost of 17% and of emissions of 29% is obtained.

Collaborative refrigerated food transport must incorporate new restrictions to guarantee the quality of the products. In [42] a case study of collaborative refrigerated food transport is presented. In this context, suppliers and customers simultaneously optimize routes and inventory plans. In the case study considered, the authors found that collaboration can substantially reduce costs and CO_2 emissions.

In [29] the implantation of a fleet of electric vehicles is analyzed together with horizontal collaboration strategies between carriers to reduce costs and CO_2 emissions. The situation is modeled as a Multi Depot Vehicle Routing Problem that is solved with an Iterated Local Search based on Biased Randomization. The results show that the choice of vehicles depends on the period of use and the variation in demand.

Quintero-Araujo et al. [36] consider and analyze, in terms of economic and environmental costs, three different horizontal cooperation scenarios: non-cooperative scenario, in which companies individually solve their vehicle routing problem; semi cooperative, in which a single vehicle routing problem is jointly solved; fully cooperative, in which a general location-route problem is jointly solved. The problems are solved using a hybrid metaheuristic algorithm combining Biased Randomization and VNS.

In [30] the impact of the location of urban consolidation centers and horizontal cooperation on CO_2 emissions and costs is analyzed. The computational results show that the greater the degree of cooperation, the greater the benefits. The work proposes a metaheuristic algorithm that combines Biased Randomization and Iterated Local Search to solve the different analyzed scenarios.

In [43] a three-step methodology is proposed to assess the benefits of collaboration. A procedure for allocating costs and CO_2 emissions among the coalition participants is also described. In the first step, eco-efficient solutions are obtained by solving a bio-objective Green Inventory Routing Problem. The costs and emissions of the eco-efficient solutions are then calculated. Finally, these values are assigned between partners. The bio-objective problem is solved using the weighting method by Romero and Rehman [38]. In the case study, all the proposed solutions result in less CO_2 emissions than the current solution. In addition, several of them correspond to solutions of lower cost than the current one.

In [18] an inventory location model is proposed to evaluate and analyze the reductions in CO_2 emissions achieved when companies cooperate in a joint supply network. A detailed analysis of this model reveals that cooperation reduces CO_2 emissions by 16% on average.

To analyze the benefits of collaboration, a Collaborative VRP with time windows, heterogeneous fleet and simultaneous pickups and deliveries is considered in [20]. This problem is solved using a neighborhood search heuristic algorithm. The computational results show that average emissions and costs are reduced by 9% and 12%, respectively.

5 Future Research Lines

A logistic network is used to formalize a collaborative transport problem. In this network, each origin, exchange or destination point is represented by a vertex; the roads that connect these points are represented by edges. Freight distribution is carried out by designing routes over this network. Traditionally, it has been assumed that this network is static, which means that the exact values of the elements that define it are known. Thus, it is assumed, for example, that the travel times or the time windows in which the delivery must be made are known. However, in real practical applications these elements are subject to a high degree of uncertainty related, for example, to transport network congestion or demand for deliveries. To the best of our knowledge, the only work that includes the treatment of uncertainty in collaborative freight distribution is [35]. On the other hand, the management of uncertainty in cross-docking has only been considered in [28]. Therefore, the incorporation and management of uncertainty in the models of cross-docking and collaborative freight distribution is a future line of research that should be explored.

Despite its potential as an alternative to reduce greenhouse gas emissions, cross-docking models have not incorporated environmental criteria so far. It is, therefore, another interesting future line of research.

In a real collaborative freight distribution problem, several conflicting objectives must be considered (CO_2 emissions, energy consumed, total costs ...). Although multi-objective models to design pickup and delivery routes have been proposed, multi-objective vehicle routing problems with cross-docking considering environmental criteria have not been studied so far.

Acknowledgements This work has been partially funded by the Spanish Ministry of Science and Innovation (project PID2019-104410RB-I00).

References

1. Agustina, D., Lee, C.K.M., Piplani, R.: A review: mathematical models for cross docking planning. Int. J. Eng. Bus. Manag. **2**(2), 47–54 (2010)
2. Aloui, A., Hamani, N., Derrouiche, R., Delahoche, L.: Systematic literature review on collaborative sustainable transportation: overview, analysis and perspectives. Transp. Res. Interdiscip. Perspect. **9**, 100291 (2021)
3. Ballot, E., Fontane, F.: Reducing transportation CO_2 emissions through pooling of supply networks: perspectives from a case study in French retail chains. Prod. Plan. Control **21**(6), 640–650 (2010)
4. Bartholdi, III, J.J., Gue, K.R.: The best shape for a crossdock. Transp. Sci. **38**(2), 235–244 (2004)

5. Bektas, T., Laporte, G.: The pollution-routing problem. Transp. Res. Part B **45**(8), 1232–1250 (2011)
6. Boysen, N., Fliedner, M.: Cross dock scheduling: classification, literature review and research agenda. Omega **38**(6), 413–422 (2010)
7. Corlu, C.G., Faulin, J., Onggo, B.S., Juan, A.A., de la Torre, R.: Simulation, optimization, and machine learning in sustainable transportation systems: models and applications. Sustainability **13**(3), 1551 (2021)
8. Cruijssen, F., Cools, M., Dullaert, W.: Horizontal cooperation in logistics: opportunities and impediments. Transp. Res. Part E **43**(2), 129–142 (2007)
9. Danloup, N., Mirzabeiki, V., Allaoui, H., Goncalves, G., Julien, D., Mena, C.: Reducing transportation greenhouse gas emissions with collaborative distribution: a case study. Manag. Res. Rev. **38**(10), 1049–1067 (2015)
10. Dondo, R., Cerdá, J.: A monolithic approach to vehicle routing and operations scheduling of a cross-dock system with multiple dock doors. Comput. Chem. Eng. **63**, 184–205 (2014)
11. Dondo, R., Mández, C.A., Cerdá, J.: The multi-echelon vehicle routing problem with cross docking in supply chain management. Comput. Chem. Eng. **35**(12), 3002–3024 (2011)
12. Enderer, F., Contardo, C., Contreras, I.: Integrating dock-door assignment and vehicle routing with cross-docking. Comput. Oper. Res. **88**, 30–43 (2017)
13. Gansterer, M., Hartl, R.F.: Collaborative vehicle routing: a survey. Eur. J. Oper. Res. **268**(1), 1–12 (2018)
14. Gansterer, M., Hartl, R.F.: Shared resources in collaborative vehicle routing. TOP **28**(3), 1–20 (2020)
15. González-La Rotta, E.C., Becerra-Fernández, M.: Cross-docking with vehicle routing problem. A state of art review [Plataformas de intercambio con ruteo de vehículos. Una revisión del estado del arte]. DYNA (Colombia) **84**(200), 271–280 (2017)
16. Grangier, P., Gendreau, M., Lehuédé, F., Rousseau, L.-M.: A matheuristic based on large neighborhood search for the vehicle routing problem with cross-docking. Comput. Oper. Res. **84**, 116–126 (2017)
17. Grangier, P., Gendreau, M., Lehéudé, F., Rousseau, L.-M.: The vehicle routing problem with cross-docking and resource constraints. J. Heurist. **27**(1), 31–61 (2021)
18. Hacardiaux, T., Tancrez, J.S.: Assessing the environmental benefits of horizontal cooperation using a location-inventory model. Cent. Eur. J. Oper. Res. **28**(4), 1363–1387 (2020)
19. Juan Ángel, J., Serrano-Hernández, A., Faulin, J., Pérez-Bernabeu, E.: Horizontal collaboration in freight transport: concepts, benefits, and environmental challenges. Sort **1**(2), 1–22 (2017)
20. Konstantakopoulos, G.D., Gayialis, S.P., Kechagias, E.P., Papadopoulos, G.A., Tatsiopoulos, I.P.: An algorithmic approach for sustainable and collaborative logistics: a case study in Greece. Int. J. Inf. Manage. Data Insights **1**(1), 1000010 (2021)
21. Lee, Y.H., Jung, J.W., Lee, K.M.: Vehicle routing scheduling for cross-docking in the supply chain. Comput. Ind. Eng. **51**(2), 247–256 (2006)
22. Liao, C.-J., Lin, Y., Shih, S.C.: Vehicle routing with cross-docking in the supply chain. Expert Syst. Appl. **37**(10), 6868–6873 (2010)
23. Maknoon, Y., Laporte, G.: Vehicle routing with cross-dock selection. Comput. Oper. Res. **77**, 254–266 (2017)
24. Mason, R., Lalwani, C., Boughton, R.: Combining vertical and horizontal collaboration for transport optimization. Supply Chain Manag. **12**(3), 187–199 (2007)
25. Mavi, R.K., Goh, M., Mavi, N.K., Jie, F., Brown, K., Biermann, S., Khanfar, A.A.: Cross-docking: a systematic literature review. Sustainability (Switzerland) **12**(11) (2020)
26. Montoya-Torres, J.R., Muñoz-Villamizar A., Vega-Mejía, C.A.: On the impact of collaborative strategies for goods delivery in city logistics. Prod. Plan. Control **27**(6), 443–455 (2016)
27. Morais, V.W.C., Mateus, G.R., Noronha, T.F.: Iterated local search heuristics for the vehicle routing problem with cross-docking. Expert Syst. Appl. **41**(16), 7495–7506 (2014)
28. Mousavi, S.M., Vahdani, B., Tavakkoli-Moghaddam, R., Hashemi, H.: Location of cross-docking centers and vehicle routing scheduling under uncertainty: a fuzzy possibilistic-stochastic programming model. Appl. Math. Model. **38**(7–8), 2249–2264 (2014)

29. Muñoz-Villamizar, A., Quintero-Araujo, C.L., Montoya-Torres, J.R., Faulin, J.: Short- and mid-term evaluation of the use of electric vehicles in urban freight transport collaborative networks: a case study. Int. J. Logist. Res. Appl. **22**(3), 229–252 (2019)
30. Nataraj, S., Ferone, D., Quintero-Araujo, C., Festa, P., Juan, A.A.: Consolidation centers in city logistics: a cooperative approach based on the location routing problem. Int. J. Ind. Eng. Comput. **10**(3), 393–404 (2019)
31. Nikolopoulou, A.I., Repoussis, P.P., Tarantilis, C.D., Zachariadis, E.E.: Adaptive memory programming for the many-to-many vehicle routing problem with cross-docking. Oper. Res. **19**(1) (2019)
32. Pan, S., Trentesaux, D., Ballot, E., Huang, G.Q.: Horizontal collaborative transport: survey of solutions and practical implementation issues. Int. J. Prod. Res. **57**(15–16), 5340–5361 (2019)
33. Peetijade, C., Bangviwat, A.: Empty trucks run reduction in Bangkok area towards sustainable transportation. Int. J. Trade Econ. Finance **3**(2), 91–95 (2012)
34. Pradenas, L., Oportus, B., Parada, V.: Mitigation of greenhouse gas emissions in vehicle routing problems with backhauling. Expert Syst. Appl. **40**(8), 2985–2991 (2013)
35. Quintero-Araujo, C.L., Gruler, A., Juan, A.A., De Armas, J., Ramalhinho, H.: Using simheuristics to promote horizontal collaboration in stochastic city logistics. Prog. Artif. Intell. 275–284 (2017)
36. Quintero-Araujo, C.L., Gruler, A., Juan, A.A., Faulin, J.: Using horizontal cooperation concepts in integrated routing and facility-location decisions. Int. Trans. Oper. Res. **26**(2), 551–576 (2019)
37. Rahbari, A., Nasiri, M.M., Werner, F., Musavi, M., Jolai, F.: The vehicle routing and scheduling problem with cross-docking for perishable products under uncertainty: two robust bi-objective models. Appl. Math. Model. **70**, 605–625 (2019)
38. Romero, C., Rehman, T.: Multiple Criteria Analysis for Agricultural Decisions. Springer (2013)
39. Sanchez, M., Pradenas, L., Deschamps, J.C., Parada, V.: Reducing the carbon footprint in a vehicle routing problem by pooling resources from different companies. Netnomics **17**(1), 29–45 (2016)
40. Santos, F.A., Mateus, G.R., Da Cunha, A.S.: The pickup and delivery problem with cross-docking. Comput. Oper. Res. **40**(4), 1085–1093 (2013)
41. Soysal, M., Bloemhof-Ruwaard, J.M., Haijema, R., van der Vorst, J.G.: Modeling a green inventory routing problem for perishable products with horizontal collaboration. Comput. Oper. Res. **89**, 168–182 (2018)
42. Stellingwerf, H.M., Laporte, G., Cruijssen, F.C.A.M., Kanellopoulos, A., Bloemhof, J.M.: Quantifying the environmental and economic benefits of cooperation: a case study in temperature-controlled food logistics. Transp. Res. Part D **65**, 178–193 (2018)
43. Stellingwerf, H.M., Kanellopoulos, A., Cruijssen, F.C.A.M., Bloemhof, J.M.: Fair gain allocation in eco-efficient vendor-managed inventory cooperation. J. Clean. Prod. **231**(10), 746–755 (2019)
44. Vahdani, B., Reza, T.-M., Zandieh, M., Razmi, J.: Vehicle routing scheduling using an enhanced hybrid optimization approach. J. Intell. Manuf. **23**(3), 759–774 (2012)
45. Van Belle, J., Valckenaers, P., Cattrysse, D.: Cross-docking: state of the art. Omega **40**(6), 827–846 (2012)
46. Wen, M., Larsen, J., Clausen, J., Cordeau, J.-F., Laporte, G.: Vehicle routing with cross-docking. J. Oper. Res. Soc. **60**(12), 1708–1718 (2009)
47. World Resources Institute, World Business Council for Sustainable Development: The Greenhouse Gas Protocol. A Corporate Accounting and Reporting Standard. World Business Council for Sustainable Development, Geneva; World Resources Institute, Washington, DC (2008)
48. Yu, V.F., Jewpanya, P., Redi, A.A.N.P.: Open vehicle routing problem with cross-docking. Comput. Ind. Eng. **94**, 6–17 (2016)

Pressure Sensor Placement for Leak Location in Zones of a Water Distribution Network

Marlon J. Ares-Milián, Marcos Quiñones-Grueiro, Cristina Verde, and Orestes Llanes-Santiago

Abstract A novel sensor placement approach for leak location in large-scale water distribution networks (WDNs) is presented. Estimating the exact location of leaks in large-scale WDNs requires a set of sensors with high sensitivity to be distributed. The partition of the WDN into leak zones allows to facilitate the location task. Thus, a topological clustering method aiming to divide the WDN into zones is combined with a novel sensor placement approach for solving the leak zone location problem. The goal of this approach is to determine the leak zone such that the water company can confine the exact location later by using specialized equipment. The Modena network is used as a real-life case study with synthetically generated field data assuming that a single leak occurs. The satisfactory performance of the proposal is demonstrated under uncertain conditions and measurement noise even when the sensitivity of the sensors is limited.

Keywords Water distribution networks · Sensor placement problem · Leak location problem · Metaheuristics

1 Introduction

The reduction of losses in WDNs is a global challenge that worries the supply managers, governments, and the society in general. These losses cause waste of fuels associated with distribution, as well as leading some countries to explore alternatives like desalination, both of which result detrimental to the environment [20].

M. J. Ares-Milián · O. Llanes-Santiago (✉)
Departamento de Automática y Computación, Universidad Tecnológica de la Habana José Antonio Echeverrııa, CUJAE, La Habana, Cuba
e-mail: orestes@tesla.cujae.edu.cu

M. Quiñones-Grueiro
Institute for Software Integrated Systems, Vanderbilt University, Nashville, TN, USA
e-mail: marcos.quinones.grueiro@vanderbilt.edu

C. Verde
Instituto de Ingeniería, Universidad Nacional Autónoma de México, Mexico City, Mexico
e-mail: verde@unam.mx

J. L. Verdegay et al. (eds.), *Computational Intelligence Methodologies Applied to Sustainable Development Goals*, Studies in Computational Intelligence 1036,
https://doi.org/10.1007/978-3-030-97344-5_10

Leaks cause a deteriorated service, frequent shutdowns for long time intervals and quality contamination. Moreover, losses caused by leaks and bursts represent statistically between 20 and 40% of the total pumped water [1]. Therefore, the design of methods to detect and locate leaks based on measurement data collected from several points of the network in real time has become an important topic of interest.

The knowledge about the internal variables of a system plays an important role for obtaining a reliable monitoring system [14]. In the case of WDNs, the number and type of sensors (flow-rate, pressure heads, or both), their distribution in the network, the minimal leak size to be detected, and the tolerance error for the leak location task altogether determine the feasibility of the leak monitoring system. Hence, researchers have paid attention to the sensor placement problem (SPP) for designing leak monitoring systems [4, 6, 19]. In most works, pressure head measurements from the network nodes are used because this type of sensors are cheaper and easier to install and maintain than flow meters. Therefore, the sensor placement strategy proposed in this work only considers pressure head sensors.

Overall, the solution to the SPP is related to the leak detection and location methods. Some placement methods in which the design criteria is the minimization of the number of non-detectable leaks have been proposed [10, 12, 23]. Meanwhile, other placement methods in which the design criteria is the minimal location error have also been proposed [2, 4, 6, 19]. This paper focuses on the latter. Therefore, the task is formulated as an optimization problem in which the leak location error is minimized given a number of distributed sensors.

Although leaks can occur at any point of the network, the localization problem is usually simplified by only considering the nodes as the potential locations [4, 6]. This simplifies the optimization problem. However, the effect of uncertainties in the consumer demand, the leak magnitude, and measurement noise compromises the reliability of the leak monitoring application. On the one hand, a possible solution to this problem is to increase the number of sensors to capture the deviations caused by leaks in real time [6]. On the other hand, recent approaches propose to improve the location reliability by modifying the dimension of the leak search space [5, 19, 22, 25, 26]. The dimension of the search problem can be reduced by assigning the position of the leak to the nearest node in the network [22]. Additionally, the search space can be further reduced by partitioning the network into non-overlapping sets of nodes such that the goal of the leak location task consists, therefore, in the identification of the partition where the leak occurs.

Network partitioning can be achieved by clustering together the nodes where the deviations from the nominal value of the hydraulic variables caused by a leak are indistinguishable from a mathematical standpoint [26]. Nonetheless, when the clustering task is performed by using the hydraulic variables of the network, such as pressure heads at the nodes, the solution obtained will depend on the conditions of the network. Thus, knowledge about operating conditions of the network in real-life scenarios (i.e. the consumers' demand) is required. It is difficult, however, to characterize all the possible operating conditions of a network in advance. Therefore, the topological parameters of the network have been used for network clustering such that nodes are grouped according to their geospatial proximity [19].

Another relevant aspect for model-based leak location methods is that the mathematical model is limited. Thus, the success of model-based leak localization methods deployed in a real network highly depends on how the uncertainty of the system is considered for designing the monitoring system [2]. Consequently, the effect of uncertainty in demands, leak size, and sensor measurement noise must be taken into account, along with the effect of limited sensor sensitivity.

The previous outcomes motivate a new proposal where the general goal is to design a sensor placement methodology for reliable leak location based on partitioned WDNs. The main contribution is the formulation of a new methodology that combines two search problems in order to achieve a tolerable error by assuming a given number of available sensors for deployment. One of the problems is associated with the partition configuration of the network given by K areas, and the other is related with the sensor placement such that a leak location criterium is minimized for the given configuration. The particular algorithm developed is based on tools of artificial intelligence and its basic functions are: partitioning the network into K different zones based on clustering tools where only topological information is available, and placing the sensors for leak zone location based on pattern recognition methods.

The structure of this paper is the following: Sect. 2 sets the theoretical basis for the rest of the paper. In Sect. 3, the methodology followed to partition the WDN, placing the sensors, and detecting and locating the leaks is presented. Section 4 presents the case study and the computational tools used. An analysis of the results obtained is presented in Sect. 5. Finally, conclusions are issued and recommendations for future works are made.

2 Preliminaries

2.1 Hydraulic Modeling of Water Distribution Networks and Leak Scenarios Generation

A WDN fed by $R_e \geq 1$ reservoirs is formed by a set B of interconnected pipes through a set of N nodes. The normal steady-state behavior of the network can be described according to first principles as follows: For any node $n_i \in N$, connected to b_i pipes, with their respective flows $q_{i,j}$, and the consumer water demand d_i at the node, relation $\sum_{j=1}^{b_i} q_{i,j} = d_i$, is satisfied; where $q_{i,j}$ is the flow rate between the nodes n_j and n_i. The sum of pressure heads for any closed loop of the network formed by p components equals zero. Thus, if there are ps source components (h_s) in the loop and pr drop components (h_r), the relation (1) is satisfied with $p = ps + pr$:

$$\sum_{s=1}^{ps} h_s + \sum_{r=1}^{pr} h_r = 0. \tag{1}$$

The flow rate $q_{i,j}$ and the pressure loss $h_{i,j}$ in any component is related according to $h_{i,j} = \theta q_{i,j}^{\alpha}$, where the parameter θ depends on the specific component and the exponent α generally has a value close to 2 [13].

Although leaks can occur indistinctly in the pipelines or the nodes, they are generally modeled as unknown aggregated demands in the nearest nodes for simplicity [4, 6, 10, 19]. Thus, a leak with outflow l_i in node i is modeled by $\sum_{j=1}^{b} q_{i,j} = d_i + l_i$; where $l_i = C_i h_i^{\gamma}$, $\gamma = 0.5$, and C_i is the discharge coefficient, which depends on the leak cross section area and the gravitational acceleration [21].

Obtaining real data from leaks of different sizes in every node of interest of a medium to large WDN is difficult because it is unlikely that they occur. A solution to this issue is to use a model of the network for generating synthetic leak scenarios. If the scenarios used in the optimization problem are not representative of the operating conditions of the network and the leak sizes that can occur, then the configuration obtained will perform poorly in real-life situations. Therefore, the feasibility of using synthetic leak scenarios for sensor placement depends on considering the stochastic conditions of the real system in the simulations. Further details on how the leak scenarios are generated in this work can be found in Sect. 4. From the generated leak scenarios, a set of residual vectors are stacked in a matrix R. A residual vector is defined as follows:

Definition 1 A residual vector $\mathbf{r} \in \mathbb{R}^M$ is defined as

$$\mathbf{r} = \mathbf{h} - \hat{\mathbf{h}}, \tag{2}$$

where M is the number of sensors distributed in the network, $\mathbf{h} \in \mathbb{R}^M$ is obtained from stochastic simulations and $\hat{\mathbf{h}} \in \mathbb{R}^M$ does not consider any uncertainty.

2.2 *Partitioned WDN and Zone Definition*

As explained previously, the WDN is partitioned into non-overlapping sets of nodes for simplifying the sensor placement task. Therefore, the following definitions will be considered throughout the paper:

Definition 2 A zone Z of a WDN formed by N nodes is defined as a subset of κ nodes $Z = \{n_1, n_2, \ldots, n_\kappa\}$ such that $\kappa < N$ and $n_i \in \mathbb{Z}_+$.

Definition 3 A partitioned WDN formed by N nodes is defined as a set of K non-overlapping zones $\mathbf{Z} = \{Z_1, Z_2, \ldots, Z_K\}$ such that $Z_k = \{n_1, n_2, \ldots, n_{\kappa_k}\}$, $\sum_{k=1}^{K} \kappa_k = N$, κ_k represents the number of nodes of zone Z_k and for any node $\{n_i \subset Z_k | n_i \not\subset Z_j, k \neq j\}$.

3 Methodology

The goal of the proposed methodology is to find the combination of network partition and sensor configuration that satisfies an error tolerance in the leak location task. The proposed methodology is outlined in Fig. 1. The first step of the methodology is to perform a topological clustering algorithm. This method considers the topological information extracted from the network, which is represented by Λ. Further details about the clustering method can be found in [17]. The parameter K represents the number of zones to divide the WDN into. This parameter defines the granularity in the leak location problem. For example, if $K = N$, the goal of the leak location task is to find the node where the leak occurs. A high number of zones (K^*) can be set initially according to the network size and number of sensors available.

The second step is to solve the SPP for K zones and M sensors to be placed, based on a set of leak scenarios to be located. The number of sensors to be considered is usually set according to the water company budget. All the matrices and leak scenarios are generated off-line before running the proposed methodology. All leak scenarios are considered in the residual matrix R and leak signature matrices S generated separately as it is explained in Sect. 3.1.

The sensor placement method is described in Sect. 3.2. A sensor configuration is obtained for the partitioned network with a leak location error (Err). If the desired error tolerance is not achieved (Err $>$ Err_{des}) then the number of zones is reduced and the two steps are repeated. If the method converges, a sensor configuration and network partitioning are obtained. Otherwise, the practitioner can select the config-

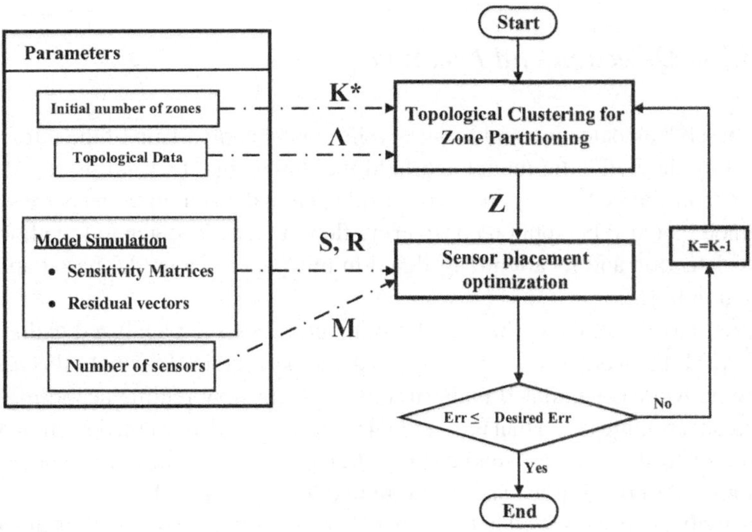

Fig. 1 Sensor placement methodology for leak zone location

uration that best suits the available number of sensors. The following assumptions are considered:

- A single leak occurs in each scenario.
- Minimum night flow (MNF) conditions: Since it is difficult to characterize all the network operating conditions, only (MNF) regime is assumed in this work during early morning hours.
- Only pressure head sensors (piezometric pressure head –mH_2O–) are considered in the placement problem. As stated during the introduction, these sensors are cheaper and easier to install and maintain than flow meters.
- The leak detection problem is assumed to be solved by using other techniques not discussed in this paper based on network input flow data [9, 16], or pressure head measurements from different nodes [15, 18].
- Steady state conditions: The sampling time usually considered by water companies ranges from 15 min to 1 h. Therefore, the dynamic behavior of the variables cannot be captured, and thus, it is not analyzed here.
- Existing errors in the calibration of some network model parameters, such as pipe friction coefficients, has a minimal or null impact on the leak localization task. This assumption has been made in previous proposals which have been successfully applied to real scenarios [22].

Overall, the proposed methodology is appealing to companies because it allows to obtain a solution that accommodates both the budget and the desired error in the leak location tool. The methodology is designed to be used in off-line settings by considering a large range of leak scenarios.

3.1 Leak Detection and Location

Model-based leak detection and location relies in the comparison of the information provided by the hydraulic model and field measurements. Pressure head measurements are considered here, as previously explained in the assumptions. Nevertheless, the method can also be applied considering flow-meters. The general methodology for leak detection and location considered in this paper is a model based approach presented in Fig. 2.

As previously explained, the leak detection problem is not analyzed in this work. Therefore, all the leaks simulated are considered detectable. To satisfy this assumption, the network is simulated under minimum night flow regime according to the different uncertainties such that the residual vectors exceed a certain detection threshold. Moreover, it must be remarked that strategies for leak detection under MNF conditions have proven to be successful in real life conditions [9].

Sensitivity analysis is used in this paper for leak location. The relationship among leaks and changes in the pressure head at different nodes of the network is modeled with a sensitivity matrix $S \in \Re^{M \times L}$ as follows [4]:

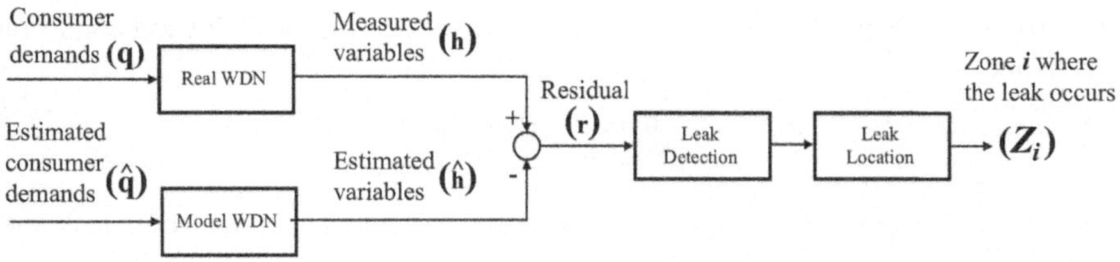

Fig. 2 Leak detection and location scheme

$$S_u = \begin{bmatrix} \mathbf{s}_1 \ \mathbf{s}_2 \cdots \mathbf{s}_L \end{bmatrix}; \quad \mathbf{s}_j = \begin{bmatrix} s_{1,j} \\ s_{2,j} \\ \vdots \\ s_{M,j} \end{bmatrix}; \quad s_{i,j} = \frac{h_{i,j} - h_i}{l_j}; \tag{3}$$

where L is the number of leaks to be located in the network nodes, $\mathbf{s}_j$ is the signature vector for leak scenario occurring in node j given M sensors, h_i and $h_{i,j}$ are the pressure heads estimated in node i without leak and with leak occurring in node j respectively, and l_j is the leak outflow. It is assumed that leaks occur only in $L \leq N$ nodes. Here, the index u denotes the operating conditions and leak magnitudes under which the matrix S has been obtained.

Leak location relies on evaluating the similarity between a residual vector and each leak signature vector, and it depends on how the residuals are generated and the similarity metric used. The simplest approach will be the one adopted in this work, where the difference is used straightforwardly. Different similarity metrics have been empirically compared in previous works, yet the one based on the correlation function given by the inner product of the residual vector and the leak signature vector (equivalent to the evaluation of the cosine angle between the vectors) provided the best results [3, 6]. The similarity measure is defined by

$$\psi_j(\mathbf{r}) = \frac{\mathbf{r}^{\mathrm{T}}\mathbf{s}_j}{|\mathbf{r}||\mathbf{s}_j|}, \tag{4}$$

where ψ_j represents the projection of the residual into the leak signature vector.

The matrix S_u is defined for specific leak magnitude and operating conditions of the network, therefore it is not unique. Consequently, the operating conditions and leak magnitude must be estimated to select an appropriate matrix that solves the leak location problem. On the one hand, it is difficult to determine all the possible operating conditions in advance, and a large number of sensitivity matrices would be required. However, under minimum night flow regime, assumed in this paper, the operating conditions do not change and a single sensitivity matrix is enough. On the other hand, it is difficult to estimate the leak size. Therefore, a set of sensitivity matrices $\mathbf{S} = \{S_1, S_2, \ldots, S_U\}$ is defined for different leak sizes. Therefore, the leak location problem in nodes of the WDN is formulated as follows:

Definition 4 A leak location is estimated to be in node j_u if the following condition is satisfied:

$$\boldsymbol{\psi}_{j_u}(\mathbf{r}) = \max_{u\in[1,U]}\left(\max_{l_u\in[1,L_u]}(\boldsymbol{\psi}_{l_u}(\mathbf{r}))\right); \tag{5}$$

where $\boldsymbol{\psi}_{l_u}(\mathbf{r})$ represents the similarity between the residual vector $\mathbf{r}$ and a leak signature vector $\mathbf{s}_{l_u}$ of matrix S_u.

3.2 Sensor Placement Problem

The objective of sensor placement is to select the sensor positions that minimize the leak location error given M available sensors. A configuration of sensors is defined by the following binary vector:

$$\mathbf{x} = \{x_1, x_2, \ldots, x_N\}; \quad \sum_{i=1}^{N} x_i = M; \tag{6}$$

where $x_i = 1$ if pressure head in node i is measured, and $x_i = 0$, otherwise. The leak signature vector and residual can be obtained as follows:

$$\mathbf{s}_j(\mathbf{x}) = X(\mathbf{x})\mathbf{s}_j^*, \quad \mathbf{r}(\mathbf{x}) = X(\mathbf{x})\mathbf{r}^*; \tag{7}$$

where $\mathbf{s}_j^* \in \Re^N$ and $\mathbf{r}^* \in \Re^N$ consider all the nodes, and $X(\mathbf{x}) = \text{diag}\{x_1, x_2, \ldots, x_N\}$. Accordingly, the similarity metric presented in Eq. (4) can be reformulated as

$$\boldsymbol{\psi}_j(\mathbf{r}^*, \mathbf{x}) = \frac{\mathbf{r}^{*\mathrm{T}} X(\mathbf{x})\mathbf{s}_j^*}{|X(\mathbf{x})\mathbf{r}^*||X(\mathbf{x})\mathbf{s}_j^*|}. \tag{8}$$

Given I leak scenarios, a set of $\mathbf{R} = \{\mathbf{r}_1, \mathbf{r}_2, \ldots, \mathbf{r}_I\}$ residual vectors are generated for testing the quality of a sensor configuration. Thus, the placement of the sensors is a combinatorial optimization problem formalized as follows [4]:

$$\mathbf{x} = \arg\min_{\mathbf{x}\in\mathbb{Z}_+{}^N} \sum_{i=1}^{I} \frac{e_i(\mathbf{x})}{I}, \quad \text{subject to} \quad \sum_{i=1}^{N} x_i = M; \tag{9}$$

$$e_i(\mathbf{x}) = \begin{cases} 0 & \text{if } j_u = j_i \mid \boldsymbol{\psi}_{j_u}(\mathbf{r}_i^*, \mathbf{x}) = \max_{u\in[1,U]}\left(\max_{l_u\in[1,L_u]}(\boldsymbol{\psi}_{l_u}(\mathbf{r}_i^*, \mathbf{x}))\right) \\ 1 & \text{otherwise} \end{cases}; \tag{10}$$

where j_i and j_u are the true and identified node locations for scenario i respectively, and L_u is the candidate set of potential nodes where the leaks can occur for sensitivity matrix S_u. The formulated problem penalizes all errors equally, independently of the topological distance from the real leak to the identified location. In addition, it only considers the exact location of the leak node as a successful result. Moreover, the performance depends on the sensitivity matrix used for generating the leak signature vectors. Thus, it is not a feasible approach for leak location in large-scale networks where the node distribution is not uniform. Below, the problem is reformulated for increasing the robustness of the sensor configuration by using the location function in (5), and reducing the dimension of the search space as follows:

$$e_i(\mathbf{x}) = \begin{cases} 0 & \text{if } j_u \subset Z_i \mid \boldsymbol{\psi}_{j_u}(\mathbf{r}_i^*, \mathbf{x}) = \max_{u\in[1,U]} \left(\max_{l_u\in[1,L_u]}(\boldsymbol{\psi}_{l_u}(\mathbf{r}_i^*, \mathbf{x}))\right) \\ 1 & \text{otherwise} \end{cases} ; \quad (11)$$

where Z_i is the true zone location ($j_i \subset Z_i$) such that the identified node belongs to the same region that the node corresponding to the ith scenario. Hence, the new candidate leak locations are zones $\mathbf{Z} = \{Z_1, Z_2, \ldots, Z_K\}$, where each one contains a different set of nodes $Z_k = \{n_1, n_2, \ldots, n_{K_k}\}$. For the sake of this analysis, a crisp partition of the search space into K clusters must be performed before solving the SPP.

4 Computational Experiments

4.1 Case Study

A large-scale network that represents a version of the distribution system of Modena city is selected for assessing the leak location capacity achieved with the sensor placement approach presented in this paper. The looped structure of the network with 317 pipes and 268 demand nodes is shown in Fig. 3. The pipe diameters are set according to Wang et al. [24]. The network is gravity-fed by four reservoirs.

4.2 Simulation Framework

EPANET [21] package, coupled with $MATLAB^{®}$, is used for steady-state simulation of the network. Since a WDN model is not perfect in real life scenarios, various forms of uncertainty are included in the simulations. A uniformly sampled distribution is used for generating the uncertain values of the demand and leak size within their respective range, while measurement noise is sampled from a Gaussian distribution. The following assumptions are made for each experiment:

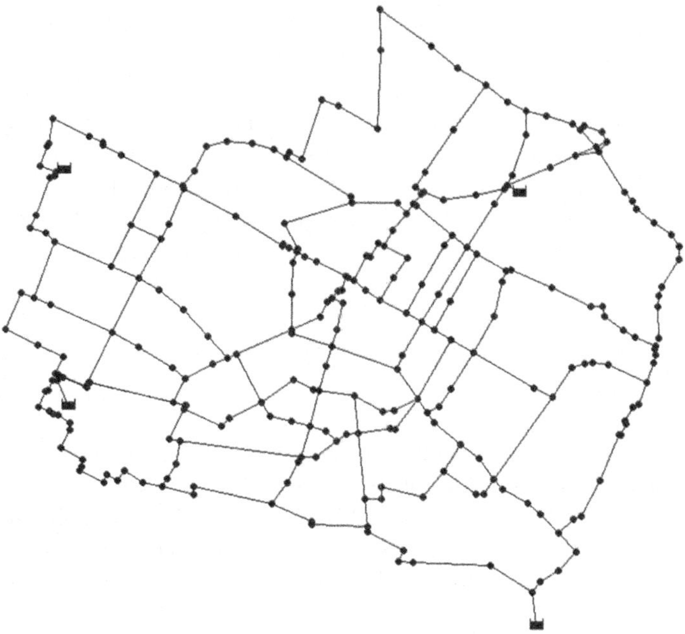

Fig. 3 Modena network

1. A single leak occurs with a magnitude related to the emitter coefficient size selected from a uniformly sampled distribution within the range $C_e \in [0.1, 2]$. The leak outflow represents between 0.1 and 3% of the network total demand. This range is selected according to the pressure head drop caused by leaks. On the one hand, a smaller emitter coefficient causes leaks which cannot be well detected given the demand uncertainty. Moreover, smaller leaks may not cause quality service problems because the pressure head is above the minimum required for this network (20 mH_2O). On the other hand, a larger emitter coefficient provokes leak outflows which are probably first reported by customers.
2. Pressure head measurements are corrupted with additive white noise (zero mean and 0.025 mH_2O of variance).
3. The leak monitoring is performed during minimum night flow conditions.
4. The demand has an uncertain value sampled from a uniform distribution within the range $[-0.05d_n, 0.05d_n]$, being d_n the nominal consumption at node n. Thus, the demand variation is scaled with respect to the nominal node consumption such that high uncertainty is associated to nodes of high consumption. The variance threshold of 5% is selected because minimum night flow regime is assumed. Otherwise, the variance could be higher.
5. A sampling period of 15 min is considered and hourly average values of the measurements are used aiming to reduce the effect of uncertainties.

A total of 10 leaks with random magnitude in the range $C_e \in [0.1, 2]$ at each node will be generated for testing the performance of each configuration (2680 leak scenarios). The following experimental studies will be performed for assessing the robustness and reliability of the proposal: (1) Relationship between performance—number of zones—number of sensors. (2) Performance degradation caused by uncertainties. (3) Influence of the sensors' sensitivity.

4.3 Solution Framework for Sensor Placement

The combinatorial optimization problem proposed in Sect. 3.2 cannot be solved with exhaustive search in a reasonable time for large-scale networks even in off-line settings. Therefore, three meta-heuristic methods: Genetic Algorithms (GAs), Particle Swarm Optimization (PSO), and Ant Colony Optimization (ACO) are compared to find a near-optimal sensor placement solution. In all three cases, the fitness function used is defined in Eq. (9), with evaluation of error according to Eq. (11). As one of the stopping criteria, a maximum of 50 stall iterations was selected. An iteration is considered stall if the best fitness function value obtained has changed less than a tolerance of 10^{-10} compared to the previous iteration. Since the initial candidate solutions are randomly initialized, the results might be conditioned by these initial values. Thus, the three algorithms are run five times and the best results are selected. A set of 20 sensitive matrices were used by generating leaks with an emitter coefficient step size of 0.1 given the previously defined range. An overview of all three methods is presented below.

GAs are optimization methods founded on the principles of genetics and natural selection [11]. GAs are population-based methods where each individual represents a possible solution to a given problem. For the SPP, a solution is a configuration represented with bit string type data (binary vectors representing $\mathbf{x}$). Three mathematical operators are applied over the population of each generation in search for the fittest solution: selection, crossover and mutation. Also, elitism is applied for retaining the best solution of each generation in the next one. The stopping conditions are to achieve a tolerance of 10^{-10} for the fitness function or a maximum of 400 generations. A population size of 50 members is selected.

ACO is a metaheuristic method for solving combinatorial optimization problems. It is inspired on the behaviour of groups of ants when exploring for food [7]. For its implementation, a set of probabilistic model parameters is defined, which are called *pheromone trails*. For the sensor placement optimization problem, a pheromone parameter was defined for every node in the network, which was updated with each iteration of the method considering the value of the fitness function for each one of 40 ants defined as the candidate solutions to the optimization problem. Since all network nodes are equally probable for a sensor to be installed in, no heuristic parameters were considered when implementing the ACO method. The stopping conditions are also to achieve a tolerance of 10^{-10} for the fitness function, or a maximum of 300 iterations.

PSO is an optimization method that presents a set of candidate solutions called particles, which are randomly initialized for a first iteration and then modified with every subsequent iteration. Inspired by the behaviour of social animals such as some types of flocking birds and school fish, as well as particle swarm theory, every particle moves through the search space considering the best fitness function scores achieved by itself and by the particle set (of 50 particles) as a whole for the previous iterations [8]. In order to adapt PSO to the combinatorial optimization problem, a continuous version $\tilde{\mathbf{x}}$ of the sensor placement vector $\mathbf{x}$ is searched within the interval $\tilde{\mathbf{x}} \in [0; 1]$. Then, for every candidate solution, binarization is effected by assigning a value of 1 to the M features in $\tilde{\mathbf{x}}$ with higher values and the value 0 to the rest, generating the corresponding $\mathbf{x}$ vector. A maximum of 400 iterations is set as a stopping criterion.

5 Results and Discussion

Reduction of the dimension of the search space through the topological clustering method is first applied. The results for 10, 15, 20 and 25 zones are shown in Fig. 4. The network partitioning is well defined from a topological point of view since all nodes of each cluster are connected. The dimension of the leak search space increases with the number of zones making the SPP more difficult to solve because of the influence of uncertainties. The number of nodes in each zone varies according to the network topology from 4 to 19 when 25 zones are defined.

The performance measure used in this work is based on the confusion matrix $A = [A(i, j)]$ generally used in pattern recognition problems. The element (i, j) of A represents the number of leaks that occur in zone i, which are located in zone j. Thus, the percentage of leaks that has been correctly located set the overall good performance (Ac). Given z zones, Ac is computed according to $\text{Ac} = \frac{1}{m} \sum_{i=1}^{z} A(i, i)$ where m is the number of leak scenarios (2680 in this case).

The dependency between location performance, number of sensors, and number of zones without any type of uncertainty is shown in Fig. 5. When uncertainty is not considered, a performance of 98% can be attained with 5 sensors, even for 25 zones, for the three optimization methods used. In the case of ACO, a minimum performance of over 99% is achieved. The leak location is still not perfect because there are nodes of the network which present a similar leak signature given the sensor positions. If more sensors than zones are used, the leak location performance is practically 100%. Note that, when 20 sensors are placed, the leak location performance also achieves almost 100% even for 25 zones because the sensor configuration allows to distinguish all leaks in the leak signature space.

In real life scenarios, there are different sources of uncertainty that make the leak location task more complicated. The performance under uncertain conditions is shown in Fig. 6. Note that, when 25 sensors are used, a perfect leak location is not achieved because uncertainties make some leak signatures indistinguishable even when only 10 zones are considered. As the number of zones increases, the location performance deteriorates for each number of sensors. If a reasonable desired per-

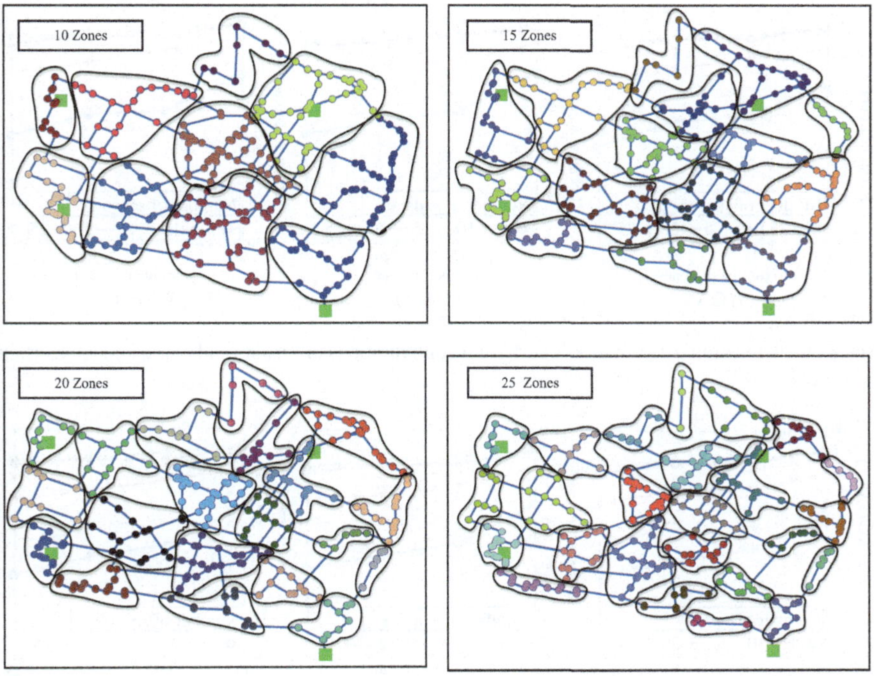

Fig. 4 Reduction of the dimension of the search space through topological clustering

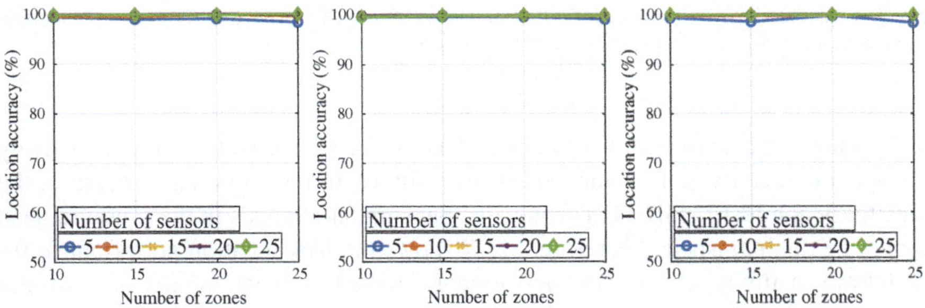

(a) GA (b) ACO (c) PSO

Fig. 5 Leak location performance for different configurations of sensors and zones

formance of approximately 90% is expected for 25 zones, then only 15 sensors are required when using ACO or PSO as a solution framework for the SPP. GA on the other hand, requires 20 sensors to achieve a performance over 90%, since, for the 25 zones and 10 sensors case, only 88.7% performance is achieved. Nevertheless, this represents a good balance between cost and performance overall. Consider also that, when the number of sensors increases, the performance does not change meaningfully. For instance, when 25 sensors are used, the performance for the 25 zone case barely reaches 94% using PSO and 93% with the other two methods.

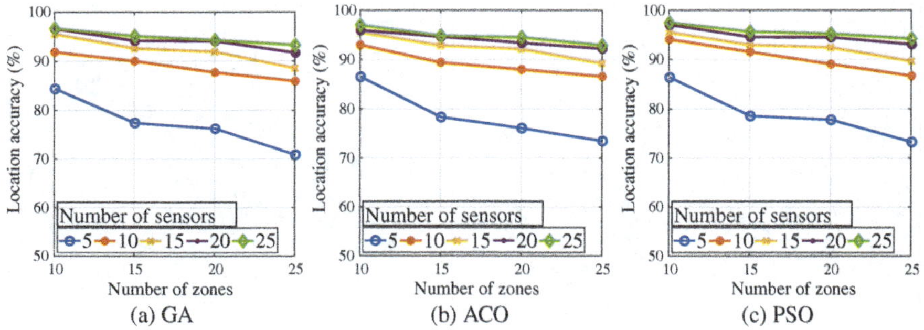

Fig. 6 Leak location performance for different configurations of sensors and zones under uncertain conditions

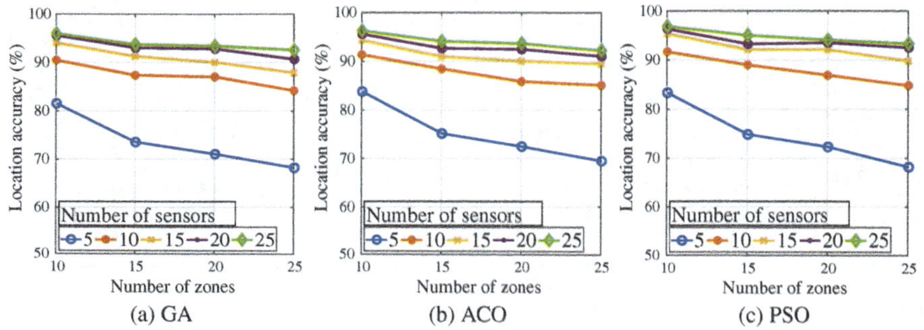

Fig. 7 Leak location performance for different configurations of sensors and zones under uncertain conditions and pressure head sensitivity of 0.1 mH_2O

Pressure head sensors are not perfect. One of the sensor features that negatively impacts the location performance is the sensitivity. In the worst-case scenario, the sensitivity can be 0.1 mH_2O. This means that smaller changes in the pressure head cannot be distinguished. Therefore, Fig. 7 shows the leak location performance for such low sensitivity sensors. The performance in the leak location presents a similar behavior to the one presented in Fig. 6 for the 15, 20, and 25 sensors case. However, performance for the 5 and 10 sensor configurations suffered significantly. This means that the number of sensors required also depends on the features of the sensors that will be distributed through the network. From a practical point of view, no significant difference in performance was presented between the three metaheuristic methods used for solving the sensor placement optimization problem.

6 Conclusions

In this paper, a novel sensor placement methodology is presented. The proposal aims to obtain a sensor configuration solution that accommodates the budget of water companies. In order to reduce the leak location problem, a clustering algorithm is proposed based on the topological features of the network. The output of this network partitioning are different sets of nodes called zones. The SPP is formulated for leak location in the predefined zones and solved using three metaheuristic optimization methods. A zone and sensor placement configuration is achieved given a number of sensors to be placed and a desired performance. Therefore, an error tolerance in the leak location task can be satisfied. The simulated experiments with the Modena network benchmark under uncertain operating conditions and low sensor sensitivity demonstrate the feasibility of the proposed approach. The leak location task can be solved with 90% performance by defining 25 zones and using 15 sensors for the network of 317 pipes and 268 demand nodes.

References

1. Agency, E.: Review of the Calculation of Sustainable Economic Level of Leakage and Its Integration with Water Resource Management Planning Contract. Technical Report, Environment Agency, OFWAT and DEFRA. Strategic Management Consultants, Bristol (2012)
2. Blesa, J., Pérez, R.: Modelling uncertainty for leak localization in water networks. IFAC-PapersOnLine **51**(24), 730–735 (2018)
3. Casillas Ponce, M.V., Garza-Castañón, L.E., Puig, V.: Model-based leak detection and location in water distribution networks considering an extended-horizon analysis of pressure sensitivities. J. Hydroinform. **16**(3), 649–670 (2013)
4. Casillas, M.V., Puig, V., Garza-Castañón, L.E., Rosich, A.: Optimal sensor placement for leak location in water distribution networks using genetic algorithms. Sensors **13**, 14984–15005 (2013)
5. Chen, J., Xin, F., Xiao, S.: An iterative method for leakage zone identification in water distribution networks based on machine learning. Struct. Health Monit. (2020)
6. Cugueró-Escofet, M.À., Puig, V., Quevedo, J.: Optimal pressure sensor placement and assessment for leak location using a relaxed isolation index: application to the Barcelona water network. Control Eng. Pract. **63**, 1–12 (2017)
7. Dorigo, M., Blum, C.: Ant colony optimization theory: a survey. Theor. Comput. Sci. **344**, 243–278 (2005)
8. Eberhart, R., Kennedy, J.: A new optimizer using particle swarm theory. In: MHS'95. Proceedings of the Sixth International Symposium on Micro Machine and Human Science, pp. 39–43 (1995)
9. Farah, E., Shahrour, I.: Leakage detection using smart water system: combination of water balance and automated minimum night flow. Water Resour. Manag. **31**(15), 4821–4833 (2017)
10. Forconi, E., Kapelan, Z., Ferrante, M., Mahmoud, H., Capponi, C.: Risk based sensor placement methods for burst/leak detection in water distribution systems. Water Sci. Technol. **17**, 1663–1672 (2017)
11. Goldberg, D.E.: Genetic Algorithms in Search, Optimization, and Machine Learning (1989)
12. Hagos, M., Jung, D., Lansey, K.E.: Optimal meter placement for pipe burst detection in water distribution systems. J. Hydroinform. **18**(4), 741–756 (2016)

13. Houghtalen, R., Hwang, N.H.C.: Fundamentals of Hydraulic Engineering Systems. Prentice Hall (2010)
14. Isermann, R.: Fault-Diagnosis Applications. Springer (2011)
15. Kang, D., Lansey, K.: Novel approach to detecting pipe bursts in water distribution networks. J. Water Resour. Plan. Manag. **140**(1), 121–127 (2014)
16. Mazzolani, G., Berardi, L., Laucelli, D., Simone, A., Martino, R., Giustolisi, O.: Estimating leakages in water distribution networks based only on inlet flow data. J. Water Resour. Plan. Manag. **143**(6), 04017014 (2017)
17. Quiñones Grueiro, M., Verde, C., Llanes-Santiago, O.: Novel leak location approach in water distribution networks with zone clustering and classification. In: Carrasco-Ochoa, J.A., Martínez-Trinidad, J.F., Olvera-López, J.A., Salas, J. (eds.) Pattern Recognition, pp. 37–46. Springer International Publishing, Cham (2019)
18. Quiñones-Grueiro, M., Verde, C., Llanes-Santiago, O.: Demand model in water distribution networks for fault detection. IFAC-PapersOnLine **50**(1), 3263–3268 (2017)
19. Raei, E., Nikoo, M.R., Pourshahabi, S., Sadegh, M.: Optimal joint deployment of flow and pressure sensors for leak identification in water distribution networks. Urban Water J. 1–10 (2019)
20. Rogers, D.: Leaking water networks: an economic and environmental disaster. Procedia Eng. **70**, 1421–1429 (2014)
21. Rossman, L.: Water supply and water resources division. National Risk Management Research Laboratory. In: EPANET 2 User's Manual. Technical Report. United States Environmental Protection Agency (2000)
22. Sophocleous, S., Savic, D., Kapelan, Z.: Leak localization in a real water distribution network based on search-space reduction. J. Water Resour. Plan. Manag. **145**(7), 04019024 (2019)
23. Venkateswaran, P., Han, Q., Eguchi, R.T., Venkatasubramanian, N.: Impact driven sensor placement for leak detection in community water networks. In: ACM/IEEE 9th International Conference on Cyber-Physical Systems (ICCPS), pp. 77–87 (2018)
24. Wang, Q., Guidolin, M., Savic, D., Kapelan, Z.: Two-objective design of benchmark problems of a water distribution system via MOEAs: towards the best-known approximation of the true Pareto front. J. Water Resour. Plan. Manag. **141**(3), 1–14 (2015)
25. Xie, X., Hou, D., Tang, X., Zhang, H.: Leakage identification in water distribution networks with error tolerance capability. Water Resour. Manag. **33**(3), 1233–1247 (2019)
26. Zhang, Q., Wu, Z.Y., Zhao, M., Qi, J.: Leakage zone identification in large-scale water distribution systems using multiclass support vector machines. J. Water Resour. Plan. Manag. **142**(11), 04016042 (2016)

Forecasting Tropospheric Ozone Using Neural Networks and Wavelets: Case Study of a Tropical Coastal-Urban Area

Mirella Lima Saraiva Araujo, Yasmin Kaore Lago Kitagawa, Davidson Martins Moreira, and Erick Giovani Sperandio Nascimento

Abstract Air quality improvement is directly associated with the Sustainable Development Goals (SDGs), established by the United Nations in 2015. To reduce potential impacts from air pollution, computational intelligence by supervised machine learning, using different artificial neural networks (ANNs) techniques, shows to be a promising tool. To enhance their abilities to predict air quality, ANNs have been combined with data preprocessing. The present work performs short-term forecasting of hourly ground-level ozone using long short-term memory (LSTM), a type of recurrent neural network, with the discrete wavelet transform. The study was performed using data from a tropical coastal-urban site in Southeast Brazil, highly influenced by intense convective weather with complex terrain. The models' performance was carried out by comparing statistical indices of errors and agreement, namely: mean squared error (MSE), normalized mean squared error (NMSE), mean absolute error (MAE), Pearson's r, R^2 and mean absolute percentage error (MAPE). When comparing the statistical metrics values, it is shown that the combination of artificial neural networks with wavelet transform enhanced the model's ability to forecast ozone levels compared to the baseline model, which did not use wavelets.

Keywords Ozone forecasting · Air quality · Neural networks · Deep learning

---

M. L. S. Araujo · D. M. Moreira · E. G. S. Nascimento (✉)
Manufacturing and Technology Integrated Campus (SENAI CIMATEC), Salvador, Bahia, Brazil
e-mail: erick.sperandio@fieb.org.br

D. M. Moreira
e-mail: davidson.moreira@fieb.org.br

Y. K. L. Kitagawa
Department of Environmental Engineering, Federal University of Espírito Santo (UFES), Vitória, Espírito Santo, Brazil

J. L. Verdegay et al. (eds.), *Computational Intelligence Methodologies Applied to Sustainable Development Goals*, Studies in Computational Intelligence 1036,
https://doi.org/10.1007/978-3-030-97344-5_11

1 Introduction

Air pollution is considered one of the main environmental risks since it has been associated with harmful effects on human health and the environment [27, 29]. The United Nation 17 Sustainable Development Goals (SDGs), adopted in 2015, have at least three goals that are directly linked to air pollution: Goal 3 focuses on good health and well-being, Goal 11 targets sustainable and safe cities, and Goal 13 combats climate change and its impacts. Some of the most obvious linkages are related to the SDGs targets 3.9 and 11.6 that aim to reduce the number of deaths and illnesses, and to mitigate the adverse per capita environmental impact of cities caused by air pollution by the year 2030.

Ambient air pollution consists of a mixture of substances that include multiple gases and particles with different sizes suspended in the air, such as tropospheric ozone (O_3), nitrogen oxides (NO_x), carbon monoxide (CO), sulfur dioxide (SO_2), fine particulate matter ($PM_{2.5}$), volatile organic compounds (VOCs), and other substances. O_3 is a short-lived trace gas (at about 23 days) [31], generated by reactions involving hydrocarbons, NO_x, VOCs, and solar radiation. Its concentrations are also highly dependent on local emission sources, topography, and meteorological conditions. In addition, long-range transboundary transport can play an important role in O_3 urban background concentration [19]. Epidemiological studies indicate that short- and long-term exposures to O_3 provoke oxidative stress and inflammatory responses of the lungs and airways, affecting lung function and inducing cardiorespiratory diseases [4, 7, 14, 15]. Along with harming human health, O_3 also has economic consequences because it affects agricultural crop yields, increases health expenditures to individuals, and addresses the climate emergency [11, 23].

To prevent the acute impact of air pollution in human lives, several countries already provide real-time air quality forecasting information that helps citizens plan their daily activities, mitigating the effects of air pollution on their health. Early air quality alerts are usually emitted using information from chemical transport models (CTMs) that are capable of describing the interaction among various pollutants and linkages between emissions and results in air pollution [6]. However, to run these models, computational power and multiple information are needed (emission inventories, initial and boundary conditions, parameterization schemes, meteorological fields, and others), making their use sometimes unfeasible [24]. Not to mention their ability to emulate atmospheric dynamics over complex terrains and regions with intense convective characteristics, like in tropical coastal-urban areas [18]. To overcome those limitations, computational intelligence arose as a promising tool and became a key point in the development of healthy and sustainable cities.

Numerous studies have successfully employed artificial intelligence applications on air quality forecasting worldwide either to predict air pollutants concentrations and to improve the performance of CTMs accuracy, working in an architecture of a hybrid model. In addition, the use of computational intelligence showed to be useful to prognosticate air contaminants that are not routinely monitored by environmental agencies because they are not regulated yet [1], or have uncertainties regarding the CTMs' outputs [28]. Computational intelligence consists of multiple categories, being the artificial neural networks (ANNs) the most commonly used technique in air pollution forecasting, and usually classified as: (1) feedforward multilayer perceptron (MLP); (2) recurrent neural network (RNN), such as long short-term memory (LSTM) and gated recurrent unit (GRU); and (3) convolutional neural network (CNN).

Recently, to enhance the short-term forecasting robustness, ANNs have been associated with signal processing techniques such as the wavelet transform (WT), adapted from the Fourier transform, which decomposes the original signals into subsequences of different frequency bands, without losing the time scale information [21, 30, 33]. Overall, the studies pointed out that air quality forecasting accuracy combined with the WT is greater than without the combination [5, 8, 13]. Additionally, for air quality studies, temporal variation has a fundamental role, for instance, O_3 displays strong diurnal patterns with higher concentrations in the afternoon. Therefore, the objective of the present study is to use the wavelet transform to preprocess ground-level ozone data, and then, to use the LSTM model to perform the short-term air quality forecasting (up to 24 h) in the tropical coastal-urban area of Vitória, in the Southeast of Brazil. Previous works [3, 17] evaluated ANNs and deep learning (MLP, LSTM, GRU, and CNN models) to predict ground-level O_3 for the next 24 $_h$ in the same study area and showed that the LSTM model had the most suitable results. However, to fill the research gaps regarding the application of the wavelets, which was not considered in the previous works, the present study aims at exploring it by analyzing, developing and evaluating the usage of WT in deep neural networks for enhancing ground-level O_3 concentrations forecasting for urban regions up to 24 h ahead.

2 Related Works

Several works in the literature can be found indicating that an ANN air pollutant forecast can be highly improved when combined with WT, those results became clear when comparing the statistical metrics such as mean squared error (MSE), normalized mean squared error (NMSE), mean absolute error (MAE), mean absolute percentage error (MAPE), Root Mean Square Error (RMSE), Pearson correlation coefficient (r), and coefficient of determination (R^2).

For instance, Dunea et al. [10] evaluated the use of WT in feedforward neural networks for NO_2, O_3, PM_{10}, and $PM_{2.5}$ in urban areas of Romania. The authors showed that the use of WT also enhanced their results, Pearson's r increased from 0.88–0.94 to 0.94–0.98, and MAPE decreased from 14–20 to 13–18%. Their results showed that the non-use of WT led to Pearson's r between 0.94 and 0.88, MAPE between 14.16 and 20.03%, changing to Pearson's r between 0.98 and 0.94, MAPE between 12.76 and 18.42% after applying wavelets.

AlOmar et al. [2] also increased the performance of the ANN model with the application of wavelets for ozone forecast in Ontario, Canada. The root mean squared error (RMSE), Pearson's r, and MAE changing, respectively, from 2.902 to 0.931, 0.956 to 0.9956, and 1.991 to 0.653. Mbatha and Bencherif [25] developed and tested five LSTM models combined with empirical WT, empirical mode decomposition, and wavelet denoising for a total column of ozone in Buenos Aires, Argentina, and found that LSTM with empirical wavelet transform outperformed the others through statistical analyses of RMSE, MAE, MAPE, and Pearson's r.

The short-term forecasting of other pollutants using WT combined with ANNs can produce similar outcomes. For instance, Cheng et al. [8] showed that the WT improved the short-term forecasting of $PM_{2.5}$ concentration in China ($R^2 = 0.88$) when compared with ANN alone ($R^2 = 0.40$). Bai et al. [5] also used WT to improve the forecasting for SO_2, r increased from 0.85 to 0.94), PM_{10} (r from 0.93 to 0.98), and NO_2 (r from 0.77 to 0.93), and Guo et al. [13] for air pollution index (r from 0.77 to 0.89). Thus, reassuring the findings that the use of wavelets does indeed produce superior results for, not only ozone forecast but several other pollutants.

3 Basic Concepts

3.1 Artificial Neural Networks

Artificial neural networks have their name based on neurobiology since their development was based on the biological structure responsible for the living organism's capacity to learn. The biologic neural network contains cells that are called neurons, those are connected to one another by axons and dendrites, and the connections that allow information to be conveyed from one neuron to another are called synapses. In an artificial neural network, there are layers of units called neurons connected to each other by numerical "weights", which is a value by which the input is multiplied. The similarity with the biological neuron is in the way the neurons communicate with each other, but it is important to state that it is not supposed to exactly model the human brain. In an artificial neural network, the input is received through the first layer and it is propagated throughout the network using the weights as intermediate parameters to reach the desired output. The network learns when the weights are updated to generate the desired output [26].

There are different types of neural networks, each with its own architecture and way of learning, in the next subsection the neural network used in this work to predict ozone concentration will be discussed.

3.1.1 Recurrent Networks

Recurrent Networks (RNN) differ from regular ANN in the fact that it works only with sequential data. When implementing it, the input needs to have a three-dimensional shape, those being the size of the input, the number of features, and a lookback window, the former defines how many time steps the network will examine at a time. The input shape combined with a structure that contains internal loops allows it to have a working memory, making it able to remember previous inputs while creating the current neuron's output. The loop inside the structure works by sending an output of a neuron forward to the next neuron, back to itself, and possibly even to a previous neuron [9]. As an example, recurrent networks are commonly used in natural language processing, where the context of a sentence is especially important to determine the word's meaning. Although this kind of network is supposed to retain temporal information, there is a problem with a simple recurrent network called vanishing gradient, which happens when the network is not able to preserve long-term information. To solve this problem the LSTM was designed.

The LSTM utilizes a memory cell that can keep its state over a long period, and a gated mechanism with three gates: the input, output and forget gate [22]. All three gates receive the previous hidden state and the current input, the input gate passes that information through activation functions that are responsible for deciding if this input is important enough to change the memory information and update the cell state. The forget gate is responsible for deciding if a piece of information should be forgotten or not. Finally, the output gate combines the information processed by the previous gate with the hidden state and current input to decide what will be the next hidden state.

3.2 Wavelets Transform

Signal processing techniques can be combined with artificial intelligence in order to change the input in a way that is easier for an artificial network to assimilate and learn its patterns, which can be done by removing noise from the data or extracting other kinds of information from it.

Joseph Fourier found out that a periodic signal could be described by the sum of different sine and cosine wave functions, each with its own frequency. That concept is used when the Fourier transform analyses a signal in the time domain and translates it to the frequency domain, showing what frequencies were combined to generate the original signal [12]. The only drawback is that the Fourier transform loses the time reference, and while it works very well in periodic signals it becomes inadequate for non-stationary signals and real-time signal processing.

The main difference between the Fourier and the wavelet transform is that Fourier uses sine waves, that is periodic, in other words, does not change its behavior over time, while the wavelet is a wave-like signal oscillation that is preceded and followed by zero values, therefore being localized in time. That characteristic makes wavelets able to analyze a signal in time for its frequency content, and since ordering data points in time is the most intuitive way to represent data, wavelets are a powerful tool.

The process consists of a convolutional procedure between the wavelet and the original signal, which means that the wavelet will pass through the entire original signal multiplying their values. The scale determines how much the wavelet is compressed or dilated, and different scaled wavelets can be used as filters for distinct frequencies. There are two kinds of wavelet transform: the continuous wavelet transform (CWT) and the discrete wavelet transform (DWT). Applying the discrete wavelet transform (DTW), the original signal can be decomposed into two coefficients with half the length of the original signal: the approximation component that contains the low-frequency information for temporal analyses, and the detail component with the high-frequency information, for frequency analyses [32]. After each convolution, the signal will smoothen, therefore, if too many convolutions are applied to the signal, it will lose important characteristic trends.

Another significant way the wavelets differ from the Fourier is that a wavelet can have different families, each with its unique shape, varying in how compact and smooth the wave looks and its location in space, that way it is possible to choose which family is best suited for treating determinate features. Within the families, there are subclasses, varying by a number of coefficients and levels of iterations [12].

4 Methodology

4.1 Data and Statistics Description

The dataset of the present work consists of various meteorological and air quality variables collected hourly, namely: SO_2, NO_2, CO, O_3, particulate matter with diameter less than $10_{\mu m}$ (PM_{10}), nitrogen monoxide (NO), nitrogen oxides (NO_x), air temperature, humidity, scalar wind direction, and wind speed. These parameters were observed through ambient air quality monitoring stations between 2001 and 2015, and the data is provided by the Local Environmental Agency of Espírito Santo State [16]. The air quality data used in this work is registered in the city of Cariacica, located in the coastal urban area of the Metropolitan Region of Vitória (MRV), Espírito Santo State, Southeast of Brazil (Fig. 1).

For a neural network to be rightfully trained, the data needs to be divided into three batches of data, those being a train set, a validation set, and a test set. That is to help the network learn the data's intrinsic patterns instead of just memorizing the input, thus, being able to realize predictions with data never seen before. Most of the data is used to train the network, while the validation set is used to assess whether the trained ANN is well generalizing the problem, or is suffering from under or over fitting. Finally, the model will be quantitatively and qualitatively evaluated using the test set, which will never be presented to the ANN during its training and validation phases. The total amount of samples was 30,492, of which 16,090 data points from the years 2001 to 2005 were used for training, and 6897 for validation and 7505 data points from the years 2006, 2008, 2011, 2012, 2014, and 2015 were used for testing.

The models' performance was quantitatively evaluated in the test set using statistical indices, which were: MSE, NMSE, MAE, MAPE, Pearson's r, and R^2. MSE, NMSE, MAE, and MAPE are indices related to errors and deviations of the model,

a) b)

Fig. 1 **a** Location of Espírito Santo State in Brazil and **b** the map of the Metropolitan Region of Vitória in Espírito Santo State with the IEMA's meteorological stations displayed as blue markers

being the closer to zero the better. Meanwhile, the indices r and R^2 measure the degree of association and agreement, with a value of zero indicating the absence of correlation and a value close to one indicating strong correlation.

4.2 Feature Augmentation with Discrete Wavelet Transform

The original data passes through a process of feature augmentation, where the five levels of DWT are applied to the data using different families. For this end, the PyWavelets package for wavelet analysis was used [20]. The DWT performs a high-pass filtering resulting in the detail coefficients, and a low-pass filtering resulting in the approximation coefficients.

The first step is finding the right wavelet family to use for each feature. Usually this process is done by performing empirical tests, however Zucatelli et al. [33] proposed an innovative technique that is able to choose the wavelet family for a time series in a more objective and automated way. The idea is to evaluate and choose the best wavelet family even before using it in the ANN to generate predictions. For that, the method proposed by Zucatelli et al. [33] analyses whether approximation coefficients are able to maintain the original signal primary characteristics, by observing the RMSE statistical metric for the signal reconstructed by the coefficient and the original values, as shown in Fig. 2. The family most suitable for that signal is the one that generates the lowest value for the RMSE, i.e. the lowest error presented during the reconstruction of the signal. This process is repeated for each feature in the dataset aiming at finding the wavelet family that best fits for that particular feature. In the present work, this method was used to select the best DWT family for each feature.

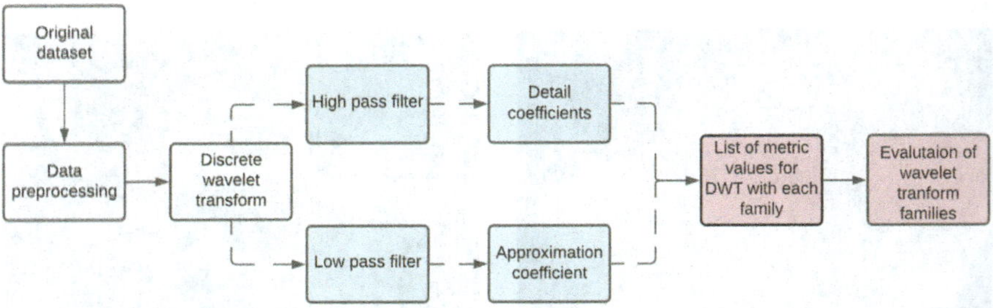

Fig. 2 Flowchart for wavelet family selection

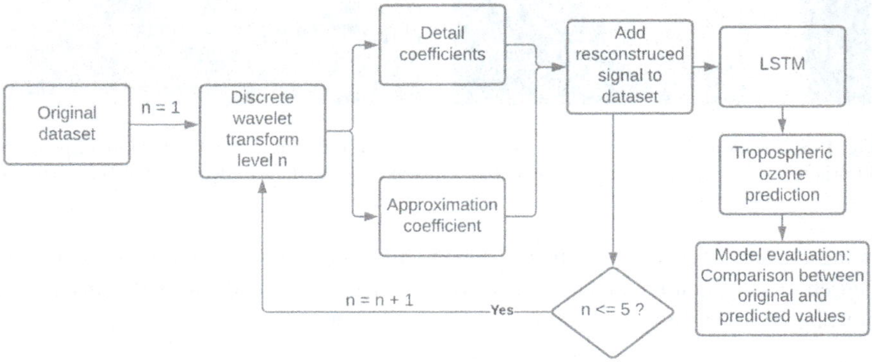

Fig. 3 Flowchart illustrating the strategy for feature augmentation, training and evaluation of the construction of LSTM model

After that, begins the process of feature augmentation, by iteratively applying the best DWT family previously chosen at each feature in the original dataset, generating the approximation and detail coefficients for each level of iteration. To enable those new values to be used as input, it needs to go through a reconstructive process being upsampled to the same size as the original data. Hence, a new dataset is constructed, being each original feature generating ten new features, corresponding to the approximation reconstructed signal and the detail reconstructed signal for five levels of iteration for the selected wavelet, as shown in Fig. 3. The last step is to scale the data to a range between −1 and +1, enabling it to be input and processed by the ANNs.

4.3 Implementing LSTM

This work proposes and evaluates the usage of wavelets to improve the ability of a long short-term memory artificial neural network to predict tropospheric ozone levels for every hour in the next 24 h. In Junior et al. [17], the LSTM was proven

Table 1 LSTM architecture

Layer	LSTM topology		
	Layer type	Neurons	Activation function
1st hidden layer	LSTM	94	ReLU
2nd hidden layer	LSTM	55	ReLU
3rd hidden layer	LSTM	36	ReLU
4th hidden layer	LSTM	48	ReLU
Output	Dense	24	Linear

to be the architecture that generated the most suitable results. Thus, for the sake of comparability, this work is using the same ANN architecture, detailed in Table 1, as well as the same dataset and periods, as in Junior et al.'s [17] work.

After implementing the architecture, other parameters need to be defined such as the loss function, which is how the network measures its success, and the optimizer, which defines how the network will be updated based on the loss function [9]. Those were also the same as in Junior et al. [17], being the loss function the MSE, and the optimizer the Adadelta function, with a learning rate set for approximately 0.194.

Thus, the LSTM is trained five times, each time adding new features that correspond to the iteration level, as shown in Fig. 3. Finally, the models' performance is evaluated by comparing the statistical metrics values for the MSE, NMSE, MAE, r, R^2, and MAPE, which was the same used in Junior et al. [17].

5 Results and Discussion

The results show that the wavelet transform application was able to significantly enhance the network's ability to predict tropospheric ozone concentration levels, when compared to the baseline scenario reported in Junior et al. [17]. Table 2 displays the average statistical indices, comparing the model's results against the observed data, over 24 h for the baseline forecast and those forecasts using the WT (for levels 1, 2, 3, 4, 5).

Table 2 Statistical metrics for each model iteration using the ANN LSTM with WT. The Baseline Scenario is reported in Junior et al. [17]

Model	MSE	MAE	r	R^2	MAPE (%)
Baseline scenario	97.675	7.477	0.775	0.604	70.108
Level 1	31.146	4.788	0.974	0.949	33.443
Level 2	11.472	2.643	0.982	0.965	19.677
Level 3	10.63	2.523	0.983	0.966	17.916
Level 4	11.797	2.694	0.979	0.959	20.413
Level 5	11.656	2.586	0.979	0.959	16.605

Table 2 shows a significant decrease of the errors. It can be noticed that Level 1 decreased the errors by approximately 68% of the MSE, 39% of the MAE, and 53% of the MAPE, when compared to the baseline scenario, in which the WT was not applied [17].

In a more intuitive way, Fig. 4 shows a graphical representation of the variation of the metric values for the forecast each time horizon after the WT application. It was not necessary to create it for every metric since MSE had similar behavior with MAE and NMSE, while Pearson's r is similar to R^2.

The first iteration level has notably higher errors and poorer correlation while also displaying the values more distant from the others, meaning that a significant change in the model's efficiency occurs between levels 1 and 2. The MSE and MAPE indicate similarities between the other levels, however, the Pearson's r graph is able to emphasize the superior results for levels 3 and 4, confirming that the third level of iteration has provided the most reliable ozone predictions for the next 24 h ahead.

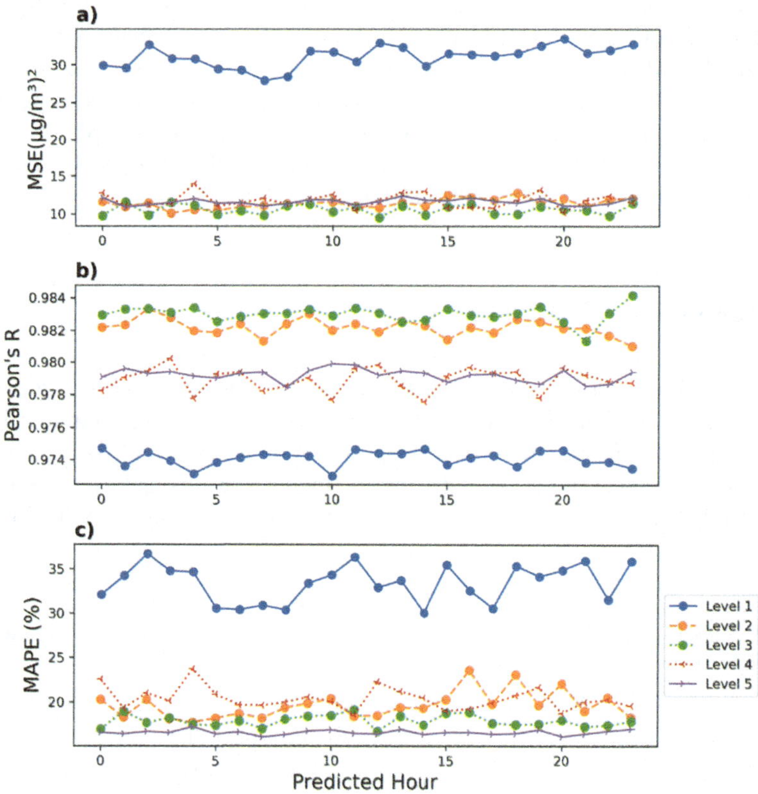

Fig. 4 Comparison between the statistical metrics for each predicted hour for the LSTM models with WT, **a** MSE in $(\mu g/m^3)^2$, **b** Pearson's r, **c** MAPE in percentage. The first level of DWT in blue, second in orange, third in green, fourth in red, and fifth in purple

In a similar way, when comparing the first to the second level, there is a relevant decrease in those metrics. The MSE of the second level was 63% lower than the first level's and the MAE and MAPE were 44% and 41% lower, respectively. Level 3 did not present high variance when compared to the results of the second level, however, it presented the lowest errors and the highest agreement indices, meaning that it had the best performance. The comparison between the third level and the baseline showed a difference of 89% for the MSE, 66% for MAE, and 74% for MAPE.

The indices of agreement for the baseline configuration (without the WT) were Pearson's r = 0.775 and $R^2 = 0.604$; meanwhile, the WT in the first and third levels were Pearson's r = 0.974, $R^2 = 0.949$, and Pearson's r = 0.983 and $R^2 = 0.966$, respectively. This shows an increase of concordance between the model and observed data. In addition, it is important to notice that the statistical metrics values showed that the results are slightly inferior after the third level of iteration, which can be associated with the fact that each wavelet transform iteration tends to smoothen the signal and after a while, the signal starts to lose important characteristic information. In that case, for optimum results, the WT should be applied to the signal until the third level, however, the WT in the fourth and fifth levels still presented better performance than the baseline scenario, which did not use WT.

Figures 5 and 6 show the qualitative comparison between the proposed hybrid models predicted values for the 24th predicted hour and the observed data for a period of 7 days for two different years, 2006 and 2014, respectively, in a more intuitive way to observe the model's level of accuracy. The qualitative analysis of

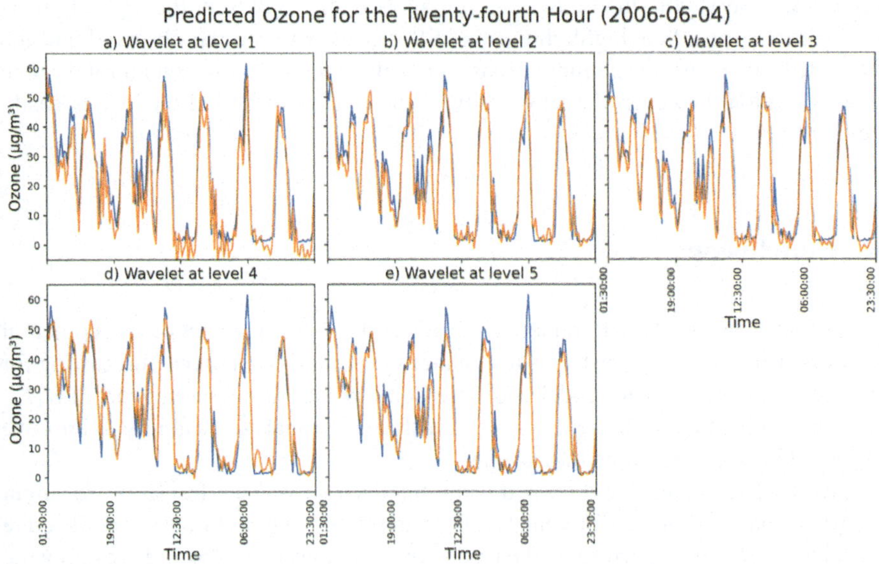

Fig. 5 Comparison between the hourly observed and predicted ozone in June 2006 for the 24 h for the LSTM models, **a** wavelets at level 1, **b** wavelets at level 2, **c** wavelets at level 3, **d** wavelets at level 4, **e** wavelets at level 5. Observed and predicted values are in blue and orange, respectively

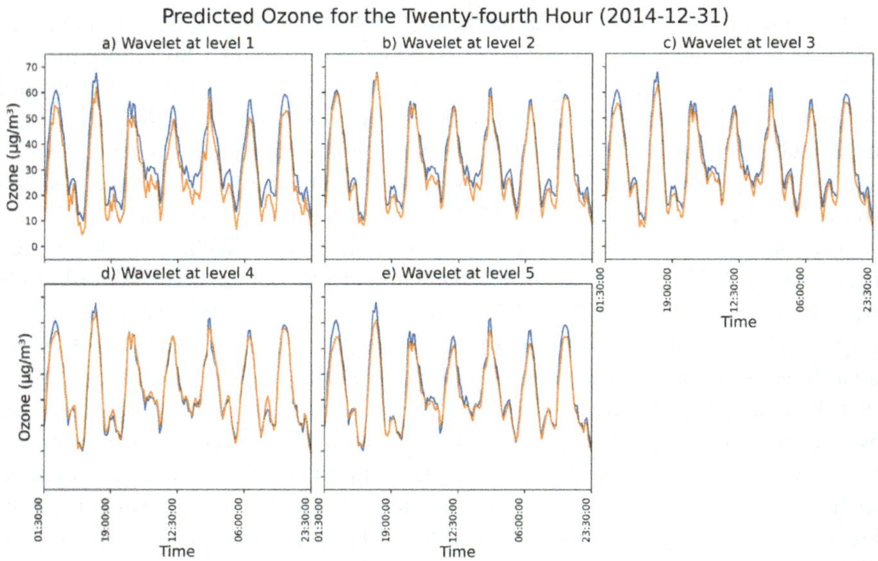

Fig. 6 Comparison between the hourly observed and predicted ozone in December 2014 for the 24 h for the LSTM models, **a** wavelets at level 1, **b** wavelets at level 2, **c** wavelets at level 3, **d** wavelets at level 4, **e** wavelets at level 5. Observed and predicted values are in blue and orange, respectively

the 24 h was chosen since it is the farthest prediction horizon into the future for each reference observation, therefore being the most likely to present the highest errors on its predictions, thus highlighting the differences between the observed and the predicted values. All the configurations were able to replicate the trend for the ozone daily variation, however, the superiority of the models that used the WT is easily perceived.

6 Conclusions

In this paper, we analyzed, trained and evaluated the construction of a deep neural network using wavelet transforms to enhance the ground-level ozone forecasting for the next 24 h ahead. The model was based on a recurrent neural network, LSTM, with the application of discrete wavelet transforms, to analyze its ability to improve air pollutant concentrations forecasting.

Five LSTM models were trained using the discrete wavelet transform for input signal decomposition in different levels of iteration. The addition of the discrete wavelet transform outperformed the results of a previous work that did not apply the use of wavelets. The indices of agreement (Pearson's r and R^2) were greater, and the errors (MSE, MAE, and MAPE) were lower, when compared to the baseline. The wavelet transform reached its maximum potential at the third level iteration.

Enhancing the ability of computational models to better predict the urban air quality is one of the biggest challenges for big cities. Raising awareness of air pollution levels can aid citizens to reduce their exposure, and consequently, reducing possible adverse health effects. In addition, this can also help further research that investigates potential strategies to improve air quality.

For future work, other artificial neural network architectures and techniques should be developed and tested in order to confirm that those findings with the use of wavelet transform in preprocessing the data for input in neural networks can be generalized. Besides, other air pollutants and air quality stations for different regions can be tested.

Acknowledgements We thank the Reference Center on Artificial Intelligence (CRIA), the Supercomputing Center for Industrial Innovation (CS2i), both at the Manufacturing and Technology Integrated Campus—SENAI CIMATEC for their support, and the Fundação de Amparo à Pesquisa do Estado da Bahia (FAPESB) for its financial aid.

References

1. Al-Dabbous, A.N., Kumar, P., Khan, A.R.: Prediction of airborne nanoparticles at roadside location using a feed–forward artificial neural network. Atmos. Pollut. Res. **8**(3), 446–454 (2017). https://doi.org/10.1016/j.apr.2016.11.004
2. AlOmar, M.K., Hameed, M.M., AlSaadi, M.A.: Multi hours ahead prediction of surface ozone gas concentration: robust artificial intelligence approach. Atmos. Pollut. Res. **11**(9), 1572–1587 (2020)
3. Alves, A., Nascimento, E.G.S., Moreira, D.M.: Hourly tropospheric ozone concentration forecasting using deep learning. WIT Trans. Ecol. Environ. **236**, 129–138 (2019). https://doi.org/10.2495/AIR190131
4. Arjomandi, M., Wong, H., Donde, A., Frelinger, J., Dalton, S., Ching, W., Power, K., Balmes, J.R.: Exposure to medium and high ambient levels of ozone causes adverse systemic inflammatory and cardiac autonomic effects. Am. J. Physiol. Heart Circ. **308**, H1499–H1509 (2015). https://doi.org/10.1152/ajpheart.00849.2014
5. Bai, Y., Li, Y., Wang, X., Xie, J., Li, C.: Air pollutants concentrations forecasting using back propagation neural network based on wavelet decomposition with meteorological conditions. Atmos. Polut. Res. **7**(3), 557–566 (2016). https://doi.org/10.1016/j.apr.2016.01.004
6. Baklanov, A., Zhang, Y.: Advances in air quality modeling and forecasting. Glob. Transit. **2**, 261–270 (2020). https://doi.org/10.1016/j.glt.2020.11.001
7. Bruyn, S., Vries, J.: Health costs of air pollution in European cities and the linkage with transport. Available at https://www.cedelft.eu/en/publications/2534/health-costs-of-air-pollution-in-european-cities-and-the-linkage-with-transport (2020).Accessed on 18 Apr 2021
8. Cheng, Y., Zhang, H., Liu, Z., Chen, L., Wang, P.: Hybrid algorithm for short-term forecasting of PM2.5 in China. Atmos. Environ. **200**, 264–279 (2019). https://doi.org/10.1016/j.atmosenv.2018.12.025
9. Chollet, F.: Deep Learning with Python. Manning, New York (2018)
10. Dunea, D., Pohoata, A., Iordache, S.: Using wavelet–feedforward neural networks to improve air pollution forecasting in urban environments. Environ. Monit. Assess. **187**(7), 1–16 (2015)
11. Feng, Z., Marco, A., Anav, A., Gualtieri, M., Sicard, P., Tian, H., Fornasier, F., Tao, F., Guo, A., Paoletti, A.: Economic losses due to ozone impacts on human health, forest productivity and crop yield across China. Environ. Int. **131**, 104966 (2019). https://doi.org/10.1016/j.envint.2019.104966

12. Graps, A.: An introduction to wavelets. IEEE Comput. Sci. Eng. **2**(2), 50–61 (1995)
13. Guo, Q., He, Z., Li, S., Li, X., Meng, J., Hou, Z., Liu, J., Chen, Y.: Air pollution forecasting using artificial and wavelet neural networks with meteorological conditions. Aerosol Air Qual. Res. **20**, 1429–1439 (2020). https://doi.org/10.4209/aaqr.2020.03.0097
14. Hu, X., He, L., Zhang, J., Qiu, X., Zhang, Y., Mo, J., Day, D.B., Xiang, J., Gong, J.: Inflammatory and oxidative stress responses of healthy adults to changes in personal air pollutant exposure. Environ. Pollut. **263**(A), 114503 (2020). https://doi.org/10.1016/j.envpol.2020.114503
15. Huang, J., Song, Y., Chu, M., Dong, W., Miller, M.R., Loh, M., Xu, J., Yang, D., Chi, R., Yang, X., Wu, S., Guo, X., Deng, F.: Cardiorespiratory responses to low-level ozone exposure: the inDoor ozone study in childrEn (DOSE). Environ. Int. **131**, 105021 (2019). https://doi.org/10.1016/j.envint.2019.105021
16. IEMA (Instituto Estadual de Meio Ambiente e Recursos Hídricos): Ambient Air Quality Monitoring System data. Available at https://iema.es.gov.br/qualidadedoar/dadosdemonitoramento/automatica (2021)
17. Junior, A.S.R., Nascimento, E.G.S., Moreira, D.M.: Assessing recurrent and convolutional neural networks for tropospheric ozone forecasting in the region of Vitória, Brazil. WIT Trans. Ecol. Environ. **244**, 101–112 (2020). https://doi.org/10.2495/AIR200091
18. Kitagawa, Y.K.L., Nascimento, E.G.S., Souza, N.B.P., Zucatelli, P.J., Kumar, P., Albuquerque, T.T.A., Moraes, R.M., Moreira, D.M.: Evaluation of the WRF-ARW model during an extreme rainfall event: subtropical storm Guará. Atmósfera (2021). (in press). https://doi.org/10.20937/ATM.52977
19. Langford, A.O., Senff, C.J., Alvarez, R.J., Banta, R.M., Hardesty, R.M.: Long-range transport of ozone from the Los Angeles Basin: a case study. Geophys. Res. Lett **37**, L06807 (2010). https://doi.org/10.1029/2010GL042507
20. Lee, G., Gommers, R., Waselewski, F., Wohlfahrt, K., O'Leary, A.: PyWavelets: a Python package for wavelet analysis. J. Open Source Softw. **4**(36), 1237 (2019)
21. Liu, H., Yin, S., Chen, C., Duan, Z.: Data multi-scale decomposition strategies for air pollution forecasting: a comprehensive review. J. Clean. Prod. **277**, 124023 (2020). https://doi.org/10.1016/j.jclepro.2020.124023
22. Lu, Y., Salem, F.M.: Simplified gating in long short-term memory (LSTM) recurrent neural networks. In: 2017 IEEE 60th International Midwest Symposium on Circuits and Systems (MWSCAS) (2017). https://doi.org/10.1109/mwscas.2017.8053244
23. Maji, K.J., Ye, W.-F., Arora, M., Shiva, S.M.N.: Ozone pollution in Chinese cities: assessment of seasonal variation, health effects and economic burden. Environ. Pollut. **247**, 792–801 (2019). https://doi.org/10.1016/j.envpol.2019.01.049
24. Mallet, V., Sportisse, B.: Air quality modeling: from deterministic to stochastic approaches. Comput. Math. Appl. **55**(10), 2329–2337 (2008). https://doi.org/10.1016/j.camwa.2007.11.004
25. Mbatha, N., Bencherif, H.: Time series analysis and forecasting using a novel hybrid LSTM data-driven model based on empirical wavelet transform applied to total column of ozone at Buenos Aires, Argentina (1966–2017). Atmosphere **11**(5), 457 (2020)
26. Nielsen, M.A.: Neural Networks and Deep Learning, vol. 25. Determination Press, San Francisco (2015)
27. Prüss-Ustün, A., Wolf, J., Corvalán, C., Bos, R., Neira, M.: Preventing disease through healthy environments: a global assessment of the environmental burden of disease. Toxicol. Lett. **259**, S1 (2016). https://doi.org/10.1016/j.toxlet.2016.07.028
28. Sayeed, A., Lops, Y., Choi, Y., Jung, J., Salman, A.K.: Bias correcting and extending the PM forecast by CMAQ up to 7 days using deep convolutional neural networks. Atmos. Environ. **253**, 118376 (2021). https://doi.org/10.1016/j.atmosenv.2021.118376
29. WHO (World Health Organization): Ambient air pollution: a global assessment of exposure and burden of disease (2016)
30. Yafouz, A., Ahmed, A.N., Zaini, N., El-Shafie, A.: Ozone concentration forecasting based on artificial intelligence techniques: a systematic review. Water Air Soil Pollut. **232**, 79 (2021). https://doi.org/10.1007/s11270-021-04989-5

31. Young, P.J., Archibald, A.T., Bowman, K.W., Lamarque, J.-F., Naik, V., Stevenson, D.S., Tilmes, S., Voulgarakis, A., Wild, O., Bergmann, D., Cameron-Smith, P., Cionni, I., Collins, W.J., Dalsøren, S.B., Doherty, R.M., Eyring, V., Faluvegi, G., Horowitz, L.W., Josse, B., Lee, Y.H., MacKenzie, I.A., Nagashima, T., Plummer, D.A., Righi, M., Rumbold, S.T., Skeie, R.B., Shindell, D.T., Strode, S.A., Sudo, K., Szopa, S., Zeng, G.: Pre-industrial to end 21st century projections of tropospheric ozone from the atmospheric chemistry and climate model intercomparison project (ACCMIP). Atmos. Chem. Phys. **13**, 2063–2090 (2013). https://doi.org/10.5194/acp-13-2063-2013
32. Zhen, Z., Wan, X., Wang, Z., Wang, F., Ren, H., Mi, Z.: Multi-level wavelet decomposition based day-ahead solar irradiance forecasting. In: 2018 IEEE Power & Energy Society Innovative Smart Grid Technologies Conference (ISGT) (pp. 1–5). IEEE, Feb 2018
33. Zucatelli, P.J., Nascimento, E.G.N., Santos, A.A.B., Moreira, D.M.: Nowcasting prediction of wind speed using computational intelligence and wavelet in Brazil. Int. J. Comput. Methods Eng. Sci. Mech. **21**(6), 343–369 (2020). https://doi.org/10.1080/15502287.2020.1841335

Power Curve Estimation of Wind Farms with Imprecise Data by Fuzzy Quadratic Programming

Ricardo Coelho

Abstract Quadratic programming can be applied in a variety of the real-world problems. As ambiguity and vagueness are natural and ever-present in real-life situations requiring solutions, it makes perfect sense to attempt to address them using Fuzzy Quadratic Programming (FQP) problems. This way of problem modeling is applied to an increasing variety of practical fields especially those with renewable energy generation. In this work, we present a parametric approach that solve quadratic programming problems under different kind of uncertainties in its data. In order to show the efficiency of this approach, it is used in a fuzzy regression analysis problem, which try to find out a relationship between variables, i.e, to describe how a dependent variable is related with independent variables. Here, we consider regression analysis with imprecise parameters which are natural in real-life situations.

Keywords Fuzzy logic · Fuzzy quadratic optimisation · Fuzzy regression analysis · Renewable energy

1 Introduction

One of the Sustainability Development Goals (SDG) is Affordable and Clean Energy, which encompass renewable energy that is generated by natural resources like sunlight, waves, geothermal heat and wind. Owing to that goal, some initiatives around the world have worked to ensure the use and development of clean energy. However, some people on the planet does not have access to electricity and the generation of renewable energy has to grow substantially in the face of increased demand.

From the several options, energy generated by wind turbines is very promising due to the wind is a renewable resource and consistent throughout the year. In the past, the wind force was used to sail and transport water to supply and irrigation.

R. Coelho (✉)
Departamento de Estatística e Matemática Aplicada, Centro de Ciências, Universidade Federal do Ceará, Av. Mister Hull, s/n, Campus do Pici, Bloco 910, 60440-900 Fortaleza, CE, Brazil
e-mail: rcoelhos@dema.ufc.br

J. L. Verdegay et al. (eds.), *Computational Intelligence Methodologies Applied to Sustainable Development Goals*, Studies in Computational Intelligence 1036,
https://doi.org/10.1007/978-3-030-97344-5_12

Nowadays, this force is transformed into electricity, which can be incorporated to the power system. Wind turbines are being installed in different types of environments and their proper operation has to take into account air humidity, air density, altitude, seasonal weather conditions and terrain complexity. Thus, it is important to model the relationship between wind speed and generated power by wind turbines.

Regression analysis, which includes any techniques for modeling and analyzing several variables, is a statistical tool that focus on finding a relationship between a dependent variable and one or more independent variables. When this relationship is found, some values of parameters are determined that help a function to best fit in a set of data observations. In regression analysis, it is also interesting to characterize the variation of the depend variable around the independent ones. However, it is more natural that the given data are imprecise and vague because there are observations that are subjective, as described in [12]. In this context, some technique to deal with these imprecise data must be used, and one of these techniques is fuzzy regression analysis, which is more considered in systems where human estimations is influential.

With this in mind, the goal of this paper is to apply our proposed method, described in [19], in the fuzzy regression analysis. This method is used to solve quadratic programming problems with fuzzy costs in the objective function.

The paper is organized as follows: Sect. 2 presents a brief description of a quadratic programming problem; Section 3 presents a description of parametric ideas to solve fuzzy quadratic programming problems with uncertainties in the relation order and/or coefficients; Section 4 briefly shows about fuzzy linear regression and how transforming a regression problem into mathematical programming problem with fuzzy costs. In this section, a novel approach is shown that solves quadratic programming problems with fuzzy costs where these are transformed into parametrical quadratic multi-objective programming problems. To clarify the above developments, numerical examples are analyzed in Sect. 5. Finally, conclusions are presented in Sect. 6.

2 Quadratic Programming

Even though Quadratic Programming (QP) is classified as a particular case of Non-linear Programming (NLP), which includes optimisation problems with a non-linear objective function subject to a set of nonlinear constraints, QP can also be viewed as a generalization of Linear Programming (LP), which only uses linear functions to formulate its optimisation problems. Clearly, QP is concerned to solve with optimisation problems modelled by a quadratic objective function subject to a set of linear constraints. Taking this into account, this set of QP problems can be modelled in the following way

$$\begin{aligned} &\text{Minimize } f(\mathbf{x}) = \mathbf{c}^t\mathbf{x} + \tfrac{1}{2}\mathbf{x}^t\mathbf{Q}\mathbf{x} \\ &\text{subject to } \mathbf{A}\mathbf{x} \leq \mathbf{b} \\ &\qquad\qquad\; \mathbf{x} \geq \mathbf{0} \end{aligned} \tag{1}$$

where $\mathbf{x}$ belongs to $\mathbb{R}^n$ space, the $n \times n$-dimensional real matrix $\mathbf{Q}$ is symmetric, $\mathbf{A}$ is any $m \times n$-dimensional real matrix, $\mathbf{c}$ is any n-dimensional real vector, and $\mathbf{b}$ is any m-dimensional real vector.

From its special features, some methods developed to non-linear programming problems are firstly applied to solve quadratic problems in order to know its efficiency and performance. In addition, there are several classes of problems that are naturally expressed as quadratic problems. Examples of such problems can be found in finance and economics, agriculture, facility allocation, structural analysis, production and operations, marketing, public policy, power generation among others.

3 A General Model for Quadratic Programming Under Fuzzy Environment

The mathematical modelling from real-world problems has to deal with uncertainty and vagueness of some parameters, which are seldom known exactly and have to be estimated by the decision maker. These uncertainties can be found in the order relation, constants, decision variables or in all parameters of the problem.

There are different methodologies to deal with uncertainty, vagueness and the theory of Fuzzy Sets, which belongs to Soft Computing (SC), is one of them. It is worth noting that some authors have applied Soft Computing methodologies to solve quadratic programming problems. In [1] is described a method to solve a quadratic programming problem with fuzzy coefficients, which divides the original fuzzy quadratic problem into two classical quadratic problems. An extended model to the fuzzy quadratic problem described in [3] uses the approach proposed by Zimmermann [24, 25]. This approach solves a classic problems that is not violated and another problem that is totally violated. In [14, 15] is discussed as transforming a fuzzy quadratic programming problem in a two-level mathematical programming problem for finding the bounds of the fuzzy objective values. In [5] is proposed a parametric approach which provides a fuzzy solution to the problem, that involves all particular-punctual-solutions provided by their method.

In this section will be presented some proposed methods to solve quadratic programming problems under fuzzy environment. The fuzzy environment encompasses the parameters that can be presented in the coefficients of the set of constraints or the costs of the objective function or in the relation of the constraints. This kind of problem under fuzzy environment can be formulated on the following way

$$\begin{aligned} \min\ & f(\mathbf{x}) = \tilde{\mathbf{c}}^t\mathbf{x} + \tfrac{1}{2}\mathbf{x}^t\tilde{\mathbf{Q}}\mathbf{x} \\ \text{s.t. } & \tilde{\mathbf{A}}\mathbf{x} \lesssim \tilde{\mathbf{b}} \\ & \mathbf{x} \geq \mathbf{0}. \end{aligned} \tag{2}$$

where $\tilde{\mathbf{c}}$ and $\tilde{\mathbf{b}}$ are an $n \times 1$ vector and an $m \times 1$ vector, respectively, with fuzzy numbers, $\tilde{\mathbf{Q}}$ and $\tilde{\mathbf{A}}$ are respectively an $n \times n$ matrix and an $m \times n$ matrix with fuzzy components, and $\lesssim$ represents the fuzzy relation order in the set of constraints. The vagueness are represented for membership functions defined by decision maker.

3.1 Parametric Idea to Solve Quadratic Programming with Fuzzy Relation Order in the Set of Constraints

The fuzzy parameters of the quadratic problem (2) are defined with fuzzy nature, that is, some violations in the accomplishment of such objective function and restrictions are permitted. Therefore, this fuzzy parameters can be determined by the decision maker.

It is clear that each membership function will give the membership (satisfaction) degree such that any $\mathbf{x} \in \mathbb{R}^n$ accomplishes the corresponding fuzzy objective function and constraint upon which it is defined. These membership functions can be formulated as follows

$$\mu_i : \mathbb{R} \rightarrow (0, 1], \quad i \in I$$

where μ is an linear membership function (and formally can also be a non linear one), and $\mathbf{I}$ is the set that contains all fuzzy parameters.

In order to solve this problem in a two-phase method, first let us define for each fuzzy constraint, $i \in I$

$$X_i = \left\{ \mathbf{x} \in \mathbb{R}^n \mid \tilde{A}_i \mathbf{x} \lesssim \tilde{b}_i, \mathbf{x} \geq 0 \right\}.$$

If $\mathbf{X} = \bigcap_{i \in I} X_i$ then the former fuzzy quadratic problem can be addressed in a compact form as

$$\min \{ f(\mathbf{x}) \mid \mathbf{x} \in \mathbf{X} \}$$

It is clear that $\forall \alpha \in (0, 1]$, an α-cut of the fuzzy constraint set will be the classical set

$$X(\alpha) = \left\{ \mathbf{x} \in \mathbb{R}^n \mid \mu_X(\mathbf{x}) \geq \alpha \right\}$$

where $\forall \mathbf{x} \in \mathbb{R}^n$,

$$\mu_X(\mathbf{x}) = \inf \mu_i(\mathbf{x}(1 - \alpha)), i \in I$$

Hence an α-cut of the i-th constraint will be denoted by $X_i(\alpha)$. Therefore, if $\forall \alpha \in (0, 1]$,

$$S(\alpha) = \left\{ \mathbf{x} \in \mathbb{R}^n \mid f(\mathbf{x}) = \min \ f(\mathbf{y}), \ \mathbf{y} \in X(\alpha) \right\}$$

the fuzzy solution to the problem will therefore be the fuzzy set defined by the following membership function

$$S(\mathbf{x}) = \begin{cases} \sup\{\alpha : \mathbf{x} \in S(\alpha)\} & \mathbf{x} \in \bigcup_\alpha S(\alpha) \\ 0 & \text{otherwise.} \end{cases}$$

Provided that $\forall \alpha \in (0, 1]$,

$$X(\alpha) = \bigcap_{i \in I} \{\mathbf{x} \in \mathbb{R}^n \mid (Ax)_i \leq r_i(\alpha), \mathbf{x} \geq 0, \mathbf{x} \in \mathbb{R}^n\}$$

with $r_i(\alpha) = b_i + d_i(1 - \alpha)$. The operative solution to the former problem can be found, α-cut by α-cut. Thus, the fuzzy quadratic programming problem was parameterized at the end of the first phase. In the second phase the parametric quadratic programming problem is solved for each of the different α values using conventional quadratic programming techniques.

The obtained results for each α value generate a set of solutions $S(\mathbf{x})$ and then the Representation Theorem can be used to integrate all these specific α-solutions. It is then demonstrated that the outlined solution to the parametric method is a valid solution to the fuzzy quadratic problem.

3.2 *Parametric Idea to Solve Quadratic Programming with Fuzzy Costs in the Objective Function*

As has already been said, the uncertainty can be present in different ways in modelling of a mathematical programming problem in particular case, as it is seen in Problem (2). Also has been seen, the parametric approach transforms the mathematical programming problem into a classical one with a parameter that represents the satisfaction level, which belongs to the interval [0, 1].

The proposed approach is transforming a mathematical programming problem with uncertain costs in the single objective function into a parametric mathematical programming problem with several objective functions. In [7, 8, 11], a multi-objective approach is developed to solve a mathematical programming problem with fuzzy cost in the objective function, which can be extended to solve quadratic programming problems with single or several objetive functions.

The uncertain costs represent the values that can

$$\begin{array}{l} \widetilde{\min}\ f(\tilde{\mathbf{c}}; \mathbf{x}) \\ \text{s.a}\ \ \mathbf{x} \in \mathbb{R}^n \end{array} \tag{3}$$

where $\tilde{\mathbf{c}} \in \mathbb{F}(\mathbb{R}^{n \times m})$ represents a vector with fuzzy parameters.

A set of membership functions defines the satisfaction level of a solution $\mathbf{x} \in \mathbb{R}^n$. These membership functions can be modelled as follows

$$\mu_i : \mathbb{R} \to (0, 1], \quad i \in I$$

such that μ is a membership function and I is the set that contains all of fuzzy parameters.

So, the fuzzy multi-objective problem can be defined by the following compact way

$$\widetilde{\min}\left\{F(\tilde{\mathbf{c}};\mathbf{x}) \mid \mathbf{x} \in \mathbb{R}^n\right\}$$

In the literature there are many ways to represent fuzzy numbers by a classic number, for instance the Yager's indexes, as described in [4]. In this work, the fuzzy number is transformed into a interval that is defined by the lower and upper bounds from the fuzzy number for a specific α-cut level.

So, the compact way of this transformation can be rewritten as

$$\widetilde{\min}\left\{F([\mathbf{c}^{\mathbf{L}}, \mathbf{c}^{\mathbf{U}}]_\alpha;\mathbf{x}) \mid \mathbf{x} \in \mathbb{R}^n\right\}$$

Thus, the fuzzy solution of the original fuzzy problem is a set of optimal solution of the parametric problem for each $\alpha \in (0, 1]$. This set of solutions can be defined in the following way

$$S(\alpha) = \min\left\{F(\mathbf{c}_\alpha;\mathbf{x}) \mid \mathbf{x} \in \mathbb{R}^n\right\}$$

where $c_\alpha \in [c^L, c^U]_\alpha$ is a real value obtained by a linear combination of the boundaries of the interval formed by the α-cut level, $\forall\alpha \in (0, 1]$.

The solution to the previous problem ban be found, α-cut by α-cut, by solving the auxiliar parametric multi-objective programming problem.

Therefore, the fuzzy optimisation problem was parameterized in the end of the first phase. In the second phase, the parametric problem is solved for each α-cut level by using classical methods to multi-objective optimisation. So, the optimal solution obtained for each α satisfies the Pareto optimality conditions and this point describes a efficient solution to the original fuzzy problem.

The obtained results for each α generate a set of α-solutions and then, according to the Representation Theorem, can be defined by

$$\tilde{S} = \bigcup_\alpha (1-\alpha)S(\alpha),$$

Thus, it is shown that the solutions, in general lines, are obtained by the parametric model, which determines a fuzzy solution $\tilde{S}$, valid for the fuzzy unrestricted multi-objective programming problem.

3.2.1 Methods to Solve Quadratic Programming Problems with Fuzzy Coefficients in the Objective Function

In this work, the uncertain parameters are regarded as fuzzy numbers and a mathematical programming problem with fuzzy costs in the objective function can be formulated as

$$\begin{array}{l} \min \left[f_1(\tilde{\mathbf{c}}_1; \mathbf{x}), f_2(\tilde{\mathbf{c}}_2; \mathbf{x}), \ldots, f_m(\tilde{\mathbf{c}}_m; \mathbf{x}) \right] \\ \text{s.a } \mathbf{x} \in \Omega \end{array} \tag{4}$$

with $\mathbf{x}$ is a n-dimensional real vector, $\tilde{\mathbf{c}}$ is a vector with fuzzy numbers. The fuzzy numbers are characterized by membership functions that are defined by decision makers. The membership functions can be defined as

$$\mu_j : \mathbb{R} \rightarrow (0, 1], \quad j \in \mathbb{J} = \{1, 2, \ldots, n\}$$

In particular these membership functions will be supposed as:

$$\mu_j(y) = \begin{cases} L_j(y) & \text{if } c_j^L \leq y \leq c_j \;\; j \in \mathbb{J} \\ R_j(y) & \text{if } c_j \leq y \leq c_j^U \end{cases} \tag{5}$$

where $\mathbf{L}(\cdot)$ and $\mathbf{R}(\cdot)$ are assumed to be strictly increasing and decreasing continuous functions, respectively, $L_j(c_j) = R_j(c_j) = 1, \; j \in \mathbb{J}$.

Multi-objective approach by permutation of the costs

The problem considered in [20] uses trapezoid membership function to describe the fuzzy parameters. However, in this work, the fuzzy parameters are described in (5). So, considering the $(1-\alpha)$-cut level for all costs, for $\alpha \in [0, 1]$,

$$\forall \mathbf{x} \in \mathbb{R}, \; \mu_j(\mathbf{x}) \geq 1 - \alpha \Leftrightarrow L_j^{-1}(1-\alpha) \leq \mathbf{x} \leq R_j^{-1}(1-\alpha), \; j \in \mathbb{J}.$$

Thus, according to the parametric transformations showed above, a fuzzy solution can be obtained from a parametric solution of a equivalent parametric multi-objective programming problem, which can be formulated as

$$\begin{aligned} \min \; & \left[f_1(\mathbf{c}_1^1; \mathbf{x}), f_1(\mathbf{c}_1^2; \mathbf{x}), \ldots, f(\mathbf{c}_1^{2^{p_1}}; \mathbf{x}), \ldots, f_2(\mathbf{c}_2^1; \mathbf{x}), f_2(\mathbf{c}_2^2; \mathbf{x}), \ldots, \right. \\ & \left. f_2(\mathbf{c}_2^{2^{p_2}}; \mathbf{x}), \ldots, f_m(\mathbf{c}_m^1; \mathbf{x}), f_m(\mathbf{c}_m^2; \mathbf{x}), \ldots, f_m(\mathbf{c}_m^{2^{p_m}}; \mathbf{x}) \right] \\ \text{s.a} \; & \mathbf{x} \in \Omega, \;\; \mathbf{c}^k \in \mathbb{E}(1-\alpha), \;\; \alpha \in [0, 1], \;\; k = 1, 2, \ldots, 2^{p_j}, \end{aligned} \tag{6}$$

where $\mathbb{E}(1-\alpha)$, $\forall \alpha \in [0, 1]$, is a set of vectors in $\mathbb{R}^{p_j}$, such that p_j describes the amount of fuzzy numbers that represent the uncertain costs in the objective function. Each one of its components is in the lower bound, $L_j^{-1}(1-\alpha)$, or in the upper bound, $R_j^{-1}(1-\alpha)$, of the respective $(1-\alpha)$-cut level, i.e., $\forall k = 1, 2, \ldots, 2^{p_j}$,

$$\begin{array}{l} \mathbf{c}^k = (c_1^k, c_2^k, \ldots, c_m^k) \in \mathbb{E}(1-\alpha) \Leftrightarrow c_j^k = L_j^{-1}(1-\alpha) \\ \text{ou } R_j^{-1}(1-\alpha), \;\; \forall \, j = 1, \ldots, m \end{array}$$

Note that the parametric solution obtained to (6) .

A fuzzy quadratic programming problem can be formulated as

$$\begin{aligned} \min\ & \tilde{\mathbf{c}}^t x + \tfrac{1}{2}\mathbf{x}^{\mathbf{t}}\tilde{\mathbf{Q}}\mathbf{x} \\ \text{s.t.}\ & \mathbf{A}\mathbf{x} \le \mathbf{b} \\ & \mathbf{x} \ge 0 \end{aligned} \tag{7}$$

where the fuzzy numbers are characterized by membership functions that are defined by decision makers. The membership functions can be defined as

$$\mu_j, \mu_{ij} : \mathbb{R} \to [0, 1], \quad i, j \in \mathbb{J} = \{1, 2, \dots, n\}$$

In particular these membership functions will be supposed as:

$$\mu_j(y) = \begin{cases} 0 & \text{if } c_j^U \le y \text{ or } y \le c_j^L \\ L_j(y) & \text{if } c_j^L \le y \le c_j \\ R_j(y) & \text{if } c_j \le y \le c_j^U \end{cases} \quad j \in \mathbb{J} \tag{8}$$

and

$$\mu_{ij}(y) = \begin{cases} 0 & \text{if } q_{ij}^U \le y \text{ or } y \le q_{ij}^L \\ L_{ij}(y) & \text{if } q_{ij}^L \le y \le q_{ij} \\ R_{ij}(y) & \text{if } q_{ij} \le y \le q_{ij}^U \end{cases} \quad i, j \in \mathbb{J} \tag{9}$$

where $\mathbf{L}(\cdot)$ and $\mathbf{R}(\cdot)$ are assumed to be strictly increasing and decreasing continuous functions, respectively, $L_j(c_j) = R_j(c_j) = 1,\ j \in \mathbb{J}$ and $L_{ij}(q_{ij}) = R_{ij}(q_{ij}) = 1,\ i, j \in \mathbb{J}$.

A multi-objective approach to solve a linear programming problem with imprecise costs is described in [7, 8]. As the linear problem is a particular case of quadratic problem, this approach can be extended to solve quadratic programming problems with fuzzy costs.

The quadratic objective function can be divided into two parts, where the first one is a linear term and the second one is a quadratic term. According to this, the fuzzy costs can only be in the first part or only the second part or in both. In this section, these three ways will be presented separately.

The linear problem considered in [7] used trapezoid membership functions for the costs but here, for the sake of simplicity, they will be supposed to be like (8) and (9).

Then, by considering the $(1 - \alpha)$-cut of every cost, $\alpha \in [0, 1]$,

$$\begin{aligned} \forall\, x \in \mathbb{R},\ \ \mu_j(x) \ge 1 - \alpha \Leftrightarrow L_j^{-1}(1 - \alpha) \le x \le R_j^{-1}(1 - \alpha), \\ \forall\, x \in \mathbb{R},\ \ \mu_{ij}(x) \ge 1 - \alpha \Leftrightarrow L_{ij}^{-1}(1 - \alpha) \le x \le R_{ij}^{-1}(1 - \alpha), \end{aligned}$$

where $i, j \in \mathbb{J} = \{1, 2, \dots, n\}$.

Thus, according to the parametric transformations shown above, a fuzzy solution to (7) may be found from the parametric solution of the multi-objective parametric quadratic programming problem

$$\begin{aligned}
\min \ & \Big[(\mathbf{c}^1)^t\mathbf{x} + \tfrac{1}{2}\mathbf{x}^t\mathbf{Q}^1\mathbf{x}, (\mathbf{c}^2)^t\mathbf{x} + \tfrac{1}{2}\mathbf{x}^t\mathbf{Q}^1\mathbf{x}, \ldots, \\
& \ldots, (\mathbf{c}^{2^n})^t\mathbf{x} + \tfrac{1}{2}\mathbf{x}^t\mathbf{Q}^1\mathbf{x}, (\mathbf{c}^1)^t\mathbf{x} + \tfrac{1}{2}\mathbf{x}^t\mathbf{Q}^2\mathbf{x}, \ldots, \\
& \ldots, (\mathbf{c}^{2^n})^t\mathbf{x} + \tfrac{1}{2}\mathbf{x}^t\mathbf{Q}^2\mathbf{x}, \ldots, (\mathbf{c}^{2^n})^t\mathbf{x} + \tfrac{1}{2}\mathbf{x}^t\mathbf{Q}^{2^{n^2}}\mathbf{x}\Big] \\
\text{s.t. } & \mathbf{Ax} \leq \mathbf{b}, \quad \mathbf{x} \geq \mathbf{0}, \\
& \mathbf{c}^k, \mathbf{Q}^p \in \mathbb{E}(1-\alpha), \quad \alpha \in [0,1], \\
& k = 1, 2, \ldots, 2^n \text{ and } p = 1, 2, \ldots, 2^{n^2},
\end{aligned} \tag{10}$$

where $\mathbb{E}(1-\alpha)$, for each $\alpha \in [0, 1]$, is the set of vectors in $\mathbb{R}^n$ such that each of its components is either in the lower bound, $L_j^{-1}(1-\alpha)$, or in the upper bound, $R_j^{-1}(1-\alpha)$, of the respective $(1-\alpha)$-cut, that is, $\forall\, k = 1, 2, \ldots, 2^n$, and $\forall\, i, j \in \mathbb{J}$

$$\mathbf{c}^k = (c_1^k, c_2^k, \ldots, c_n^k) \in \mathbb{E}(1-\alpha) \Leftrightarrow c_j^k = \begin{cases} L_j^{-1}(1-\alpha) & \text{or} \\ R_j^{-1}(1-\alpha). \end{cases}$$

and

$$\mathbf{Q}^k = (q_{11}^k, \ldots, q_{1n}^k, \ldots, q_{nn}^k) \in \mathbb{E}(1-\alpha) \Leftrightarrow q_{ij}^k = \begin{cases} L_{ij}^{-1}(1-\alpha) & \text{or} \\ R_{ij}^{-1}(1-\alpha). \end{cases}$$

The obtained parametrical solutions for any of multi-objective models above, to the different α values, generate a set of solutions and then we use the Representation Theorem to integrate all of these particular alpha-solutions.

4 Fuzzy Regression Analysis

Regression analysis is widely used for prediction and forecasting, where its use has substantial overlap with the field of machine learning. It is also used to understand, which among the independent variables are relates to the dependent one and to explore the forms of these relationships. The performance of regression analysis in proactive depends on the form of the data-generating process and how it relates to the regression approach being used.

In [21], fuzzy regression analysis is classified in two categories that are possibilistic regression analysis, which is based on possibility concepts and least squares method, which minimizes errors between the given outputs and the estimated ones. The first work about possibilistic regression analysis was proposed by Tanaka et al. [22], in which is used a fuzzy linear system as a regression model. In [20] and [21], a interval regression analysis, which is a simplest version of possibilistic regression, based on quadratic programming is shown. It is an approach that unify the possibility and necessity regression analysis. Other proposals based on the use of possibility concepts can be found in [2, 16, 17, 23]. Another direction of fuzzy regression is fuzzy least squares approaches and some approaches can be found in [9, 18]. In

Table 1 Input-output data

Samples	Output	Inputs
1	y_1	$x_{11}, \ldots, x_{1n}$
.	.	.
.	.	.
.	.	.
m	y_m	$x_{m1}, \ldots, x_{mn}$

[10], an approach that uses quadratic programming is shown and it integrates the two categories from fuzzy regression analysis, where it reconciles the minimization of estimated deviations of the central tendency with the minimization of estimated deviations in the spreads of membership functions. A regression problem can be formulated as a mathematical programming problem, where the objective is to minimize the difference between the estimated values and the observed values. These data lack this kind of exact knowledge, and only approximate, vague and imprecise values are known. Moreover, these imprecise values can be dealt with fuzzy logic. In this case, the concept of fuzzy mathematical programming emerges when it is used. In this work, the input and output data are real numbers and the coefficients of these data are fuzzy numbers.

In this section, a brief introduction about fuzzy linear regression, which is used extensively in practical applications, is shown that has been based on linear programming when introduced by Tanaka et al. [22]. A approach that solves quadratic programming problems with fuzzy costs, where these are transformed into parametrical quadratic multi-objective programming problems, is also described in this section.

4.1 Fuzzy Linear Regression

Fuzzy linear regression was the first type of regression analysis to be studied because models, which depend linearly on their unknown parameters, are easier to fit the models, which are non-linearly related to their parameters. Fuzzy linear regression models are often fitted using the least squares approach, which is used in this work, or possibilistic regression analysis.

In fuzzy linear regression, the model is a linear combination of the parameters, which are represented by fuzzy numbers, and independent variables that determines the dependent variable. In Table 1, the independent variable, x_{ij}, represents the jth input for the ith sample, when the dependent variable, $\hat{y}_i$, is an estimated value of the output, y_i, that is an observation for the ith sample.

The goal is to find the fuzzy parameters out that obtain the best estimation of the regression problem. Thus, the problem of fuzzy regression can be formulated by a straight line as

$$\tilde{y}_i = \tilde{A}_0 + \sum_{i=1}^{m} \tilde{A}_i x_i \tag{11}$$

where $\tilde{A}_0$ is a fuzzy coefficient and $\tilde{A}_i (i = 1, \ldots, m)$ is a vector of fuzzy coefficients. Each fuzzy coefficient is characterized by a function membership defined by decision maker. The membership function can be defined as

$$\mu_{\tilde{A}_j} : \mathbb{R} \rightarrow [0, 1], \quad j \in \{0, 1, \ldots, m\}$$

In particular these membership functions will be supposed as:

$$\mu_{\tilde{A}_j}(y) = \begin{cases} 0 & \text{if } A_j^U \leq y \text{ or } y \leq A_j^L \\ h_{\tilde{A}_j}(y) & \text{if } A_j^L \leq y \leq A_j \\ g_{\tilde{A}_j}(y) & \text{if } A_j \leq y \leq A_j^U \end{cases} \quad j \in \mathbb{J} \tag{12}$$

As the target is to minimize the difference between the observations and the estimations,

$$y_i - \tilde{y}_i = \epsilon_i \tag{13}$$

where ϵ_i is the residual, which is a random variable with zero mean. Thus, according to [22] the fuzzy regression analysis by using least square model can be transformed into a fuzzy quadratic programming problem as

$$\begin{aligned} &\min \textstyle\sum_{i=1}^{m} \left(y_i - \tilde{A}_0 - \sum_{j=1}^{n} \tilde{A}_j x_{ij} \right)^2 \\ &\text{s.t. } \mathbf{A}^t \mathbf{x}_j + \mathbf{A}^{\mathbf{U}^t} |\mathbf{x}_j| \geq y_j \\ &\quad\;\; \mathbf{A}^t \mathbf{x}_j - \mathbf{A}^{\mathbf{L}^t} |\mathbf{x}_j| \leq y_j \\ &\quad\;\; c_i \geq 0, \quad i = 0, 1, \ldots, n, \quad j = 1, \ldots, m \end{aligned} \tag{14}$$

5 Numerical Example

In this section, a kind of datasets that represents the wind turbines is analyzed and is available in [13]. For this reference, six wind turbines were selected, WT1 to WT6, where WT1 to WT4 are inland wind turbines, while WT5 and WT6 are offshore ones. The data at wind farms are collected in blocks for each 10 min because wind speed are considered stationary, and other environmental factors nearly constant, over 10 min duration. Some sensors on a meteorological mast are used to collect and the environmental data was used to generate the power output, which is estimated by the environmental factors like wind velocity, wind direction, air density, temperature, air pressure, humidity among others. For this work, the factors used are wind velocity and direction, which are the averages of the recordings in a 10 min duration, and air density that can be computed by air temperature and pressure. However, in order to show the estimation, the figures depict the relation between wind speed and normalized power outputs.

The tests were all performed on a Macbook Pro with 2.7 GHZ Intel® Core™ i7 processor, 16 GB RAM running MacOS High Sierra operational system. All the problems presented in this work were resolved using NSGA-II evolutionary algorithm [6] which was implemented in MATLAB® 9.7.0 program. The evolutionary algorithm parameter are 1000 generations and 20 individuals in the population, while the crossover and mutation index are 0.6 and 0.1, respectively.

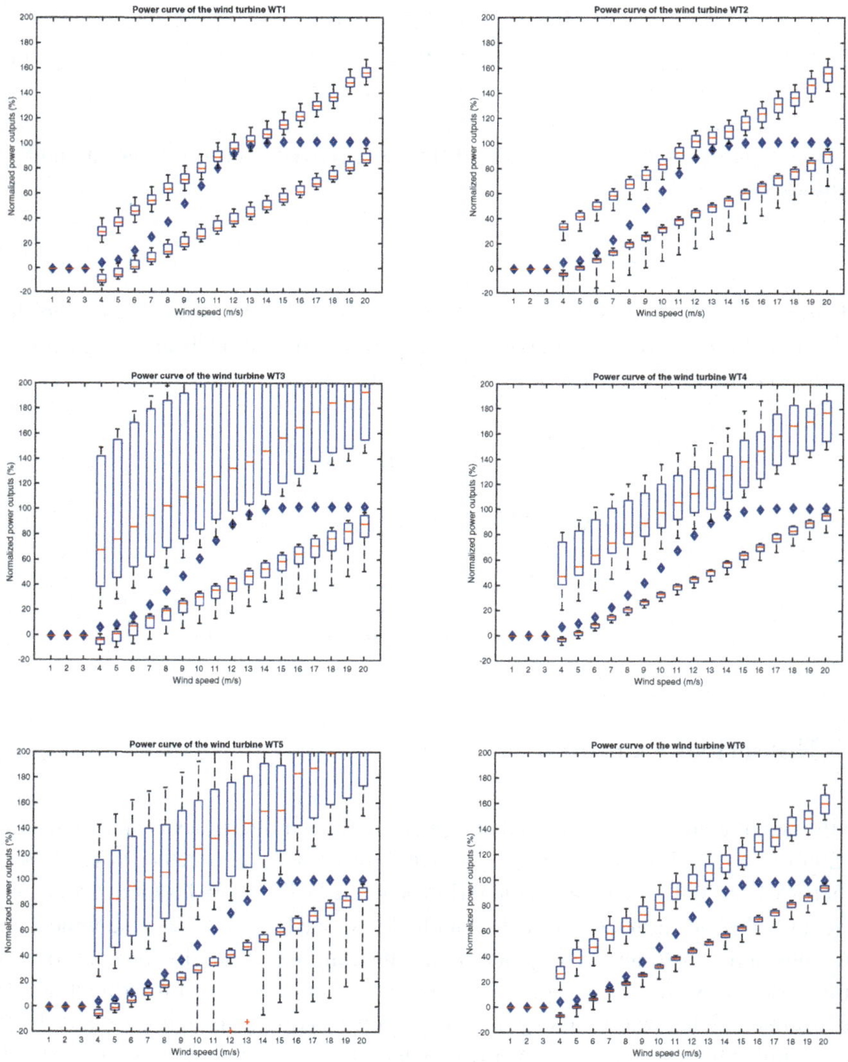

The pictures show that the observations, which are represented by diamonds, are also inside of the interval formed by the obtained solutions for the quadratic formulation. The upper and lower box-and-whisker diagram, which shows the spread of the data, represent upper and lower values, respectively, of each one fuzzy number of the population of obtained optimal solutions. The box lies between the upper and lower quartiles, and the median can also be indicated by dividing the box into two. The whiskers are straight line extending from the end of the box to the maximum and minimum extreme values. It is easy to see that the most of the lower values are more concentrate and some minimum extreme values are more distant from the lower quartile. Therefore, the quadratic formulation obtains parameters that fit better to the observed values.

6 Conclusion

Quadratic Programming problems are very important in a variety of both theoretical and practical areas. When real-world applications are considered, the vagueness appears in a natural way and then it makes perfect sense to think in Fuzzy Quadratic Programming problems. In contrast to what happens with Fuzzy Linear Programming problems, unfortunately until now no solution method has been found for this important class of problems. In this context this paper has presented an operative and novel method for solving Fuzzy Quadratic Multi-Objective Programming problems which is carried out by performing two phases which finally provide the user with a fuzzy solution. The obtained solutions allow the authors to follow along this research line trying to solve real problems in practice, in such a way that oriented Decision Support Systems involving Fuzzy Quadratic Programming problems can be built.

An evolutionary algorithm called NSGA-II was used and it produces a sequence of points according to a prescribed set of instructions, together with a termination criterion. Usually we look for a sequence that converges to a set of efficient solutions, but in many cases the sequence converges to less favorable solutions. Then the procedure may stop either (1) if a point belonging to a prefixed set (the solution set) is reached, or (2) if some prefixed condition for satisfaction is verified. In any case, assuming that a solution set is prefixed, the algorithm would stop if a point in that solution set is reached. Frequently, however, the convergence to a point in the solution set is not easy due to the existence of local optimum points, for example. Thereupon, we must redefine some rules to finish the iterative procedure.

Besides, the control rules of the algorithms to solve convex programming problems could be associated to the solution set, and to the criteria for terminating the algorithm. As it is clear, fuzziness could be introduced in both points, not assuming it as inherent in the problem, but as help for obtaining, in a more effective way, some solution for satisfying the decision-maker's wishes. This mean that the decision maker might be more comfortable obtaining a solution expressed in terms of satisfaction instead of optimisation, as it is the case when fuzzy control rules are applied to the processes.

References

1. Ammar, E., Khalifa, H.A.: Fuzzy portfolio optimization a quadratic programming approach. Chaos Solitons Fractals **18**, 1045–1054 (2003)
2. Bardossy, A.: Note on fuzzy regression. Fuzzy Sets Syst **37**, 65–75 (1990)
3. Bector, C.R., Chandra, S.: Fuzzy mathematical programming and fuzzy matrix games. In: Volume 169 of Studies in Fuzziness and Soft Computing. Springer, Berlin (2005)
4. Bortolan, G., Degani, R.: A review of some methods for ranking fuzzy subsets. Fuzzy Sets Syst **15**, 1–19 (1985)
5. Cruz, C., Silva, R.C., Verdegay, J.L.: Extending and relating different approaches for solving fuzzy quadratic problems. Fuzzy Optim. Decis. Making **10**, 193–210 (2011)
6. Deb, K.: Multi-Objective Optimization Using Evolutionary Algorithms. Wiley, LTD, Chichester (2001)
7. Delgado, M., Verdegay, J.L., Vila, M.A.: Imprecise costs in mathematical programming problems. Control Cybern. **16**(2), 113–121 (1987)
8. Delgado, M., Verdegay, J.L., Vila, M.A.: Relating different approaches to solve linear programming problems with imprecise costs. Fuzzy Sets Syst. **37**, 33–42 (1990)
9. Diamond, P.: Fuzzy least squares. Inform. Sci. **46**, 141–157 (1988)
10. Donoso, S., Marín, N., Vila, M.A.: Fuzzy regression with quadratic programming: an application to financial data. Lect. Notes Comput. Sci. **4224**, 1304–1311 (2006)
11. Jiménez, F., Cadenas, J.M., Sánchez, G., Gómez-Skarmeta, A.F., Verdegay, J.L.: Multi-objective evolutionary computation and fuzzy optimization. Int. J. Approx. Reason. **43**(1), 59–75 (2006)
12. Klir, G.J.: Where do we stand on measures of uncertainty, ambiguity, fuzziness, and the like. Fuzzy Sets Syst. **24**, 141–160 (1987)
13. Lee, G., Ding, Y., Genton, M., Xie, L.: Power curve estimation with multivariate environmental factors for inland and offshore wind farms. J. Am. Stat. Assoc. **110**, 56–67 (2015)
14. Liu, S.T.: Quadratic programming with fuzzy parameters: a membership function approach. Chaos Solitons Fract. **40**(1), 237–245 (2009)
15. Liu, S.T.: Solving quadratic programming with fuzzy parameters based on extension principle. In: FUZZ-IEEE 2007. IEEE International Conference Fuzzy Systems, pp. 1–5, London, UK (2007)
16. Peters, G.: Fuzzy linear regression with fuzy interval. Fuzzy Sets Syst. **63**, 45–55 (1994)
17. Redden, D.T., Woodall, W.H.: Properties of certain fuzzy linear regression methods. Fuzzy Sets Syst **64**, 361–375 (1994)
18. Savic, D., Pedrycz, W.: Evaluation of fuzzy linear regression models. Fuzzy Sets Syst. **39**, 51–63 (1991)
19. Silva, R.C., Cruz, C., Yamakami, A.: A Parametric Method to Solve Quadratic Programming Problems with Fuzzy Costs. IFSA (2009)
20. Tanaka, H., Koyama, K., Lee, H.: Interval regression analysis based on quadratic programming. IEEE 325–329 (1996)
21. Tanaka, H., Lee, H.: Interval regression analysis by quadratic programming approach. IEEE Trans. Fuzzy Syst. **6**(4), 473–481 (1998)
22. Tanaka, H., Uejima, S., Asai, K.: Linear regression analysis with fuzzy model. IEEE Trans. Syst. Man Cybern. **12**(6), 903–907 (1982)
23. Tanaka, H., Watada, J.: Possibilistic linear systems and their application to the linear regression model. Fuzzy Sets Syst **27**, 275–289 (1988)
24. Zimmermann, H.J.: Fuzzy programming and linear programming with several objective functions. Fuzzy Sets Syst. **1**, 45–55 (1978)
25. Zimmermann, H.J.: Fuzzy mathematical programming. Comput. Oper. Res. **10**(4), 291–298 (1983)

Effect of the Growth of Carbon Emissions as a Function of GDP Per Capita and Its Relationship with the SDG in American and European Countries: An Application of Neural Networks

Anderson Quintero, Mercedes Gaitán-Angulo, Melva Inés Gómez-Caicedo, Álvaro Luis Mercado Suárez, and Rosa Eugenia Reyes Gil

Abstract The repercussions of progressive environmental deterioration in the world have generated studies such as the one presented through the Kuznet environmental curve, which theorizes phenomena that relate the use of natural resources with the increase in the Gross Domestic Product (GDP) of economies. Thus, the case analyzed in this chapter makes it possible to identify the effect of the growth of carbon emissions as a function of per capita GDP in American and European countries, its behavior and the results generated by it. The above responds to the provisions of Sustainable Development Goals 8 "Decent Work and Economic Growth" and 13 "Climate Action", which correspond to economic growth and climate action, respectively. The methodological development of this study is based on the construction of a deep neural network, which allows concluding on the results of the economies of America and Europe and identifying those that have implemented environmentally friendly processes and, at the same time, sustainable economic growth.

Keywords SDG · Environment · GDP · Kuznets' curve · Neural networks

A. Quintero · M. Gaitán-Angulo (✉)
Fundación Universitaria Konrad Lorenz, Bogotá, Colombia
e-mail: mercedes.gaitana@konradlorenz.edu.co

M. I. Gómez-Caicedo · Á. L. M. Suárez · R. E. R. Gil
Fundación Universitaria los Libertadores, Bogotá, Colombia
e-mail: migomezc@libertadores.edu.co

Á. L. M. Suárez
e-mail: alvaro.mercado@libertadores.edu.co

R. E. R. Gil
e-mail: rereyesg@libertadores.edu.co

J. L. Verdegay et al. (eds.), *Computational Intelligence Methodologies Applied to Sustainable Development Goals*, Studies in Computational Intelligence 1036,
https://doi.org/10.1007/978-3-030-97344-5_13

1 Introduction

The United Nations World Commission on Environment and Development, in the document entitled "Our Common Future", often referred to as the Brundtland Report, popularized the concept of sustainable development based on the following statement: "It is up to humankind to make development sustainable, that is, to ensure that it meets the needs of the present without compromising the ability of future generations to meet their own needs" [1].

This definition reconciles two elements that seemed incompatible for decades: economic growth and environmental protection. It also recognizes the need and responsibility to comply with intergenerational equity, legitimized at the First World Conference on the Human Environment in Stockholm (1972), as well as the recognition of long-term damage and the limits of the natural environment [2].

In this way, the concept of sustainable development is associated with elements of economic, social, and environmental character [3], which promote the participation of all actors in the implementation of policies, actions and processes [4] that are enduring in time.

Thus, Sustainable Development facilitates the interaction of three complex systems: the global economy, global society and the physical environment, within an environment of good governance [5]. In other words, it consists of the challenge of integrating economic, social and environmental objectives in the long term, while preserving biological, cultural and economic diversity [6].

For this reason, the United Nations Conference on Sustainable Development in 2012, better known as Rio + 20, reaffirmed and revitalized the commitment of nations to Sustainable Development, ratified the Rio 1992 principles and defined the priority thematic axes necessary to achieve Sustainable Development, thus giving rise to the formulation of the Sustainable Development Goals (SDGs) and the 2030 Agenda, allowing developed and developing countries to commit to living sustainably.

From the above, it can be inferred that Sustainable Development has become the new paradigm of development of nations, in which all countries must commit to sustainability, i.e., it requires that economic, social and environmental aspects are aligned, in addition to the commitment of all members of society.

In this sense, this chapter is divided into three parts: the first presents a general context of sustainable development, the SDGs and the 2030 Agenda; the second part mentions the model to be used to analyze the relationship between economic growth and the environment; and in the third part, the results of the research are related and finally the main conclusions of the research are presented.

2 Sustainable Development, SDG and Agenda 2030

The Sustainable Development Goals (SDGs) were born at the United Nations Conference on Sustainable Development in Rio de Janeiro in 2012. These goals are intended

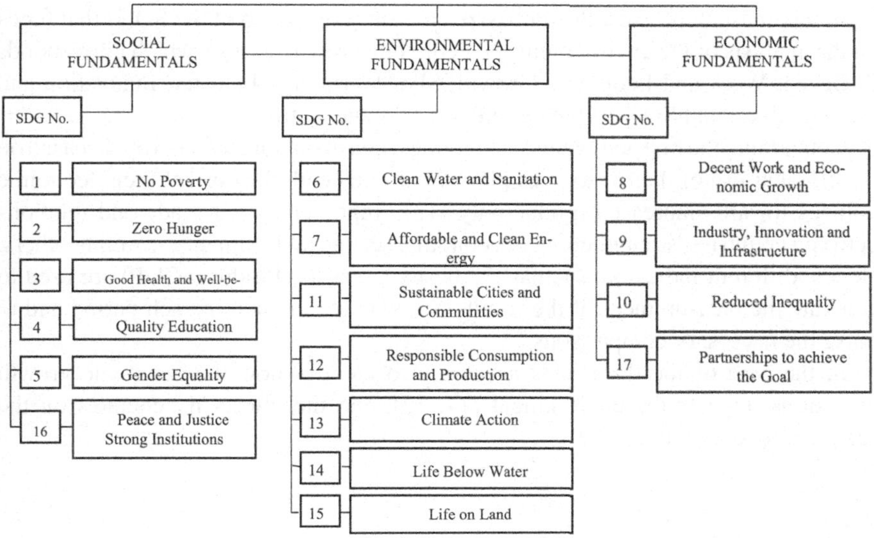

Fig. 1 SDGs Pillars. *Source* Perea Hinestroza [8] and Gil (2018)

to achieve a set of targets and a globally agreed agenda that would address the world's environmental, political and economic challenges [7]. The SDGs are the reformulation and evolution of the Millennium Development Goals, originally established in 2000 to combat extreme poverty, hunger, deadly diseases, and expand access to education. The 17 SDGs are based on the five pillars that make up the United Nations 2030 Agenda: People, Planet, Prosperity, Peace and Partnerships, with a significant interrelationship among them.

Its conceptualization and articulation in the countries allow the inclusion of three pillars that are the foundation for the design and development of policies and actions aimed at solving social, environmental and economic problems [8]. Figure 1 identifies their relationship with the SDGs.

Hence, the 194 signatory countries of Sustainable Development in September 2015, agreed on the importance of declaring that "the main objective is to ensure the protection of people, the planet and prosperity" [9]; and established the importance of improving living conditions, based on the premise that the environment, society and the economy must interact together for optimal development of coexistence and respect for natural capital centered on strategic ecosystems.

Likewise, with the Paris Agreement (2015) and the Pathway Framework for Disaster Risk Reduction (2015–2030), the 2030 Agenda provides a set of common standards and timely and achievable goals to achieve mitigate climate change, its effects on communities and natural disasters. These issues involve all members of the global society and the only way to achieve them is through joint work where all sectors of society (state, business, education and civil society), act towards their fulfillment to achieve a more humane, sustainable and safe planet.

In this sense, this research emphasized the analysis of two of the SDGs that focus on the analysis of the environment and economic growth: SDG 8 which corresponds to Decent Work and Economic Growth, SDG 9 which is Industry, Innovation and Infrastructure, and SDG 13 which refers to Climate Action.

Concerning the first and second, economies seek to obtain greater growth, based on the efficient use of their resources [10–12]. To this end, they orient their economic policies towards increased productivity [13], improvements in trade and relations between countries, strengthening of economic sectors and companies, among others. Hence, different indicators such as the Gross Domestic Product (GDP) are used to generate the measurement of the results obtained by a nation in each period and to make the necessary comparisons.

In the case of the third, it is necessary to identify how the economic growth of nations impacts the environment [14, 15], and the effects it generates on the surrounding community.

3 Methodology

Although the theoretical basis of the Kuznets'Curve generates a hypothetical relationship between the gross earnings of a country or region and its pollutant emissions [16], the contrast that can be established with the real data can generate highly unstable dynamics due to mismatches between the study regions, inflation and the advance of technology.

Thus, changes in environmental policies, capital investment and other factors can generate triggering changes in the evolution of the domestic economy and subsequently in the associated pollutant emissions [17].

Therefore, the focus of this methodological development will be to establish a deep neural network model that allows identifying the annual progressive trend of environmental deterioration as a function of the Gross Domestic Product per capita in the European and Latin American region, calculated from the results obtained from the application of the Kuznets' Curve.

Thus, the characteristics of this study explore the quantitative quantities used to calibrate a deep neural network with which a relationship between Gross Domestic Product and environmental deterioration will be implicitly established.

Organism studied

Two phenomena were used for the development of this study. To quantify the monetary benefit produced by a country we used the Gross Domestic Product (GDP) per capita [18], which refers to the total gross earnings of a country divided by its total population.

For the second phenomenon, environmental deterioration was quantified based on CO_2 emissions [19] or carbon dioxide, which measures the number of tons that are emitted when using fossil fuels and greenhouse gases.

These emissions are an indicator of environmental deterioration because regions exposed to these gases eventually become biologically unproductive areas needed to absorb this CO_2. In addition to being the standard unit dependent on human economic production and technological development.

Data description

The data explored in this study were taken from a common repository, the World Bank Data [20, 21]. This repository stores a wide and varied set of data.

For this specific analysis, data about GDP per capita and CO_2 emissions measured in kilotons were used. Both records contained the information per country for each metric and were divided annually from 1960 to 2018.

The processing of the database consisted of isolating the countries of Latin America, as well as Europe and Western Asia, to have only the records of these specific regions.

The countries used for the Latin American region were: Argentina Belize, Bolivia, Brazil, Colombia, Costa Rica, Cuba, Dominica, Dominican Republic, Ecuador, Grenada, Guatemala, Guyana, Honduras, Haiti, Jamaica, St. Lucia, Mexico, Nicaragua, Panama, Peru, Paraguay, El Salvador, Suriname, St. Vincent and the Grenadines. For the European and Western Asia region, the following countries were considered: Albania, Andorra, Armenia, Austria, Azerbaijan, Belgium, Bulgaria, Bosnia and Herzegovina, Belarus, Switzerland, Cyprus, Czech Republic, Germany, Denmark, Spain, Estonia, Finland, France, United Kingdom, Georgia, Gibraltar, Greece, Greenland, Croatia, Hungary, Ireland, Iceland, Italy, Liechtenstein, Lithuania, Luxembourg, Latvia, Monaco, Republic of Moldova, North Macedonia, Montenegro, Netherlands, Norway, Poland, Portugal, Romania, San Marino, Serbia, Slovak Republic, Slovenia, Sweden, Turkey, Ukraine and Kosovo.

Development of the study

For the development of this study, a preliminary analysis was initially generated from the available information to identify trends between countries and regions, irregular distributions and atypical data.

The most relevant characteristics of each region in terms of GDP per capita and CO_2 emissions were also considered for the design of the automated adjustment method.

Once the initial information bases were established, a deep neural network was designed and trained to take as inputs the CO2 values and as output the value of the gross domestic product in the same year. The network specifications were established from two hidden layers, each with 8 neurons, as shown in Fig. 2. The activation function used was ELU and was designed using Python, specifically from the tensor library Flow and Keras.

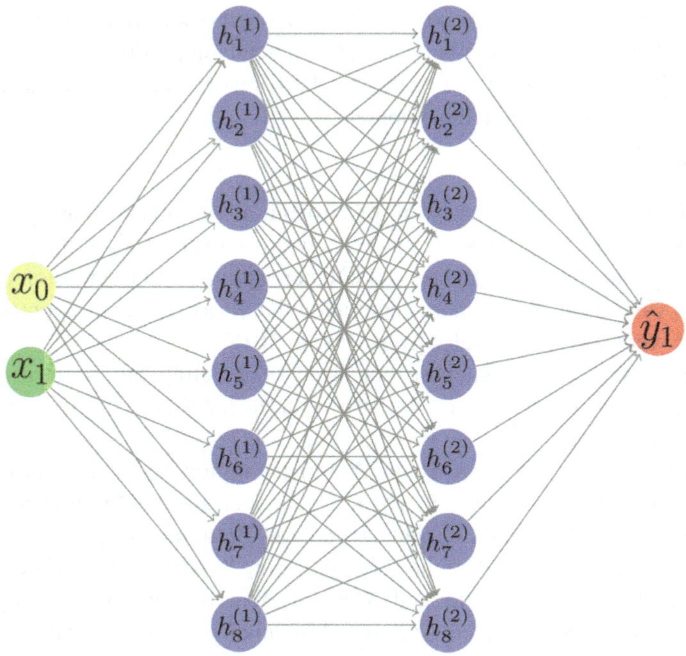

Fig. 2 Neural network model applied to establish the relationship between GDP and CO_2 emissions. *Source* Own elaboration

4 Results

The distribution of GDP per capita indicated that the overall global per capita GDP of the Latin American countries does not have a common distribution, indicating the inequality of the contexts among the territories. Although this could generate disadvantages in the development of trends, it will be observed later how the overall average in fact maintains a close relationship between Latin American regions.

Likewise, the pollutant emissions of which Argentina, Brazil and Mexico stand out as having a much larger scale than the general average, as well as a much larger annual deviation. The distribution of annual emissions indicates that countries are generating more and more carbon dioxide and cases such as Colombia have a very accelerated growth rate.

On the other hand, the distribution of GDP in Europe indicates that Bosnia, Czech Republic, Georgia, Lithuania, Latvia, Moldova, Montenegro, Ukraine, Kosovo, Albania, Armenia, Bulgaria, Gibraltar, Hungary, Macedonia, Netherlands, Romania, Serbia and Turkey have a lower distribution than the most representative countries, generating a relative bias among the regions of the study, while countries with higher GDP per capita generate a huge annual variation.

Unlike European GDP, CO_2 emissions do not have a high amount of outlier data, indicating that the characteristics of this phenomenon are much more frequent and invariant over time, likewise, the variation of emissions does not describe highly unstable behaviors. Aspects that are significantly analyzed by the countries and measured through the policies implemented since SDG 13.

In the case of Europe, there is great stability in its annual distribution, which indicates that eventually, the countries have maintained a constant consumption of carbon dioxide, except for Germany, whose annual evolution seems to generate enormous amounts in comparison to the other territories.

Thus, when contrasting Carbon Dioxide as a function of Gross Domestic Product in each of the countries, the results shown in Fig. 3 are obtained.

Figure 3 develops the desired behavior, as linear, pseudo-linear and logarithmic relationships between GDP can be identified. Ideally, an explicit relationship can be constructed that can be fitted by parameter-adjusted models.

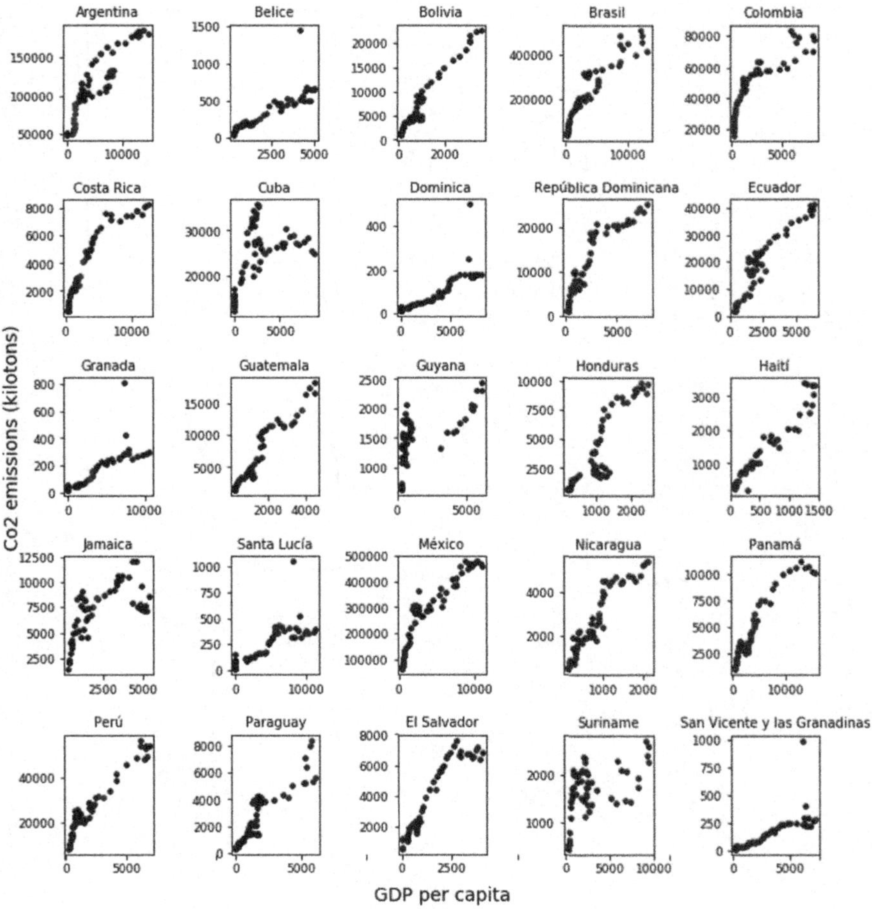

Fig. 3 Carbon dioxide emissions as a function of GDP per capita Latin America. *Source* Own elaboration

Figure 4 also describes CO_2 emissions as a function of GDP for each country in West Asia and Europe as a whole. Visually, in contrast to the Latin American territories, there does not seem to be an explicit relationship between the distribution of countries and the GDP/CO_2 ratio.

Therefore, the general characteristics of the territories seem to indicate that the conglomerate of these regions could adjust the relationships between GDP per capita and CO_2 emissions.

To adequately visualize this possible relationship, a comparison was generated between average GDP and CO_2 emissions, whose annual evolution is shown in Fig. 5.

Figure 5 made it possible to establish a hypothetical relationship between both behaviors because growth and evolution over time are monotonous and seem to follow an analogous behavior.

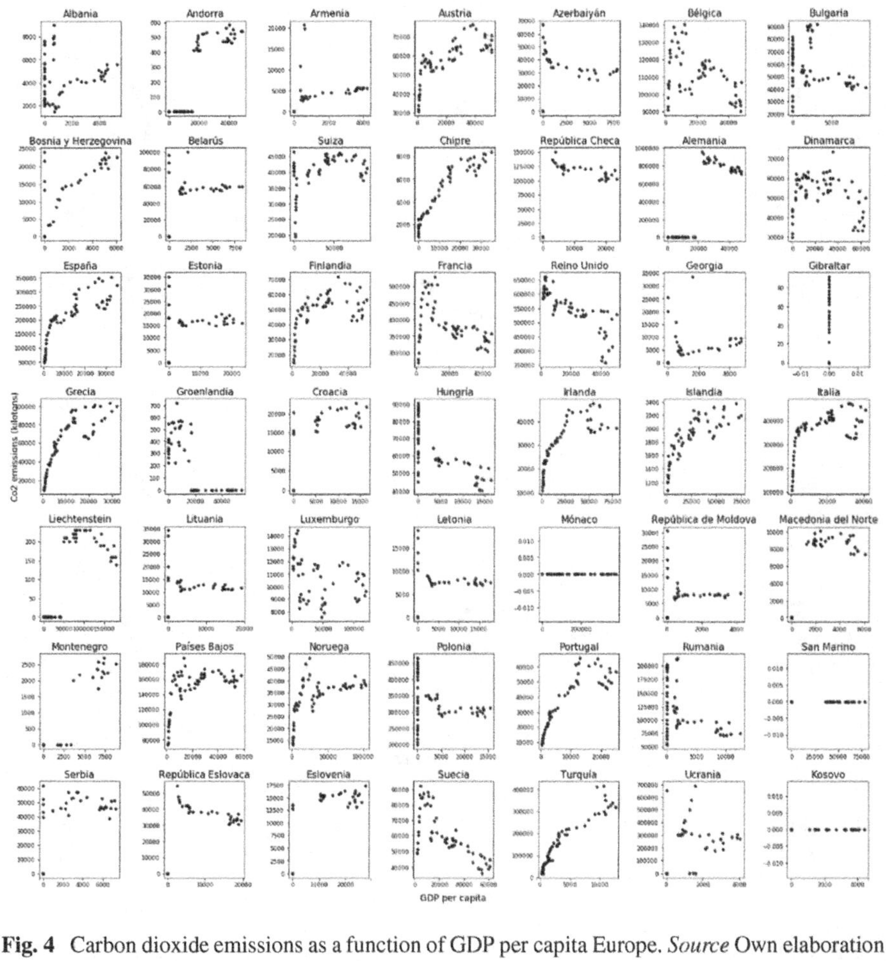

Fig. 4 Carbon dioxide emissions as a function of GDP per capita Europe. *Source* Own elaboration

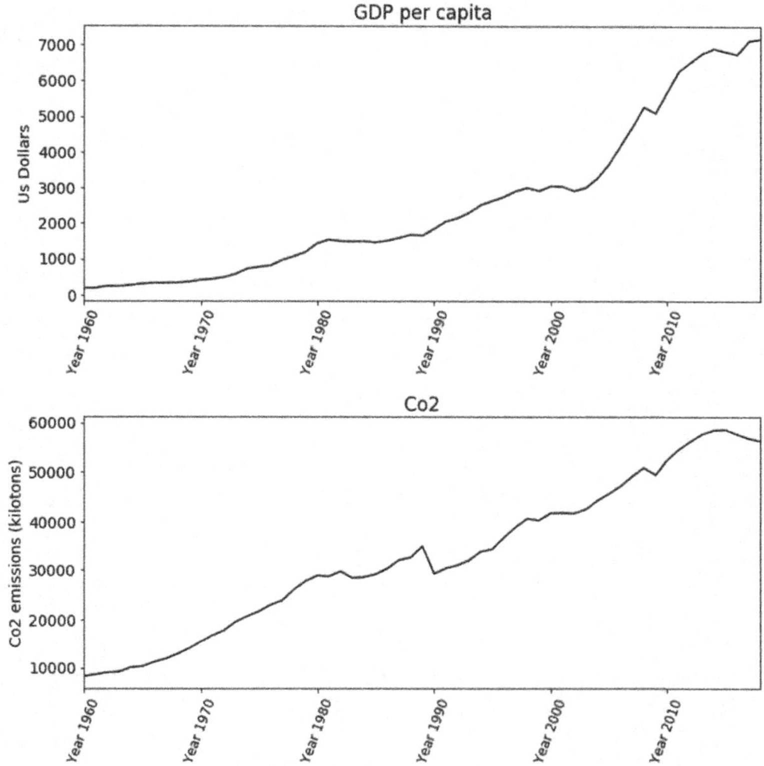

Fig. 5 Average annual evolution of gross domestic product and carbon emissions in Latin America. *Source* Own elaboration

The opposite is the case in Europe, whose distribution is totally different from one another, as shown in Fig. 6.

The independence of these behaviors makes it difficult to establish a relationship of proportionality between the two variables. This makes it more difficult to analyze the impact that may arise in the implementation of measures that seek to respond to SDGs 8, 9 and 13.

To confirm the possible existence of an implicit relationship between GDP and pollutant emissions, the neural network explained above was adapted with individual data from Latin America and Europe, the results of which are shown in Figs. 7 and 8.

As can be seen, the neural network has been able to detect a relevant trend in the Latin American conglomerate indicating that this region possesses characteristics that fit appropriately with the Kuznets' curve hypotheses.

Although the relationship does not necessarily imply causality, the contextual characteristics of this region seem to affirm the initial hypotheses of the environmental Kuznets' curve.

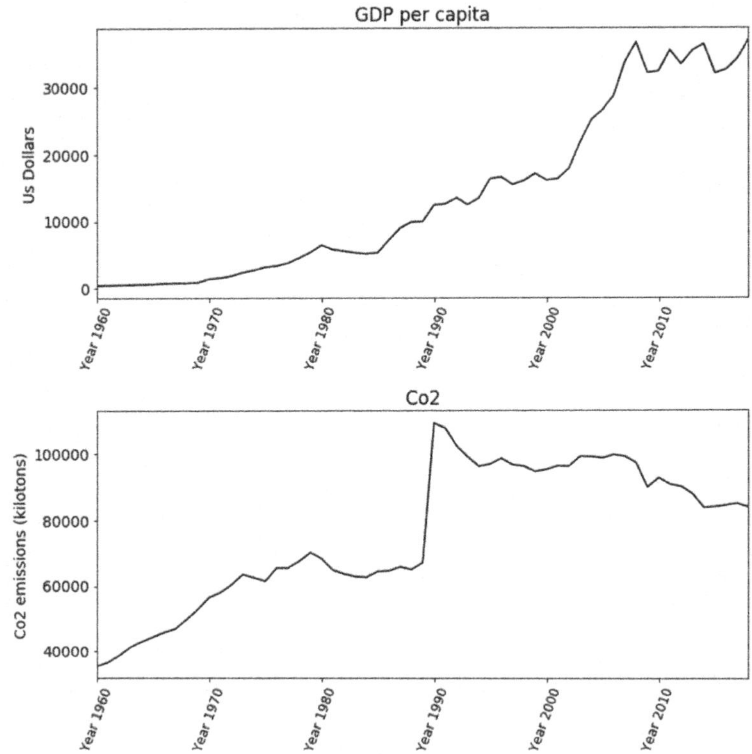

Fig. 6 Average annual change in gross domestic product and carbon emissions in Europe. *Source* Own elaboration

On the other hand, Fig. 8 illustrates the result of applying neural networks to fit the phenomenon in the European context.

The European region and Western Asia do not stand out for their high relationship based on this methodology, despite pretending to establish a relationship without the need for a concrete expression, it is not possible to generate a sufficiently precise adjustment without over-feeding the network.

Therefore, from images 7 and 8 it is possible to determine that, for the Latin American case, it does indeed adequately adjust the relationship, despite the minimal amount of data present, indicating that the relationship is clearly identified and reaffirming the existence of this implicit behavior in the region.

On the other hand, Europe does not seem to have a global case distributed throughout the region, as the perceived behavior is far from that adjusted by the neural network, however, this statement may be due to the lack of information or to the changes in trends in recent years.

In general terms, Kuznets' hypothesis of a possible implicit relationship between a country's earnings and its environmental effects seems to be more perceptible in countries with greater ecological wealth, while individually European countries

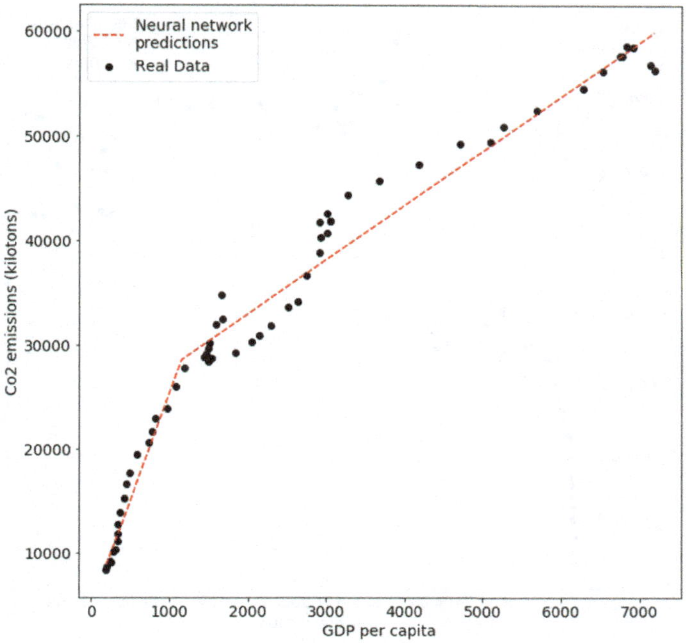

Fig. 7 Neural network adjustment to CO_2 emissions as a function of GDP per capita in Latin America. *Source* Own elaboration

maintain a minimal convergence between their annual earnings and environmental degradation, possibly due to the growth of their environmental awareness or to the increase in technology, this being one of the last phases described by Kuznet.

5 Discussion and Conclusions

The European context indicates that there is not enough evidence to determine a possible general relationship, however, the data do show a decrease in carbon dioxide emissions, without directly affecting GDP growth.

Although it was not possible to find an expression relating both variables, the evolution of carbon dioxide emissions seems to be in line with the last phase of the Kuznets' Curve, which establishes that the technological growth of territories allows the establishment of efficient and sustainable alternatives whose implementation decreases environmental degradation, generating independence between profit and wear.

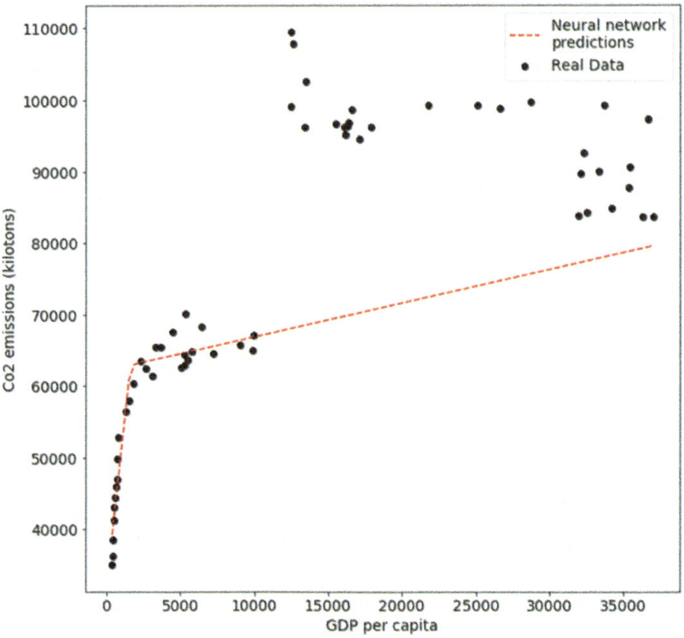

Fig. 8 Neural network adjustment to CO_2 emissions as a function of GDP per capita in Europe. *Source* Own elaboration

References

1. World Commission on Environment and Development: Our Common Future. Oxford University Press, Oxford (1987)
2. Rodríguez Becerra, M.: El Desarrollo Sostenible: Utopía o realidad para Colombia? En M. Rodríguez Becerra, La política ambiental del fin de siglo: Una agenda para Colombia, pp. 19–43, CEREC (1994)
3. Mensah, J.: Sustainable development: meaning, history, principles, pillars, and implications for human action: literature review. Cogent Soc. Sci. **5**(1) (2019). https://doi.org/10.1080/23311886.2019.1653531
4. Haider, L.J., Boonstra, W.J., Peterson, G.D., Schlüter, M.: Traps and sustainable development in rural areas: a review. World Dev. **101**, 311–321 (2018). https://doi.org/10.1016/j.worlddev.2017.05.038
5. Sachs, J.: La era del desarrollo sostenible. Paidós Empresa (2014)
6. Gallopín, G.: Sostenibilidad y desarrollo sostenible: un enfoque sistémico. CEPAL. https://repositorio.cepal.org/handle/11362/5763 (2003)
7. Halkos, G., Gkampoura, E.-C.: Where do we stand on the 17 sustainable development goals? An overview on progress. Econ. Anal. Policy **70**, 94–122 (2021)
8. Perea Hinestroza, L.: Los Objetivos de Desarrollo Sostenible y su inclusión en Colombia. Revista Producción + Limpia **14**(1), 122–127 (2019). https://doi.org/10.22507/pml.v14n1a8
9. PNUD, P.D.: Programa de las Naciones Unidas para el Desarrollo (2019)
10. Buera, F.J., Kaboski, J.P.: Scale and the origins of structural change. J. Econ. Theory **147**(2), 684–712 (2012)

11. Chavarro, D., Vélez, M.I., Tovar, G., Montenegro, I., Hernández, A., Olaya, A.: Los Objetivos de Desarrollo Sostenible en Colombia y el aporte de la ciencia, la tecnología y la innovación. Work Document **1**, 30 (2017)
12. López-Mera, S.F.: Qué tan lejos está el ODS# 8 para Colombia? Una década de medición del trabajo decente. Soc. Econ. (43) (2021)
13. Cómbita Mora, G., Pérez Rodríguez, Ó.: Crecimiento inclusivo: una estrategia integral para la obtención de los objetivos de desarrollo sostenible. J. Desarrollo y Sociedad **86**, 43–72 (2020)
14. Dantas, T.E.T., de-Souza, E.D., Destro, I.R., Hammes, G., Rodriguez, C.M.T., Soares, S.R.: How the combination of circular economy and Industry 4.0 can contribute towards achieving the sustainable development goals. Sustain. Prod. Consum. **26**, 213–227 (2021)
15. Fischer-Kowalski, M., Amann, C.: Beyond IPAT and Kuznets curves: globalization as a vital factor in analyzing the environmental impact of socio-economic metabolism. Popul. Environ. **23**(1), 7–47 (2001)
16. Gu, W., Zheng, X.: An empirical study on the impact of sustainable entrepreneurship: based on the environmental Kuznets model. J. Bus. Res. **123**, 613–624 (2021)
17. Gunawan, J., Permatasari, P., Tilt, C.: Sustainable development goal disclosures: do they support responsible consumption and production? J. Clean. Prod. **246**(118989), 118989 (2020)
18. Cambridge Dictionary: Meaning of GDP per capita in Cambridge Dictionary. Retrieved 10 Aug 2021, from: https://dictionary.cambridge.org/es/diccionario/ingles/gdp-per-capita (2021)
19. Global Footprint Network National Footprint and Biocapacity Accounts Data Footprint Network. 2021. Retrieved 10 Aug 2021, from: https://data.footprintnetwork.org/#/abouttheData
20. Ridge, O.: Emisiones de CO_2 (kt) | Data. In Banco Mundial. Retrieved 10 Aug 2021, from: https://datos.bancomundial.org/indicator/EN.ATM.CO2E.KT (2019)
21. World Data Bank: PIB per cápita (US$ a precios actuales)|Data. Banco Mundial. Retrieved 10 Aug 2021, from: https://datos.bancomundial.org/indicator/NY.GDP.PCAP.CD?end=2020&start=1960&view=chart (2019)

Transportation and Distribution Models

A Fuzzy Approach for Organizational Transportation Aligned with the Sustainable Development Goals: Health Promotion, Inequality Reduction, and Responsible Consumption

Eduardo Sánchez-Ansola, Ana Camila Pérez-Pérez, and Alejandro Rosete

Abstract In 2015 the United Nations Organization approved the Agenda for Sustainable Development. This agenda defines the Sustainable Development Goals, including Good Health and Well-being, Reduced Inequalities, and Responsible Consumption and Production. The present research introduces a new model for the School Bus Routing Problem (SBRP) to have a direct impact on the three mentioned objectives. The proposed model focuses on fuzzy modeling of maximum students' walking distance and bus capacity restrictions. Likewise, a new characteristic in the students that imply the differentiation in the distance to walk is included. Experimental results are obtained by applying a parametric linear approach to solve the fuzzy model in conjunction with metaheuristics. The analyzed results allow us to affirm that cost-saving can be achieved at the expense of slightly increasing the student's walking distance and a little overload of the bus capacity. Also, the results reveal that this saving can be obtained by reducing students' inequalities.

Keywords Fuzzy optimization · Parametric approach · School bus routing problem · Sustainable development goals

1 Introduction

On its official website, the United Nations (UN) defines sustainable development as: "...development that meets the needs of the present without compromising the ability of future generations to meet their own needs..." [1]. Consequently, in 2015

E. Sánchez-Ansola · A. C. Pérez-Pérez · A. Rosete (✉)
Department of Artificial Intelligence and Informatics Systems Infrastructure, Universidad Tecnológica de La Habana José Antonio Echeverría (CUJAE), La Habana, Cuba
e-mail: rosete@ceis.cujae.edu.cu

E. Sánchez-Ansola
e-mail: esancheza@ceis.cujae.edu.cu

A. C. Pérez-Pérez
e-mail: acperez@ceis.cujae.edu.cu

J. L. Verdegay et al. (eds.), *Computational Intelligence Methodologies Applied to Sustainable Development Goals*, Studies in Computational Intelligence 1036,
https://doi.org/10.1007/978-3-030-97344-5_14

the 2030 Agenda for Sustainable Development was approved, holding 17 Sustainable Development Goals (SDG) to stimulate actions in areas of critical importance for humanity and the planet.

The current problems that humanity faces and that have led to the definition of the SDGs need to be approached from the point of view of Sustainability Science, a field that seeks to interpret the fundamental relation between nature and society. Sustainability Science requires new approaches, tools, and techniques that allow decision-makers to make decisions based on a wide range of uncertainty. In this regard, methodologies based on Soft Computing can play a significant role in solving these goals [2].

According to Di Vaio et al. [3], Artificial Intelligence (AI) techniques have been proved to be a useful tool to work focused on SDGs. Two significant aspects where AI can contribute to SDGs are optimization (including metaheuristics) and customer satisfaction. In particular, transport planning and organization problems are influential elements in sustainable development that have received the attention of Soft Computing [4]. According to Corona [2], Verga et al. [4], metaheuristics and fuzzy modeling have been widely used in research on transport for sustainable development.

The School Bus Routing Problem (SBRP) aims to create a set of routes for the transportation of students to their schools from different locations [5]. According to Ellegood et al. [6], one of the current challenges in this type of problems is to make them as close to reality as possible, with the treatment of uncertainty being an essential element.

In classical SBRP literature, the problem is expressed in terms of schools and students, but the model can be considered in a more broad sense in terms of organizations and persons. Despite this generalization, in the rest of the paper, we use the usual concepts of SBRP related to schools.

In this work, a fuzzy model of SBRP is presented that focuses on the restrictions of the student's walking distance and the bus capacity. Likewise, a new characteristic is modeled in the students that imply the differentiation in the distance to walk, making it possible for those students (persons, in general) with some difficulty and/or disability to walk less to reach a stop than the rest of the students. The aforementioned conception implies a new version of SBRP with distinct types of persons in the organizational transportations, which may be relevant to allow different considerations, such as disabilities, ages of workers, and weight of tools that some workers must carry.

This research aims to have a direct impact on three of the SDGs:

- SDG 3: Good Health and Well-being. By allowing certain levels of flexibility in student's walking distance, some of them would have to walk a little more, which promotes physical exercise and (probably) social interaction, influencing both their health and their social well-being. In contrast, a reduced walking distance may be convenient for the health of those with problems or complex conditions.
- SDG 10: Reduced Inequalities. Allowing differences in the maximum walking distance by some students allows a differentiated treatment according to the singularities of each person, thus reducing inequalities between students.

- SDG 12: Responsible Consumption and Production. By modeling the capacity of the buses and the maximum walking distance as fuzzy elements, a reduction of the total distance traveled by the buses (and the use of fuels) is achieved, at the expense of a small overhead on the buses and/or increased walking distance. This allows a more responsible consumption of resources and a direct impact on the reduction of environmental pollution.

The rest of the document is organized as follows. Section 2 addresses the general characteristics of the SBRP. In Sects. 3 and 4, the mathematical model and its fuzzy approach for the SBRP are presented. Section 5 presents and discusses the experimental results. Finally, the conclusions and future work are presented in Sect. 6.

2 School Bus Routing Problem

The School Bus Routing Problem (SBRP) seeks to plan an efficient program for a fleet of school buses where students are transported to (or from) their schools while satisfying various constraints [7]. It was identified as a problem in [8], but only recently the contributions to the modeling and solution of this problem have increased, as can be observed in [6].

The main categories used to classify SBRP [6] are:

1. Sub-problems: Bus Stops Selection (BSS), Bus Routes Generation (BRG), Bus Route Scheduling (BRS), and School Bell Time Adjustment (SBA).
2. Number of schools: one or multiple schools.
3. Service environment: urban or rural.
4. Load type: distinguishes whether a bus can carry students from/to more than one school. Mixed loads generally allow a bus to carry students from multiple schools.
5. Fleet mix: identifies whether the buses under consideration have the same capacity (homogeneous) or different capacities (heterogeneous).
6. Objective: generally, it is the minimization of one or more of the following aspects: Number of buses (N), Total bus traveled distance or time (TBD), Total student riding distance or time (TSB), Total student walking distance (SWD), Maximum route length (MRL), etc.
7. Constraints: each SBRP model has to meet one or more of the following conditions: Bus capacity (C), Maximum riding time (MRT), School Time Window (TW), Maximum walking time or distance (MWT), etc.

Taking these characteristics into account, the present research is focused on an SBRP with: bus stop selection and bus route generation similar to [9, 10], a single school and homogeneous fleet [11, 12]; having as objective the total traveled distance by the buses [13, 14] and the considered constraints are the bus capacity and the maximum walking distance [9, 10, 15].

Also, we consider multiple classes of students. In [16], a similar procedure was presented but the model focused on the capacity that each student occupies in the bus. However, here the class distinguishes the student's maximum walking distance.

In a broad sense, several fuzzy optimization models have been presented for other transportation optimization problems like the Vehicle Routing Problem (VRP) and some of its variants. For example, the authors of [17] presented several models for variants of the VRP with uncertain travel times modeled as triangular fuzzy numbers. A model for the Truck and Trailer Routing Problem (TTRP) with imprecise capacity constraints and fuzzy treatment was proposed in [18]. More recently, some researchers focus on fuzzy optimization of eco-friendly problems like the Fuzzy Green Vehicle Routing Problem (FGVRP) with fuzzy demands presented in [19] or the Fuzzy Electric Vehicle Routing Problem with time windows and recharging stations (FEVRPTW) [20] where service times, battery energy consumption, and travel times are treated as uncertain. To the best of our knowledge, the only available fuzzy model for SBRP is the one presented in [21], which is a fuzzy extension of the SBRP model presented in [22].

3 Mathematical Model

The fuzzy model presented here is a fuzzy extension of the model proposed in [21]. The main difference is that the model presented here allows:

- a relaxation of the bus capacity, in conjunction with the students' walking distance relaxation
- the consideration of different types of students that implies that the maximum walking distance (and the corresponding tolerance) may be varied. In general, each student may be different in terms of its maximum walking distance.

The following Input Variables are considered:

c: Capacity of each bus.

B, b: Set and index of buses, $b = 1 \ldots |B|$.

P, p: Set and index of possible stops, $p = 0 \ldots |P|$, where $p = 0$ indicates the school.

E, e: Set and index of students, $e = 1 \ldots |E|$.

d_e: Maximum walking distance for student e.

V^p: A set of vectors with coordinate pairs representing the possible stops.

V^e: A set of vectors with pairs of coordinates representing the house of each student.

H_e: Maximum admissible tolerance for the distance that the student e could walk.

T: Maximum allowed tolerance to overload the bus capacity.

The following Auxiliary Variables are considered:

D: A distance function that indicates the cost between a pair of stops or between a student and a bus stop.

C^{P}_{pq}: A distance matrix between each pair of stops (p, q).

C^{E}_{ep}: A distance matrix between each pair of student-stop (e, p).

c^{p}_{i}: Coordinates of the stop located on the i index of V^{p}.

c^{e}_{j}: Coordinates of the student located on the j index of V^{e}.

$$C^{P}_{pq} = \begin{cases} D\left(c^{p}_{p}, c^{p}_{q}\right), p \neq q \\ 0, p = q \end{cases} \quad \forall p, \forall q \in P \tag{1}$$

$$C^{E}_{ep} = D\left(c^{e}_{e}, c^{p}_{p}\right) \quad e \in E, p \in P - \{0\} \tag{2}$$

The following Decision Variables are considered:

R_{bm}: Indicates the stop that is visited by bus b in the order m. $|R_{bm}|$, indicates the number of bus stops visited by the bus b.

Z_e: Indicates the stop where the student e is picked up.

Objective Function is defined as follows:

$$\mathrm{Min} \sum_{b=1}^{|B|} \left(C^{P}_{0[1]} + \sum_{m=1}^{|P|-1} C^{P}_{[R_{bm}][R_{bm+1}]} + C^{P}_{[|R_{bm}|]0} \right) \tag{3}$$

Subject to the following constraints:

$\leq^{f}$: A fuzzy comparison operator indicating the fuzzy imprecision of the constraints (5) and (6).

$$\{R_{bm} | R_{bm} = p\}| \leq 1 \quad \forall p \in P - \{0\}, \forall b \in B, \forall m \in \{1, \ldots, |P|\} \tag{4}$$

$$\{(e, p) | Z_e = p\} \subseteq \left\{(e, p) | C^{E}_{ep} \leq^{f} d_e\right\} \quad \forall e \in E, \forall p \in P \tag{5}$$

$$|\{e | \exists m\ R_{bm} = Z_e\}| \leq^{f} c \quad m \in \{0, \ldots, |P|\}, \forall b \in B, \forall e \in E \tag{6}$$

$$|\{R_{bm} | R_{bm} = Z_e\}| = 1 \quad \forall e \in E \tag{7}$$

The objective function, Eq. (3), minimizes the total distance traveled by the entire bus fleet. Equations (4), (5), (6), and (7) represent the restrictions that must be met for the solution to be feasible. Equation (4) guarantees that each stop is visited at most once, except for stop p = 0 which represents the school, the final destination of all buses. Equation (5) ensures that each student can reach their assigned bus stop. Equation (6) takes into account that the capacity of each bus is not exceeded on the route. And, finally, with Eq. (7) it is guaranteed that each stop to which at least one student is assigned is visited by a bus. Equations (8) and (9) are used to replace Eqs. (5) and (6) and the following explanation argues why.

$$\{(\mathrm{e},\mathrm{p})|\mathrm{Z_e}=\mathrm{p}\} \subseteq \left\{(\mathrm{e},\mathrm{p})|C^{E}_{ep} \leq \mathrm{d}_e + H_e(1-\alpha_w)\right\} \quad (8)$$

$$|\{\mathrm{e}|\exists \mathrm{m}\ \mathrm{R_{km}} = \mathrm{Z_e}\}| \leq \mathrm{c} + T(1-\alpha_c) \quad (9)$$

As can be seen in the previous model, restrictions (5) and (6) are the focus of the proposed fuzzy model. The new way of posing these constraints implies that the feasibility of a student reaching a stop and the bus capacity becomes fuzzy (i.e., not crisp) and therefore it is possible to satisfy these constraints with different degrees of membership.

Taking these elements into account, if the maximum walking distance for students is 200 m, then a student that walks 190 m satisfies it with a grade of 1. On the other hand, if a student walks 210 m, the degree of membership (degree of satisfaction of this constraint) may be less than 1, but it may be higher than the membership if the stop is 250 m away. Conversely, a stop situated 500 m away could be considered unreachable.

On the other hand, if the original bus capacity is 50, the load of 45 students satisfies it with a grade of 1. If 55 students are loaded, the degree of membership may be less than 1, but higher than if 60 students are loaded. Contrariwise, the load of 100 students could be considered a forbidden overload.

All these values (e.g., 200, 500 for the walking distance or 45, 50 for the bus capacity) will depend on the admissible conditions and the allowed tolerance. These values will imply that solutions will have distinct degrees of compliance with the restrictions. From the decision-making point of view, these relaxations may allow a modest increase in the distance that the students could walk and in the bus capacity, to find a relaxed solution with a better value of the objective function (reduced cost).

To model this situation, it is necessary to define a tolerance $\mathrm{H_e}$, which determines the maximum admissible tolerance for the distance that a student could walk, and a tolerance T, which determines the maximum allowed overload for the bus capacity. If T and all the values of $\mathrm{H_e}$ are equal to 0, then the problem is reduced to the crisp case. Figure 1 (left) shows the function to measure the degree of compliance with the restriction of bus capacity taking into account the capacity c and the tolerance T. Figure 1 (right) shows the function to measure the degree of compliance with the restriction of maximum walking distance taking into account the distance d and the tolerance H_e.

To understand this function, having c as the primary bus capacity and T as the maximum admissible tolerance, a bus that loads less than or equal to c has a degree of compliance of 1. On the other hand, if the bus load is between c and c + T, it has a degree of compliance in the interval [0, 1]. Finally, if the bus load is greater than c + T, the degree of compliance is 0, then it is assumed that the bus can't carry this load. A similar thought can be applied to the students walking distance restriction.

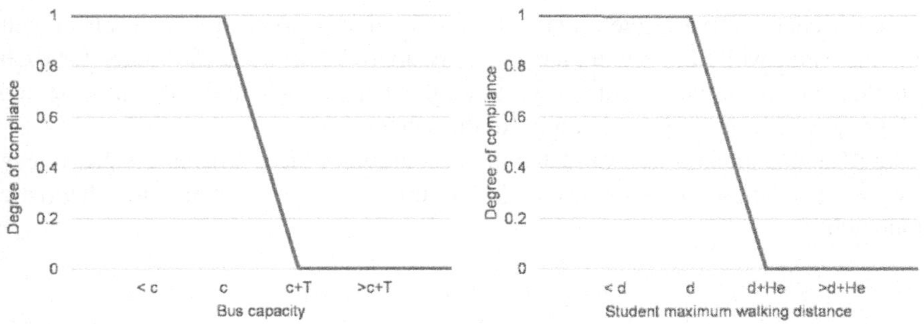

Fig. 1 Membership function of the compliance of the constraints associated with the bus capacity (left) and the student maximum walking distance (right)

These are linear functions and the parametric approximation method [23], based on the principles of parametric linear programming, and the concept of alpha-cuts [23] can be applied. The alpha-cut notion applied to this case implies that different sets of feasible solutions are associated with a particular value of alpha, i.e., those solutions with a degree of feasibility (the accomplishment of the original conditions) greater than or equal to alpha. Consequently, with small values of alpha, few relaxed solutions are considered feasible. Then Eqs. (8) and (9) are used instead Eqs. (5) and (6).

With these changes, when $\alpha_w = 1$ and $\alpha_c = 1$ the problem remains crisp, and then students can only reach those bus stops settled at the original maximum walking distance and the bus only allows the initial load, i.e., it is the most restrictive case. On the other hand, when $\alpha_w = 0$, students are allowed to reach those bus stops located at the original maximum distance plus the maximum tolerance, i.e., it is the greatest relaxation. Likewise, when $\alpha_c = 0$ the capacity of the bus can be increased up to the maximum tolerance (T).

4 Solution Approach

We use the parametric approach [23], to achieve the solution to the new fuzzy SBRP model. In this approach, a fuzzy problem ($P_\sim$) is transformed into a set of crisp problems ($\bigcup_\alpha P_\alpha$), where each problem corresponds to a different combination of values of α_w and α_c. A solution is found for each of these crisp problems (S_α) and finally, the set of these solutions ($\bigcup_\alpha S_\alpha$), each associated with an alpha value, composes the solution of the original fuzzy problem ($S_\sim$).

The metaheuristic method, presented in [22], was used to solve each crisp instance. This metaheuristic combines two heuristic algorithms for the construction of the initial solution with a local search strategy that has a probabilistic-based selection mechanism for mutation operators. The mutation operators used in the selection mechanism for the local search were Swap, Two-opt, Section swap, and Reorder.

In the construction of the initial solution, the first heuristic assigns each student to a bus stop, while the second has a greedy approach to build the routes for each bus that start from the school and go through each stop where there is at least one assigned student, to ultimately return to the school.

In [22], this method was validated in the solution of the test instances previously used in [13]. These previous results demonstrate the efficacy of this metaheuristic approach.

5 Results and Discussion

To validate the proposed model, we take 31 instances of those used in [13] to execute the solution algorithms described above. In the instances used in [13], all students were similar, so they were slightly modified here to be able to use them with the proposed model. This modification consists of maintaining 25% of the students with the initial distance, these students being the ones who require a differentiated treatment, and the remaining 75% of the students increasing their walking distance by 50%, these being standard students. In this way, in an instance where the maximum walking distance was 10, now that value corresponds to the students with special requirements and 15 is the value for the rest of the students. The characteristics of these instances are available in: https://drive.google.com/file/d/126pXCN_BIZ9y14kXBgkMODGBLbPIopnq/view?usp=sharing.

Taking into account that these instances correspond to the crisp instances (with $\alpha_w = 1$ and $\alpha_c = 1$), each one induces an instance of the proposed fuzzy model. For the creation of these fuzzy instances, the tolerance of the original walking distance was set to 20%, i.e., $H_e = 0.2 * d_e$, and the tolerance of the initial bus capacity was also set to 20%, i.e., $T = 0.2 * c$. In real-world conditions, each student (person) may define its particular settings. Following the parametric approach, 5 values of α_w and 3 values of α_c allow different relaxation for each fuzzy instance, thus allowing different degrees of membership to the crisp case, $\alpha_w \in \{0, 0.25, 0.5, 0.75, 1\}$ and $\alpha_c \in \{0, 0.4, 1\}$. Combining these alpha values, the solution of the 31 fuzzy instances derived in the solution of $31 * 5 * 3 = 465$ crisp instances. It is important to say that we selected the values for the alphas in a fitting way, in the form that the bus capacity always results in an integer value and the walking distance tolerance was explored in 5 segments.

Table 1 shows the best solution value obtained for each instance from 30 executions of the metaheuristics with 10,000 evaluations of the objective function. Each row constitutes the fuzzy solution S_α, for each instance of the fuzzy problem. The caption of each column stands for the combinations of the alpha-cut for each fuzzy element. For example, 0.75/1 represents the combination of $\alpha_w = 0.75$ and $\alpha_c = 1$. To simplify Table 1, only the integer part of the results is shown (details are available in https://drive.google.com/file/d/17ghCpEccaELsFgf67AC3TbrFagtJn4Zj/view?usp=sharing).

Table 1 Results obtained for the 31 fuzzy instances

		α_w/α_c														
#	Instance	1/1	0.75/1	0.5/1	0.25/1	0/1	1/0.4	0.75/0.4	0.5/0.4	0.25/0.4	0/0.4	1/0	0.75/0	0.5/0	0.25/0	0/0
1	Inst6	**98**	**82**	82	82	82	98	82	82	82	82	98	82	82	82	82
2	Inst9	**287**	287	287	287	287	287	287	287	287	287	**239**	239	239	239	239
3	Inst11	**182**	182	182	182	182	182	182	182	182	**176**	**176**	176	176	176	176
4	Inst15	**9**	9	9	9	9	9	9	9	9	9	9	9	9	9	9
5	Inst21	**160**	160	160	160	160	160	160	160	160	160	**148**	148	148	148	148
6	Inst24	**33**	33	**12**	12	12	**24**	**12**	12	12	12	24	12	12	12	12
7	Inst27	**250**	250	250	250	250	250	250	250	250	250	**245**	**240**	240	240	240
8	Inst32	**27**	**19**	19	19	**8**	**26**	19	**8**	8	8	**19**	19	8	8	8
9	Inst33	**392**	392	392	392	392	392	392	392	392	392	**347**	347	347	347	347
10	Inst37	**171**	171	171	171	171	171	171	171	171	171	171	171	171	171	171
11	Inst40	**31**	**10**	10	10	10	31	10	10	10	10	**21**	10	10	10	10
12	Inst42	**506**	506	506	506	506	506	**500**	500	500	500	**417**	417	417	417	417
13	Inst54	**175**	**148**	148	148	**137**	**163**	**142**	**111**	111	111	163	142	111	111	111
14	Inst55	**35**	35	**32**	**19**	19	35	**26**	**19**	19	19	35	26	19	19	19
15	Inst56	**8**	8	8	**4**	4	8	**4**	4	4	4	8	4	4	4	4
16	Inst57	**921**	**899**	**884**	884	**869**	**891**	891	884	884	869	**715**	715	**705**	705	705
17	Inst60	**463**	463	**453**	**444**	444	**450**	450	**447**	444	**438**	**432**	432	**423**	423	423
18	Inst64	**43**	43	**33**	**29**	29	43	43	**29**	29	29	**29**	29	29	29	29
19	Inst70	**329**	**327**	**320**	320	320	**292**	292	292	292	292	**279**	**276**	276	276	**260**
20	Inst72	**78**	**63**	63	63	63	78	63	63	63	63	78	**53**	53	53	53

(continued)

Table 1 (continued)

		α_w/α_c														
#	Instance	1/1	0.75/1	0.5/1	0.25/1	0/1	1/0.4	0.75/0.4	0.5/0.4	0.25/0.4	0/0.4	1/0	0.75/0	0.5/0	0.25/0	0/0
21	Inst78	**223**	223	223	**203**	**181**	223	**208**	208	203	181	223	208	208	203	181
22	Inst79	**64**	64	64	64	64	64	64	64	64	64	**59**	**53**	53	53	53
23	Inst80	**44**	**29**	**21**	21	21	44	**17**	17	17	17	**17**	17	17	17	17
24	Inst84	**772**	772	**738**	**725**	725	**726**	726	**701**	**690**	690	**677**	**665**	665	**661**	**647**
25	Inst90	**1362**	**1357**	**1335**	1335	**1332**	**1307**	**1269**	1269	**1261**	**1254**	**1167**	**1165**	**1153**	**1141**	1141
26	Inst95	**393**	393	393	393	393	**376**	**354**	354	354	354	**321**	**318**	**298**	298	298
27	Inst97	**1525**	1525	**1522**	**1475**	**1474**	**1451**	**1427**	1427	1427	**1422**	**1374**	**1349**	1349	**1342**	**1338**
28	Inst99	**1207**	1207	**1170**	**1162**	**1094**	**1157**	**1071**	1071	1071	**1034**	**1107**	**1033**	1033	**1018**	**1014**
29	Inst102	**374**	**342**	**336**	336	**301**	**366**	342	336	**314**	**300**	**353**	342	**305**	**275**	275
30	Inst108	**1201**	**1180**	**1143**	**1119**	**1108**	**1175**	**1144**	**1115**	**1073**	1073	**1067**	**1045**	**992**	992	992
31	Inst112	**116**	116	**104**	104	104	116	**105**	104	**94**	94	**110**	**104**	**85**	85	85

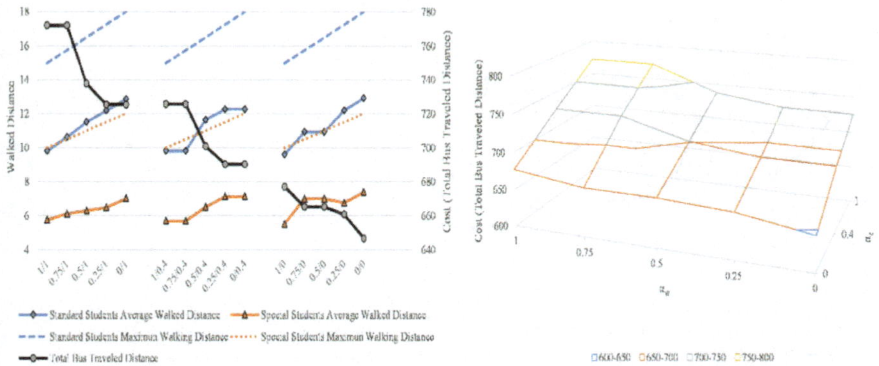

Fig. 2 Inst84. (left) The relation between average distance walked by students (left y-axis) and total bus traveled distance (right y-axis) in each combination of the alpha-cuts (x-axis). (right) Relationship between walking distance relaxation (α_w) and capacity relaxation (α_c) concerning the cost

From the point of view of a decision-maker, there is no rationale to allow a greater relaxation of the constraints if it does not imply an improvement in the cost. Thus, the solutions of interest are highlighted in bold in Table 1 because they are relevant to the decision-maker. It is worth noting that in some instances (e.g., Inst15 and Inst37) the allowed relaxations do not allow any cost-saving. However, this is not the general case.

From the results shown, instances like 84, 90, 97, and 108, stand out with more variation of the objective function in each column. In Fig. 2 (left), the relationship between the average distance that the students walk and the total buses traveled distance (cost to minimize), can be observed on Inst84.

In Fig. 2 (left), it is evident how students with special needs walk a shorter distance on average than standard students. Likewise, is noticeable that as the students' maximum walking distance is relaxed (and therefore, students walk more to reach the assigned bus stop), the cost of the objective function decreases simultaneously. This decrease is about 20% in the solution with the higher relaxation. However, the students never reach the maximum relaxation allowed in their maximum walking distance.

Figure 2 (right) shows the fuzzy solution of the instance Inst84 in a graphical way, where each axis corresponds to each allowed relaxation (α_w and α_c) with the corresponding cost. It is observable that, when combined, allowing more relaxation in both factors, the cost tends to decrease. Besides, the relaxation of individual elements also brings a cost decrease.

Taking this analysis into account, Fig. 3 (left) shows, for all the instances evaluated, the comparison between the average proportional savings in the total distance traveled by the buses and the average proportional increase in the distance walked by the students. All the data associated with these values are available in (https://drive.google.com/file/d/17ghCpEccaELsFgf67AC3TbrFagtJn4Zj/view?usp=sharing).

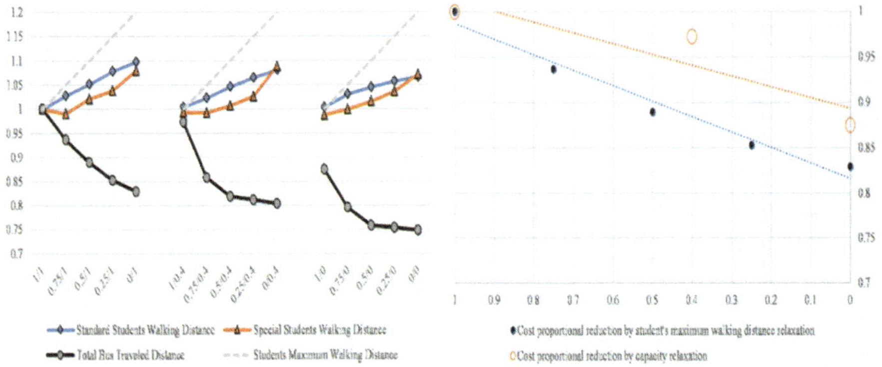

Fig. 3 (left) Comparison between the proportional savings in total bus traveled distance against the proportional increase in students' walking distance. (right) Cost proportional savings by the different relaxation levels on fuzzy elements

In this figure, it is appreciated how the total distance traveled by the buses has achieved a saving of 25% with the higher relaxation by increasing just an extra 6% of the walking distance for standard students and students with special needs. It is worth noting that the most relaxed case in terms of the increase of the walking distance (until 20%) was not necessary to achieve the best improvement in the objective function. This corresponds to the several instances (rows) in Table 1 where the column associated to $\alpha_w = 0$ is not highlighted. This analysis allows us to affirm that the differentiated treatment to the students' walking distance is effective, both in saving resources and reducing inequalities.

Another analysis that can be performed is the individual influence on the value of the objective function by the different relaxation levels of the elements treated as fuzzy (the bus capacity and the student's maximum walking distance). This influence is present in Fig. 3 (right), where the relaxation of the bus capacity yields a bit more than 10% of savings with the highest lessening allowed and the relaxation of the student's walking distance yields a bit more than 20% of savings with the highest lessening. This analysis may indicate that slightly overloading buses can bring, on average, fewer cost-savings than slightly increasing the students walking distance.

Finally, the obtained results can be analyzed guided through the main characteristics of the studied instances: the number of bus stops, the number of students, and the distance they can walk. In Fig. 4, the x-axis should be interpreted as the relationship between the alpha-cuts of the fuzzy elements, thus: 1/1 represents the solution in the original problem, 0.75/0.4, represents the solution with the alpha-cut 0.75 for the walking distance and 0.4 for the bus capacity, and 0/0 represents the solution with the highest level of relaxation in both factors. This analysis is helpful to understand which characteristics of the problem allow better exploitation of the relaxations.

In this way, Fig. 4 (top-left) shows the cost-savings ratios for maximum students walking distance. It is observable that the highest average savings occur in cases

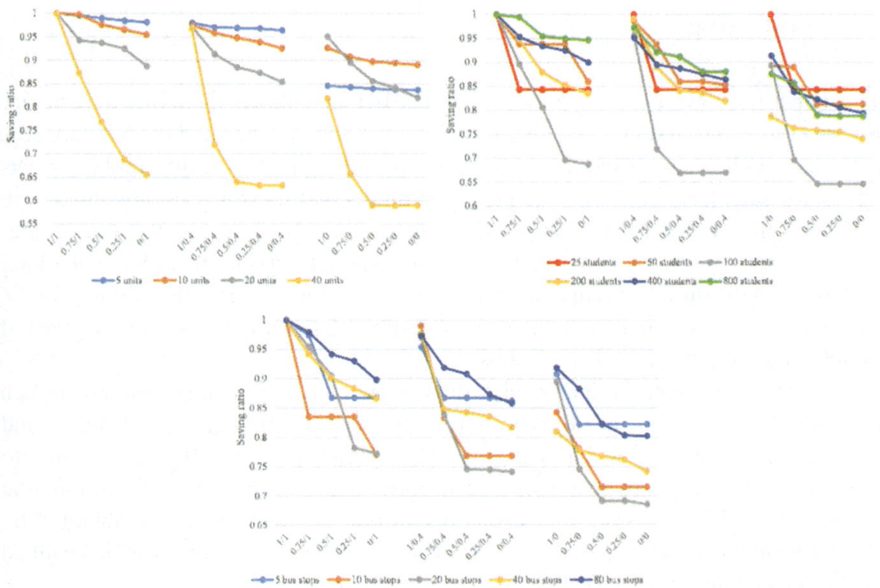

Fig. 4 Saving ratios in the instances grouped by: the maximum students walking distance (top-left), the number of students (top-right), and the number of bus stops (bottom)

where the walking distance is the greatest (40 units), although in general, in all types of instances, cost savings are obtained at the highest levels of relaxation.

Figure 4 (top-right) shows that the cases where 100 students need to be carried achieved the highest cost-saving. Likewise, situations where more students are carried (200, 400, and 800) are also noticeable. However, in all cases, cost-saving between 15 and 35% at the highest level of relaxation of both fuzzy elements is obtained.

Likewise, in Fig. 4 (bottom) it can be appreciated that in all cases grouped by bus stops count, at highest levels of relaxation, cost-saving between 18 and 30% are obtained. From this point of view, in the instances with 20 bus stops the most significant cost-saving is achieved.

In general, the results obtained in these instances indicate that greater values of the number of bus stops and students, and larger values of walking distances in the examples tend to allow a higher reduction in the cost, and thus less fuel is used to transport the students to their schools.

The trade-off observed in the solutions allows conducting a more insightful decision-making process. This is aligned with the intention of the Sustainable Development Goals to promote responsible consumption that improved health and well-being and that also reduces inequalities.

6 Conclusion

This paper introduced a fuzzy model for the SBRP that allows modeling uncertainty for the constraints associated with the student walking distance and the bus capacity. Likewise, it allows to differentiate the students and then to make different considerations for each kind of them. The proposed model and its solution allow obtaining a fuzzy solution, as a set of crisp solutions to the SBRP with interesting trade-offs between cost and accomplishment of the original maximum student's walking distance and the initial bus capacity. A main contribution of the proposal is to provide a tool to the decision-makers to analyze multiple solutions with different degrees of membership to the original conditions.

On the other hand, the analyzed results allow us to affirm that cost-saving can be achieved at the cost of slightly increasing the student's walking distance and a modest overload of the bus capacity. The cost-saving implies less distance to travel by the buses, thus less fuel is consumed and a reduction of environmental pollution is achieved. Also, the results reveal that this saving can be obtained by reducing students' inequalities because the students with special necessities walked less distance than the standards ones.

Finally, this proposal allows us to trace a way forward in fuzzy optimization of SBRP. In this way, a more extensive study to define the alpha values to get the most attractive trade-offs between cost and relaxation is required. Equally, the research on SBRP and other routing optimization problems may be conducted to achieve results aligned with others Sustainable Development Goals like Sustainable Cities and Communities.

References

1. U. Nations: *The Sustainable Development Agenda* (2021). Available: https://www.un.org/sustainabledevelopment/development-agenda/ (19/2/2021)
2. Corona, C.C.: Soft computing techniques and sustainability science, an introduction. In: Cruz Corona, C. (ed.) Soft Computing for Sustainability Science, pp. 1–10. Springer International Publishing, Cham (2018)
3. Di Vaio, A., Palladino, R., Hassan, R., Escobar, O.: Artificial intelligence and business models in the sustainable development goals perspective: a systematic literature review. J. Bus. Res. **121**, 283–314 (2020)
4. Verga, J., Silva, R.C., Yamakami, A.: Multimodal transport network problem: classical and innovative approaches. In: Cruz Corona, C. (ed.) Soft Computing for Sustainability Science, pp. 299–332. Springer International Publishing, Cham (2018)
5. Bock, A., Grant, E., Könemann, J., Sanità, L.: The School Bus Problem on Trees, pp. 10–19. Berlin, Heidelberg (2011)
6. Ellegood, W., Solomon, S., North, J., Campbell, J.: School bus routing problem: contemporary trends and research directions. Omega (2019)
7. Ezquerro Eguizábal, S., Moura Berodia, J.L., Ibeas Portilla, Á., Benavente Ponce, J.: Optimization model for school transportation design based on economic and social efficiency. Transport Policy **67**, 93–101 (2018)

8. Newton, R., Thomas, W.: Design of school bus routes by computer. Socio-Econ. Plan. Sci. **3**, 75–85 (1969)
9. Sales, L., Sousa, C., Oliveira, T., Athayde, B.: Memetic algorithm for the heterogeneous fleet school bus routing problem. J. Urban Plan. Dev. **144** (2018)
10. Pérez-Rodríguez, R., Hernández-Aguirre, A.: Probability model to solve the school bus routing problem with stops selection. Int. J. Comb. Optim. Problems Inform. **7**, 30–39 (2016)
11. Unsal, O., Yigit, T.: Using the genetic algorithm for the optimization of dynamic school bus routing problem. Broad Res. Artif. Intell. Neurosci. **9** (2018)
12. Sun, S., Duan, Z., Xu, Q.: School bus routing problem in the stochastic and time-dependent transportation network. PLoS One **13** (2018)
13. Schittekat, P., Kinable, J., Sörensen, K., Sevaux, M., Spieksma, F., Springael, J.: A metaheuristic for the school bus routing problem with bus stop selection. Eur. J. Oper. Res. (2013)
14. Caceres, H., Batta, R., He, Q.: School bus routing with stochastic demand and duration constraints. Transp. Sci. (2017)
15. Miranda, D.M., Camargo, R.S.d., Conceio, S.V., Porto, M.F., Nunes, N.T.R.: A multi-loading school bus routing problem. Expert Syst. Appl. **101**, 228–242 (2018)
16. Bowerman, R., Hall, B., Calamai, P.: A multi-objective optimization approach to urban school bus routing: formulation and solution method. Transp. Res. Part A: Policy Pract. **29A**, 107–123 (1995)
17. Brito, J., Martínez, F.J., Moreno, J.A., Verdegay, J.L.: Fuzzy approach for vehicle routing problems with fuzzy travel time. In: International Conference on Fuzzy Systems, pp. 1–8, 2010
18. Torres, I., Cruz, C., Verdegay, J.L.: Solving the truck and trailer routing problem with fuzzy constraints. Int. J. Comput. Intell. Syst. **8**, 713–724 (2015)
19. Giallanza, A., Puma, G.L.: Fuzzy green vehicle routing problem for designing a three echelons supply chain. J. Clean. Prod. **259**, 120774 (2020)
20. Zhang, S., Chen, M., Zhang, W., Zhuang, X.: Fuzzy optimization model for electric vehicle routing problem with time windows and recharging stations. Expert Syst. Appl. **145**, 113123 (2020)
21. Sánchez-Ansola, E., Pérez-Pérez, A.C., Rosete, A.: School bus routing problem with fuzzy walking distance. In: Polytechnic Open Library International Bulletin of Information Technology and Science (POLIBITS), vol. Special Issue: Emerging Challenges and Trends in Business Intelligence, pp. 69–75, July–Dec 2020
22. A.C. Pérez Pérez, E. Sánchez-Ansola, A. Rosete, Una solución metaheurística al problema de planificación de rutas de auto-buses escolares con flota homogénea y selección de paradas. Ingeniería **26** (2021)
23. A. Ebrahimnejad, J.L. Verdegay, Fuzzy Sets-Based Methods and Techniques for Modern Analytics. Springer, Cham (2018)

A VND to Solve the Bi-objective Tourist Trip Design Problem with Fuzzy Constraints and Reducing CO_2

José Ruiz-Meza, Julio Brito, and Jairo R. Montoya-Torres

Abstract This work proposes solving a bi-objective problem of planning tourist routes for multiple days, selecting the transport mode and multiple fuzzy constraints. The imprecise and non-strict constraints are time windows, the duration of the route and the budget available to the tourist.To find the optimal planning using two criteria: obtaining the most excellent satisfaction or benefit for the tourist and reducing CO_2 emissions. The problem is modeled as a Bi-objective Fuzzy Multi-Constraint Multi-Modal TOPTW and solve with a Variable Neighborhood Descent. This work contributes to the design of tourist routes with efficient use of transport modes, reducing trips, reducing CO_2 emissions, and favouring the use of less polluting transport.

Keywords Bi-objective problem · Tourist trip design problem · Fuzzy constraints · Variable neighborhood descent

1 Introduction

The Sustainable Development Goals (SDG) that refer to climate change are considered a priority issue, given the global threat it constitutes. Reducing the impact that productive activities have on CO_2 emissions is essential to mitigate climate

J. Ruiz-Meza (✉) · J. R. Montoya-Torres
Facultad de Ingeniería, Universidad de La Sabana, km 7 autopista norte de Bogotá, D.C., Chía, Cundinamarca, Colombia
e-mail: joserume@unisabana.edu.co

J. R. Montoya-Torres
e-mail: jairo.montoya@unisabana.edu.co

J. Brito
Departamento de Ingeniería Informática y de Sistemas, Instituto Universitario de Desarrollo Regional, Universidad de La Laguna, Santa Cruz de Tenerife, Spain
e-mail: jbrito@ull.edu.es

J. L. Verdegay et al. (eds.), *Computational Intelligence Methodologies Applied to Sustainable Development Goals*, Studies in Computational Intelligence 1036,
https://doi.org/10.1007/978-3-030-97344-5_15

change. Sustainable tourism can contribute, among others, to the proper use of natural resources, energy efficiency, the reduction of pollution and CO_2 emissions, and the protection of the environment. In this sense, it is essential in tourism to reduce the negative externalities of transport. Travellers are increasingly aware of the need to contribute to sustainable tourism and plan their trips with suitable criteria.

This work contributes to the design of tourist routes with efficient use of transport modes, reducing trips, reducing CO_2 emissions and favouring the use of less polluting transport.

When tourists visit a destination, they plan their visit for one or more days and select the preferred attractions or points of interest (POIs). As time is limited, it is impossible to visit everything that interests them. Organising a route by selecting the POIs to stay and sequencing the order of the visit is a complex task. The availability of applications that generate this plan with quality and efficiency considering all the tourist's constraints is of great interest. In the literature these route planning problems are called Tourist Travel Design Problems (TTDP) [19]. Each of the POIs is associated with several characteristics which are taken into account in the selection: the minimum time required for a visit, the time in which it is operational (opening and closing time), the cost of the activity and preference indicators such as the benefit or degree of satisfaction perceived when visiting the tourist. Details on each route are required, such as: information on the travel time between POIs, total distance or duration, cost or budget, the starting and ending point, and means of transport to move from one point to another. All this information is necessary and is used to make decisions about the most suitable routes.

The simplest and most widely used models to address the TTDP are the Orienteering Problem (OP) or the Team Orienteering Problem (TOP). The OP introduced by Tsiligiridis [18], unlike the TOP, only considers one route. The TOP presented by Chao, Golden and Wasil [4] is able to simultaneously plan a certain fixed number of routes, corresponding to the number of days available to tourists. When the opening hours for each POI is considered, the TOP is extended to the Team Orienteering Problem with Time Windows (TOPTW) [14].

These combinatorial problems belong to the class of NP-hard problems. They become more complex with real-world characteristics in terms of the number and size of variables. In these cases, the inability of exact approximations to solve medium and large-scale problems makes heuristics and metaheuristics considered for the search for optimal solutions. Vansteenween [21] provides an overview of existing research on OP and TOP, including solution methods.

Available information or the limitations imposed by the tourist frequently are subjective and are subject to levels of impression or vagueness. Travel time, visiting time and schedules depend on the conditions of the environment and the state of traffic, roads or the weather. Typically, the time window for some visited POIs is not strict and imprecise. Likewise, the planning expected by tourists has a high degree of flexibility in the duration of the route and budget. In all these cases, the uncertainty is associated with the linguistic terms used in its expression. In this context, it is necessary to investigate models and methodologies that support new applications in the organization and plan the user experience, incorporating imprecision and flexibility.

Fuzzy sets and systems are used to solve decision and optimization problems whose available information is vague and imprecise, as well as formulate constraints and objectives expressed by decision-makers in the same terms. Fuzzy optimisation problems are those in which some of its components are defined and formulated in fuzzy terms. The solutions to these problems are described in terms of degrees of feasibility and optimality. The fundamentals of fuzzy optimisation problems where objectives and constraints can be fuzzy and characterised with membership functions as fuzzy sets were introduced by Bellman and Zaded [1].

This work proposes solving a bi-objective problem of planning tourist routes for multiple day, with the selecting of transport mode and multiple fuzzy constraints. The fuzzy constraints are time windows (specifically the closing time), the duration of the route and the available budge of the tourist. Two criteria are considered to find the optimal planning: obtaining the highest level of satisfaction or benefit for the tourist and reducing CO_2 emissions. This proposed fuzzy model is transformed into an equivalent optimisation model [23]. This approach provides methods to solve these problems obtaining consistent solutions with their fuzzy nature. We propose a variant of the TOP in the formulation of the problem. The resolution of the problem is obtained by applying a strategy based on Variable Neighborhood Search (VNS).

The rest of the chapter is organised as follows. The next section presents a review of the state of the art of the context and the methodologies to which this work contributes. Section 3 describes the formulation of the fuzzy model and its transformation through the proposed approximation. Section 4 introduces the VND metaheuristic for finding solutions. In Sect. 5, we describe computational experimentation and its results. Finally, some conclusions are included in the last section.

2 Background

Destinations or companies in the tourism sector are interested in offering applications and services to tourists which solve route planning optimisation problems. These tools are incorporated into electronic tourist guides or planning and/or recommendation systems on websites or platforms. The literature offers some papers with applications dedicated to solving these problems. The eCOMPASS application facilitates multimodal routes using public transport and minimises waiting times at stops [6]. City Trip Planner [22] and PlanTour [3] organise routes for several days and consider multiple time windows in which POIs are available. In general, travel times between POIs are obtained from online services. Some applications use georeferenced information generated by users to get preferences or visit times, such as PersTour [10] or information from social networks, such as PlanTour.

There are some papers which use the applied metaheuristic in our work, namely the VNS. Vansteenwegen et al. [20] analyze different metaheuristics applied to the TOP and present the application of Skewed VNS (SNVS). This metaheuristic is used in problems that show separate solution spaces and near-optimal solutions. Schilde et al. [17] apply two metaheuristics to a TTDP modelled as Multi-Objective Orienteering Problem (MOOP). The first algorithm is Pareto Ant Colony Optimization (P-ACO), and the second is a Pareto-based VNS extension (P-VNS). A hybrid procedure between Genetic Algorithm (GA) and VND (GA-VND) to solve the OP with time windows and hotel selection (OPHS-TW) is developed in [5]. More recently, a Memetic Algorithm (MA) has been applied for a similar hotel selection problem. This metaheuristic adds a VNS to a Differential Evolution Algorithm (DEA) as local search methods to build the solution. Mrazovic et al. [15] developed a VNS to solve a TTDP modelled as a Mixed Orienteering Problem (MOP) to consider the arcs between POIs.

Fuzzy approaches are quite common for routing problems, but not specifically in TTDP problems and Orienteering Problems. To our knowledge, there is an initial work [11] in the planning of tourist routes, which applies fuzzy theory and considers time-varying travel time and ambiguous location value. In more recent years, Hasuike et al. present several works for the tour planning problem with uncertain travelling times and satisfaction values of sightseeing places, which they treat as fuzzy random variables [7–9]. Other authors applied fuzzy optimisation to the Orienteering Problem using the max-min formulation of the fuzzy sets obtained from the fuzzy goals [25] . Mendez [12] uses fuzzy number comparisons to treat a TOPTW with fuzzy scores. Finally, Brito et al. apply a GRASP to solve the TOP with fuzzy scores and constraints [2] and Ni et al. present the orienteering problem in a fuzzy environment and represent the scores and the travel time as fuzzy variables [16].

3 Bi-objective Fuzzy Model Formulation

The TTDP is modelled as a Fuzzy Multi-constraint, Multi-modal TOPTW. Let us consider a weighted undirected graph $G = (N, V)$, where N is the set of vertices. $N = \{1, 2, 3, .., i\}$ corresponding to the nodes and V is the set of arcs (ij), $i, j \in N$. Different modes of transport m each with different emissions CO_{2ijm} move between the arcs. The model aims to minimize the costs of CO_2 emissions and maximize the benefit of the tourist based on the benefits P_i associated with each node. The nodes must be visited in a time window $[a_i, b_i]$ within a route k. Each route has a maximum budget of time $Tmax$. Each tourist has a cost of P to complete their itinerary. The parameters and variables of the model are presented in Table 1.
Bi-objective function

$$Maximize = \gamma \sum_{i=1}^{I} \sum_{k}^{K} P_i Y_{ik} \varphi + \beta \left(\sum_{i}^{I} \sum_{j}^{I} \sum_{k}^{K} \sum_{m}^{M} ceE_{ijkm} \right) \quad (1)$$

Table 1 Notation

Notation	Description
$i, j \in I$	Set of nodes
$0 \in I^0$	Depot node
$k \in K$	Set of routes
$m \in M$	Set of transport modes
P_i	The score of nodes
S_i	Visit time
$Tmax$	Maximum time
$[a_i, b_i]$	Time windows
P	Maximum budget
CO_{2ijm}	Emissions
t_{ijm}	Travel time matrix
ce	Emissions cost
φ	Profit-to-cost conversion parameter
T_{ik}	Integer variable representing travel time
E_{ijkm}	Integer variable representing emissions calculation
X_{ijk}	Binary variable. 1 if a trip is made in the arc on the route, 0 otherwise
G_k	Binary variable. 1 if the route is enabled, 0 otherwise
Y_{ik}	Binary variable. 1 if the POI is on the route, 0 otherwise
F_{ijkm}	Binary variable. 1 if a trip is made in arc, route and transport, 0 otherwise

Subject to:

$$\sum_{j=1}^{J} X_{ijk} = \sum_{j=1}^{J} X_{jik} \qquad \forall i \in I^0, \forall k \in K \tag{2}$$

$$\sum_{j=1}^{J} X_{ijk} = G_k \qquad \forall i \in I^0, \forall k \in K \tag{3}$$

$$\sum_{k}^{K} Y_{ik} \leq 1 \qquad \forall i \in I \backslash \{0\}, \forall k \in K \tag{4}$$

$$\sum_{i=1, i \neq h}^{I} X_{ihk} = \sum_{j=1, j \neq h}^{J} X_{hjk} = Y_{hk} \qquad \forall h \in I, \forall k \in K \tag{5}$$

$$\sum_{m}^{M} F_{ijkm} \leq 1 \qquad \forall ij \in I : i \neq j, \forall k \in K \tag{6}$$

$$\sum_{m}^{M} F_{ijkm} = X_{jik} \qquad \forall ij \in I : i \neq j, \forall k \in K \tag{7}$$

$$T_{ik} + S_i + \sum_{m}^{M} t_{ijm} F_{ijkm} - T_{jk} \leq M(1 - X_{ijk}) \qquad \forall ij \in I \backslash \{0\}, \forall k \in K \tag{8}$$

$$\sum_{i}^{I} S_i Y_{ik} + \sum_{j}^{J} \sum_{m}^{M} t_{ijm} F_{ijkm} \leq_f Tmax \qquad \forall k \in K \tag{9}$$

$$a_i \leq T_{ik} \leq_f b_i \qquad \forall i \in I \backslash \{0\}, \forall k \in K \tag{10}$$

$$E_{ijkm} = \mathrm{CO}_{2ijm} F_{ijkm} \qquad \forall ij \in I, \forall k \in K, \forall m \in M \tag{11}$$

$$\sum_{i \neq j}^{I} \sum_{j \neq i}^{J} \sum_{m}^{M} F_{ijkm} V c_{ijm} \leq_f P \qquad \forall k \in K \tag{12}$$

$$\sum_{i}^{I} \sum_{k}^{K} X_{ijk} = K \qquad \forall j \in I^0 \tag{13}$$

$$X_{ijk}, Y_{ik}, F_{ijkm}, G_k \in 1, 0 \qquad \forall ij \in I, \forall k \in K, \forall m \in M \tag{14}$$

$$T_{ik}, E_{ijkm} \geq 0 \qquad \forall ij \in I, \forall k \in K, \forall m \in M \tag{15}$$

Equation (1) minimize CO_2 emissions cost and maximize profit. Constraints (2) ensure routes start and end at the same node (I^0). Constraints (3) establish the routes to be used. Constraints (4) ensure that a node must belong exclusively to a single route. Constraints (5) guarantee flow. Constraints (6) and (7) guarantee the selection of only one mode of transport on each arc (ij). Constraints (8) calculates the travel time. Constraints (9) ensure that time budget is not exceeded. Constraints (10) ensure compliance with time windows. Constraints (11) calculate the emissions at each arc (ij). Constraints (12) ensure that the travel cost budget is not exceeded. Constraints (13) denote the number of routes to be performed. Finally, constraints (14) and (15) determine the nature of the variables. Note that the $\leq_f$ denotes that the constraint is fuzzy.

4 Solution Method

4.1 Fuzzy Optimization Approach

In the previous section, we formulated the TTDP as a linear programming problem with fuzzy inequalities in some constraints. Fuzzy Linear Programming (FLP) constitutes the basis for solving fuzzy optimization problems and their solution methods have been the subject of many studies in the fuzzy context. To solve the optimization problem with fuzzy constraints, Verdegay [24] using the representation theorem for fuzzy sets, proves that the solutions for the case of linear functions can be obtained from the auxiliary model:

$$\begin{aligned} &\text{Maximize} z = cx \\ &\text{subject to} Ax \leq b + \tau(1-\alpha) \\ &x \geq 0, \alpha \in [0, 1] \end{aligned} \tag{16}$$

where $\tau = (\tau_1, \tau_2, ..., \tau_m) \in \Re^m$ is the tolerance level vector. Thus, we use that approach to obtain an equivalent model to deal with fuzzy constraints of the proposed TTDP which is obtained by replacing (10), (11) and (13) by the following equivalent constraints (17), (18), and (19):

$$\sum_i^I S_i Y_{ik} + \sum_j^J \sum_m^M t_{ijm} F_{ijkm} \leq Tmax + \tau_1(1-\alpha) \quad \forall k \in K \tag{17}$$

$$a_i \leq T_{ik} \leq b_i + \tau_2(1-\alpha) \quad \forall i \in I \backslash \{0\}, \forall k \in K \tag{18}$$

$$\sum_{i \neq j}^I \sum_{j \neq i}^J \sum_m^M F_{ijkm} V c_{ijm} \leq P + \tau_3(1-\alpha) \quad \forall k \in K \tag{19}$$

where $\tau_1, \tau_2, \tau_3 \in \Re$ are the tolerance level vectors or the maximum violations in the fulfillment of the constraints provided by the decision maker, and $\alpha \in [0, 1]$. When this model is applied a new optimal solution is obtained for each value of α. The end result is an optimal solution range varying in α. The result is consistent with the fuzzy nature of the problem.

4.2 VND Metaheuristic Procedure

The VNS is a procedure that systematically explores the search space through various neighbourhoods. It was first proposed in 1997 by Mladenović and Hansen [13]. To solve the Bi-objective Fuzzy Multi-Constraint Multi-modal TOPTW a Variable Neighborhood Descent (VND) is developed. The VND method is an extension of the VNS obtained with the changes of neighbourhoods performed in a deterministic way (i.e., l to l_{max}). The rationale behind the VND is that the local optimum defined by one neighbourhood structure is not necessarily the local optimum of another neighbourhood structure. Therefore, the search can methodically cross over different search spaces defined by different neighbourhood structures. The different neighbourhoods are usually associated with types of moves or a transformation of the current solution that states the neighbourhood structure. The use of the VND leads to a change in the neighbourhood structure when no improving move is found. The VND procedure is given in the algorithm 1.

Algorithm 1 VND algorithm

x initial solution for VND
while (Stop = False) **do**
  $l = 1$
  while $l \leq l_{max}$ **do**
    x' is the best solution in $N_l(x)$
    if $f(x') < f(x)$ **then**
      $x = x'$ **and** $l = 1$
    else
      $l = l + 1$
    end if
  end while
end while
return best solution

A description of the procedure on how to obtain initial solutions and the algorithm notation follows (Table 2).

4.3 Construction of Initial Routes

To obtain an initial feasible solution x, the construction of the routes is selected randomly. Each route ends if the budget, time or number of available POIs is exceeded. Transformed fuzzy constraints are also evaluated in this process. The following procedure consists of randomly selecting each POI, evaluating the arc ij with each mode of transport m, and analyzing the fulfillment of the model constraints for the inclusion of each POI. The POI is added to the list of selected nodes M times (i.e., the

Table 2 Algorithm notation

Input/output	Description	Input/output	Description
Sg	Initial routes	*Cmx*	Maximum number of routes
Tg	Initial transports	*bt*	Total time available
Cmx	Maximum number of routes	*bm*	Total budget
P_{im}	Travel cost	t_{im}	Travel time
S	Visit time	$[a_i, b_i]$	Time windows
POI	Nodes	*NT*	transport modes
γ	Gamma	β	Beta
α	variation range	τ	Tolerance value
Comp	Off-route nodes	*pG*	Budget
tS	order of routes	*Int*	Nodes within the route
l	Numbers of deterministic changes of the local search		

number of times corresponding to the number of transports that meet the constraints). Similarly, the modes of transport that comply are aggregated in another list to apply the bi-objective strategy.

The bi-objective adjustment strategy calculates the emissions on arc ij for each mode of transport m and ranks them from largest to smallest. The highest emissions are associated with travel time. The number of POIs on the route tend to decrease as travel time becomes longer. Selection of the transport mode to use in the ij arc, is generated random and compared with the weight of the objectives (γ and β). Finally, the POI and the selected transport mode are included to the initial route. The algorithm for generating the initial routes is shown in Algorithm 2.

Algorithm 2 Proposed VND algorithm—Step 1

Input:$(POI, S, Cmx, bt, bm, NT, \gamma, \alpha, \tau)$
Output:$(Sg, Tg, Comp, pG)$
set $(Sg = [], Tg = [], Comp = [], pG = [], tS = [], n = 0)$
$Comp = POI$
$B \leftarrow Random$
insert $(tS(B))$
for i:=0 **to** Cmx **do**
  while $n \leq$ **len** POI **and** $t_{im} \leq bt + \tau(1-\alpha)$ **do**
    $X \leftarrow Random$
    for m:=0 to NT **do**
      check-constraints for X **and** m
      insert $(LBase(X), TBase(m))$ **and update** M
    end for
    order $OrderEmissions$ $(LBase)$
    $Y \leftarrow Random(0-1)$
    if $Y \leq \gamma$ **then**
      $Tg_{[tS]} = OrderEmissions[0]$
    else
      $Tg_{[tS]} = OrderEmissions[Random : (1toNT)]$
    end if
    insert $(Sg_{[tS]}(X)$ **and update** t_{im}, p_{im}, pG
  end while
end for
$Comp = Comp - Sg$
return $x \subset Sg, Tg, Comp, pG$

4.4 Neighbourhood Change

The procedure uses a finite set of pre-selected neighbourhood structures. Neighbourhoods structures are based on a move on the solution. In our algorithm we applied two movements: *insert* and *replace* to explore a larger space of solutions. The movements are performed with a given disturbance as opposed to the random disturbance that is performed in the VNS [13].

- *Insert*: includes l POIs in an order $l = 1, \ldots, l_{max}$ that have been left out of the path.
- *Replace*: replaces l POIs not included in the route, by POIs belonging to the route. To reduce the number of inefficient routes (i.e., that do not comply with the restrictions) and reduce the number of route revisions that may arise ($Comp$ X POI with $l = 1$ and $C_{n,Comp}$ X POI with $l = l_{max}$). We define a fitness function to evaluate the lowest value time window a_i where $i \in Comp$ and a_j where $j \in Int$. The function replaces a POI if and only if, the time window previously belonging to the external POI has a tolerance $J \geq 0$ such that, $\mid a_i - a_j \mid \leq J$. The tolerance J is defined by the decision maker.

We apply two strategies for organizing the neighbourhood movements. The first strategy ($VND1$) starts with *insert* $l = l_{max}$ up to $l = 1$. Subsequently, *replace* $l = 1$ is applied to $l = l_{max}$. The second ($VND2$) starts with *replace* $l = 1$ until $l = l_{max}$ and then *insert* $l = l_{max}$ until $l = 1$.

5 Experimentation

This section shows the experiments performed to validate the proposed solution methodologies to solve the TTDP introduced in this chapter. Our objectives with this experimentation are to validate the VND strategies to solve the bi-objective problem, evaluate their performance about the exact procedure and the validity of our approach when handling fuzzy constraints.

Instances of size $N = 11$ and $N = 21$ were used in the algorithm tests. The results for $K = 2$ and $K = 3$ routes are compared with optimal results of the model. The value of α for tests is $0.2, 0.5, 0.8, 1$. The value of τ (tolerance) is 20% for all fuzzy constraints. the $Tmax$, P values were set to 600 and 40,000 for $N = 11$ and 700 and 1000 for $N = 21$, respectively. The value of γ and β is set to $1, 0.5, 0$.

The multi-objective adjustment strategy varies the maximum and minimum values in both objectives according to the weights of γ and β. In each variation of the weights, $n = 20$ executions were made, each one with a total of 1000 runs of the algorithm applying the strategy *VND*1 with $\alpha = 1$ in the instance $N = 21$ and $K = 3$. In Figs. 1, 2, and 3 the behavior of the solutions is shown in two different executions. The variations of the objectives according to the assigned weight are as follows:

- $\gamma = 1$, $\beta = 0$: highest profit = 280, lower profit = 130, highest emissions = 518.1, and lower emissions = 1145.1.
- $\gamma = 0.5$, $\beta = 0.5$: highest profit = 280, lower profit = 110, highest emissions = 1208.5, and lower emissions = 416.3.
- $\gamma = 0$, $\beta = 1$: highest profit = 280, lower profit = 60, highest emissions = 1009, and lower emissions = 191.

The procedures are executed by varying the values of the α-cuts for a certain tolerance established in each of the constraints. Consequently, with the diffuse nature of the constraints, various solutions are obtained by modifying the values.

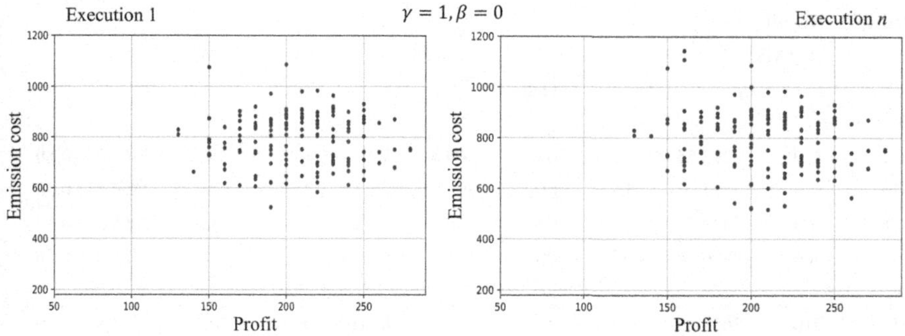

Fig. 1 Variation of results for $\gamma = 1$ and $\beta = 0$

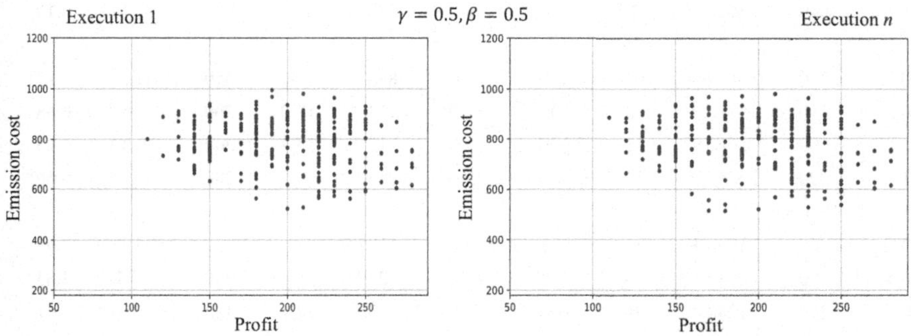

Fig. 2 Variation of results for $\gamma = 0.5$ and $\beta = 0.5$

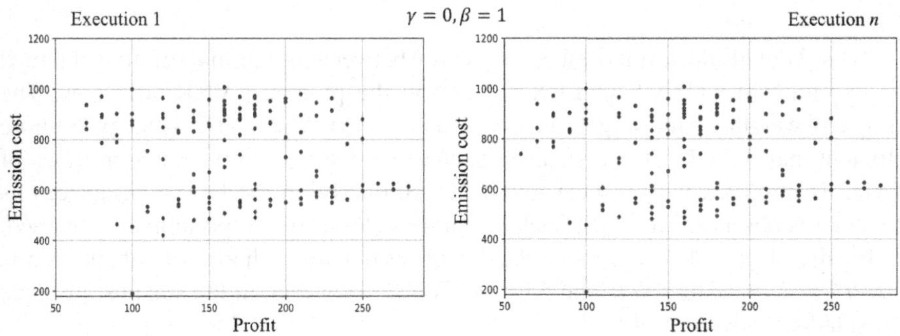

Fig. 3 Variation of results for $\gamma = 0$ and $\beta = 1$

Table 3 Results

α	GAMS				*VND* 1			*VND* 2		
	γ	β	Time	Gap	γ	β	Time	γ	β	Time
$N = 11$ $K = 2$										
0.2	509	2273.55	224.73	0%	509	2380	3.25	509	2364.3	**0.80**
0.5	509	2421.9	115.81	0%	509	2422	2.63	509	2363.5	**0.56**
0.8	509	2342.74	121.67	0%	502	2322.8	6.88	502	2265.2	**0.54**
1	505	2406.74	51.33	0%	441	2322.7	3.81	441	2322.7	**0.66**
$N = 11$ $K = 3$										
0.2	710	3651.86	3.31	0%	710	4020.2	0.26	710	3831.8	**0.19**
0.5	710	3806.18	5.31	0%	710	3731.9	0.55	710	3716	**0.18**
0.8	710	3747.69	6.13	0%	646	3630.5	0.24	646	3631.3	**0.12**
1	649	3673.19	152.82	0%	634	3533.9	0.17	634	3549.7	**0.14**
$N = 21$ $K = 2$										
0.2	320	882.59	3.63	0%	300	868.5	1.31	**310**	787	5.17
0.5	310	582.23	3.38	0%	290	574.6	1.59	**300**	666.9	**0.68**
0.8	300	576.17	3.32	0%	290	712.3	7.94	**300**	918.1	**6.73**
1	290	605.17	3.52	0%	280	754	9.22	280	749.1	**0.66**
$N = 21$ $K = 3$										
0.2	360	1438.1	4.32	0%	360	1197.9	8.6	360	1197.9	**4.99**
0.5	360	1687.49	5.76	0%	340	1239.6	18.48	**350**	1512.5	**4.83**
0.8	360	1240.42	3.96	0%	330	919.5	24.47	**350**	1316.6	**7.61**
1	350	1148.39	1647.27	2.8%	320	1055.7	2.86	**340**	1310.4	**1.82**

Table 3 details the results of the comparison between the optimal solution obtained by executing the GAMS CPLEX solver and the proposed VND strategies. The assigned weight of the targets is $\gamma = 1$ and $\beta = 0$. The *VND*2 strategy is more efficient than *VND*1 . In the smallest instance, a decrease in the solution times of *VND*2 is observed with respect to *VND*1 . Similarly, in the largest instance, this behavior is observed and additionally, solutions closer to the optimum are obtained.

Finally, the local search move that excels at obtaining the best final solution is *insert* $l = 2$ (42% in *VND*1 and 55% in *VND*2). However, in the smallest instance the standout move is *replace* $l = 1$.

6 Conclusion

In this work, a VND algorithm is presented to solve the TTDP modeled as a Bi-objective Fuzzy Multi-constraint Multi-modal TOPTW with minimization of CO_2 emissions. An adjustment strategy was developed to verify the multi-objective behavior of the problem. Two neighbourhood structures based on *insert* and *replace* movements were also applied, and the efficiency of both was verified.

The algorithm is able to solve the problem and generates a set of solutions according to the fuzzy nature of the constraints. The solutions of the algorithm compared with the solutions obtained in GAMS demonstrate high efficiency. At lower α-cut (0.2) the optimal solution is generally obtained. For future works it is proposed to add other local search movements that allow exploring the solutions space. Additionally, it is suggested to apply a methodology in order to solve the multi-objective problem based on algorithms of non-dominated solution populations (PNDs), such as VEGA, NSGA-II or MOEA/D or based on Pareto optimality.

Acknowledgements This work has been partially funded by Universidad de La Sabana (grant INGPhD-37-2020), the Colombian Ministry of Science, Innovation and Technology under a Doctoral Scholarship awarded to the first author and the Spanish Ministry of Science and Innovation (project PID2019-104410RB-I00).

References

1. Bellman, R.E., Zadeh, L.A.: Decision-making in a fuzzy environment. Manage. Sci. **17**(4), B-141 (1970)
2. Brito, J., Expósito, A., Moreno, J.A.: Solving the team orienteering problem with fuzzy scores and constraints. In: 2016 IEEE International Conference on Fuzzy Systems (FUZZ-IEEE), pp. 1614–1620. IEEE (2016)
3. Cenamor, I., de la Rosa, T., Núñez, S., Borrajo, D.: Planning for tourism routes using social networks. Exp. Syst. Appl. **69**, 1–9 (2017)
4. Chao, I., Golden, B., Wasil, E.: The team orienteering problem. Eur. J. Oper. Res. **88**(3), 464–474 (1996)
5. Divsalar, A., Vansteenwegen, P., Chitsaz, M., Sörensen, K., Cattrysse, D.: Personalized multi-day trips to touristic regions: a hybrid GA-VND approach. In: Evolutionary Computation in Combinatorial Optimisation, pp. 194–205. Springer, Berlin (2014)
6. Gavalas, D., Konstantopoulos, C., Mastakas, K., Pantziou, G., Vathis, N.: Efficient heuristics for the time dependent team orienteering problem with time windows. Comput. Oper. Res. **62**, 36–50 (2015)
7. Hasuike, T., Katagiri, H., Tsubaki, H., Tsuda, H.: Tour route planning problem for sightseeing with the multiroute under several uncertain conditions. In: 2012 IEEE International Conference on Systems, Man, and Cybernetics (SMC), pp. 715–720. IEEE (2012)
8. Hasuike, T., Katagiri, H., Tsubaki, H., Tsuda, H.: Route planning problem under fuzzy sightseeing times and satisfaction values of sightseeing places. In: 2013 Joint IFSA World Congress and NAFIPS Annual Meeting (IFSA/NAFIPS), pp. 140–145. IEEE (2013)
9. Hasuike, T., Katagiri, H., Tsubaki, H., Tsuda, H.: Flexible route planning for sightseeing with fuzzy random and fatigue-dependent satisfactions. J. Adv. Comput. Intel. Intel. Inform. **18**(2), 190–196 (2014)
10. Lim, K., Wang, X., Chan, J., Karunasekera, S., Leckie, C., Chen, Y., Tan, C.L., Gao, F.Q., Wee, T.K.: Perstour: a personalized tour recommendation and planning system. In: Extended Proceedings of the 27th ACM Conference on Hypertext and Social Media (HT'16), Demonstration Track. Halifax, NS, Canada, July 2016
11. Matsuda, Y., Nakamura, M., Kang, D., Miyagi, H.: A fuzzy optimal routing problem for sightseeing. IEEE J. Trans. Electron. Inform. Syst. **125**, 1350–1357 (2005)
12. Mendez, C.E.C.: Team Orienteering Problem with Time Windows and Fuzzy Scores. PhD thesis, National Taiwan University of Science and Technology (2016)
13. Mladenović, N., Hansen, P.: Variable neighborhood search. Comput. Oper. Res. **24**(11), 1097–1100 (1997)

14. Montemanni, R., Gambardella, L.M.: Ant colony system for team orienteering problems with time windows. Found. Comput. Decis. Sci. **34**, 287–306 (2009)
15. Mrazovic, P., Larriba-Pey, J.L., Matskin, M.: Improving mobility in smart cities with intelligent tourist trip planning. In: 2017 IEEE 41st Annual Computer Software and Applications Conference (COMPSAC). IEEE (2017)
16. Ni, Y., Chen, Y., Ke, H., Ralescu, D.A.: Models and algorithm for the orienteering problem in a fuzzy environment. Int. J. Fuzzy Syst. **20**(3), 861–876 (2018)
17. Schilde, M., Doerner, K.F., Hartl, R.F., Kiechle, G.: Metaheuristics for the bi-objective orienteering problem. Swarm Intel. **3**(3), 179–201 (2009)
18. Tsiligiridis, T.: Heuristic methods applied to orienteering. J. Oper. Res. Soc. **35**, 797–809 (1984)
19. Vansteenwegen, P., Oudheusden, D.V.: The mobile tourist guide: an or opportunity. OR Insight **20**(3), 21–27 (2007)
20. Vansteenwegen, P., Souffriau, W., Berghe, G.V., Oudheusden, D.V.: Metaheuristics for tourist trip planning. In: Lecture Notes in Economics and Mathematical Systems, pp. 15–31. Springer, Berlin (2009)
21. Vansteenwegen, P., Souffriau, W., Oudheusden, D.V.: The orienteering problem: a survey. Eur. J. Oper. Res. **209**(1), 1–10 (2011)
22. Vansteenwegen, P., Souffriau, W., Vanden Berghe, G., Van Oudheusden, D.: The city trip planner: an expert system for tourists. Exp. Syst. Appl. **38**(6), 6540–6546 (2011)
23. Verdegay, J.: Fuzzy Information and Decision Processes. North-Holland, Fuzzy Mathematical Programming (1982)
24. Verdegay, J.L.: Fuzzy optimization: models, methods and perspectives. In: Proceeding 6th IFSA-95 World Congress, pp. 39–71 (1995)
25. Verma, M., Shukla, K.K.: Application of fuzzy optimization to the orienteering problem. Adv. Fuzzy Syst. **2015**, 8 (2015)

A Multi-objective Solid Transportation Problem in Sustainable Development

Thiziri Sifaoui and Méziane Aïder

Abstract Sustainable development is regarded as the achievement of continuous economic development without damaging the environment and natural resources. In today's, competitive market, most of us are willing to pay less and earn more quickly without considering the negative externality to the environment and the quality of life of future generations. Considering this case, this paper investigates the study of the multi-objective multi-item solid transport problem with time-variant, taking into account the pollution minimization. The parameters of the multi-objective multi-item solid transport problem are taken as interval-valued trapezoidal fuzzy numbers. A new procedure is taken into consideration for converting multi-objective multi-item solid transport problem into deterministic form and then a hybrid method expressed with different membership function is used to solve the converted transport problem.

Keywords Multi-objective solid transportation problem · Carbon emission · Interval valued fuzzy numbers

1 Introduction

Sustainability is not about analyzing threats, but about analyzing systems. Specifically, it is about how environmental, economic, and social systems interact to their mutual advantage or disadvantage at various scales of spatial operations. The transportation of merchandise from one location to another is a human activity observed from ancient times to the present day.

T. Sifaoui (✉)
Faculty of Sciences, LAROMAD, UMM-Tizi-Ouzou, Algeria

Faculty of Sciences and Technology, Amine Elokkal El Hadj Moussa Eghamouk University, Tamanghasset, Algeria
e-mail: thiziri.sifaoui@gmail.com

M. Aïder
Faculty of Maths, LAROMAD, USTHB, PB 32, 16111 Bab Ezzouar, Algeria
e-mail: m-aider@usthb.dz

J. L. Verdegay et al. (eds.), *Computational Intelligence Methodologies Applied to Sustainable Development Goals*, Studies in Computational Intelligence 1036,
https://doi.org/10.1007/978-3-030-97344-5_16

The current transportation scenarios are totally under the influence of automation created by science. In transportation, several vehicles are used, which run on distinct fuels to communicate energy. For this energy transmission, there are several types of gases emitted by vehicles, creating pollution in the atmosphere. In most situations, the fact of pollution is not taken into account by the agency, person, or governing body, they are focused on maximizing profits or minimizing costs. It is time to address this particular problem, otherwise, the question will arise on the existence of our next generation.

The main idea of the transportation problem (TP) is to find the least total cost of transporting a commodity to satisfy the demand at destinations using the supplies available at the origin. Multi-Item Solid Transportation Problem (MISTP) arises when modes of conveyance and types of goods are taken into account in the objective and constraint set. Transportation problems can be applied in diverse situations such as scheduling, production, investment, facility location, and many others. Human languages involve imperfect or unknown information and lack certainty. It is impossible to accurately describe an existing state or a future outcome, several types of uncertainty have been suggested in the transportation problems. Sifaoui and Aïder [29] proposed an uncertain interval programming model for multi-objective multi-item fixed charge solid transportation problem with budget constraint and safety measure. Oheigeartaigh [23] developed an algorithm for solving transportation problems where the supplies and demands are fuzzy. Single-objective TP is not suitable for solving complex scenarios in real-life situations; Thus, we consider MOTP, which has been applied in many scientific fields, including engineering, economics, and logistics, where optimal decisions must be made in the presence of trade-offs between two or more conflicting objectives. Minimizing cost while maximizing comfort while buying a car, and maximizing performance while minimizing fuel consumption and emission of pollutants of a vehicle are examples of multi-objective optimization problems involving two and three objectives, respectively.

In practical problems, there can be more than three objectives. Singh and Yadav [30] addressed intuitionistic fuzzy multi-objective linear programming problems using intuitionistic fuzzy numbers. Rani and Gulati [25] used a fuzzy programming technique to solve a fully fuzzy multi-objective multi-item solid transportation problem. Nagarajan and Jeyaraman [22] studied the Multi-Objective Interval Solid Transportation Problem under Stochastic Environment using Fuzzy Approach. Mahapatra et al. [19] proposed a solution procedure of multi-objective transportation problems in which supplies and demands are normal random variables. Pramanik and Roy [24] presented a fuzzy goal programming approach for solving multi-objective transportation problems with fuzzy coefficients. Anukokila et al. [1] presented a goal programming approach for solving a multi-objective fractional transportation problem with fuzzy parameters. Maity et al.[20] proposed Time Variant Multi-Objective Interval Valued Transportation Problem in Sustainable Development.

None of the previous works has addressed Time-Variant Multi-Objective Multi-Item Solid Transportation Problem in Sustainable Development in terms of (γ, δ) interval-valued fuzzy numbers.

In this sense, the first contribution of this paper is that we introduce fuzzy numbers with (λ, ρ) interval values into Time-Variant Multi-Objective Multi-Item Solid Transportation Problem in Sustainable Development. To be more precise, all parameters are considered as fuzzy numbers with (λ, ρ) interval values.

The second contribution of this paper is to propose a new method based on a new ranking of (λ, ρ) interval-valued fuzzy numbers.

The third contribution of this paper is to take into consideration the Sustainable Development.

The fourth contribution of this paper is using a hybrid method expressed with different membership function to solve the model.

The rest of the paper is organized as follows. In Sect. 2, some basic definitions and arithmetic operations are reviewed. Section 3 presents Time-Variant Multi-Objective Multi-Item Solid Transportation Problem in Sustainable Development in terms of (γ, δ) interval-valued fuzzy numbers. Section 4 contains the proposed method. Section 5 gives an illustration of the proposed method. In Sect. 6, we present our conclusion.

2 Preliminaries

In this section, we give some basic definitions, the arithmetic operations and the comparison of interval- valued intuitionistic fuzzy numbers.

Definition 1 (Chiang [7]) Note a level γ-trapezoidal fuzzy number or a generalized trapezoidal fuzzy number by:

$$\widetilde{A} = (a_1, a_2, a_3, a_4; \gamma), \quad 0 < \gamma \leq 1,$$

with the membership function defined as follows:

$$\mu_{\widetilde{A}}(x) = \begin{cases} \gamma \dfrac{x - a_1}{a_2 - a_1}, & \text{if } a_1 \leq x \leq a_2, \\ \gamma, & \text{if } a_2 \leq x \leq a_3, \\ \gamma \dfrac{a_4 - x}{a_4 - a_3}, & \text{if } a_3 \leq x \leq a_4, \\ 0, & \text{otherwise.} \end{cases}$$

Let $F_{TN}(\gamma)$ be the family of all level γ-trapezoidal fuzzy numbers, that is:

$$F_{TN}(\gamma) = \{\widetilde{A} = (a_1, a_2, a_3, a_4; \gamma), a_1 \leq a_2 \leq a_3 \leq a_4\},$$

with $0 < \gamma \leq 1$.

Definition 2 (Yao and Lin [32]) Let $\widetilde{A}^L \in F_{TN}(\gamma)$ and $\widetilde{A}^U \in F_{TN}(\delta)$. A level (γ, δ)-interval-valued trapezoidal fuzzy number $\underset{\sim}{\widetilde{A}}$, denoted by:

$$\widetilde{\widetilde{A}} = [\widetilde{A}^L, \widetilde{A}^U] = \langle (a_1^L, a_2^L, a_3^L, a_4^L; \gamma), (a_1^U, a_2^U, a_3^U, a_4^U; \delta) \rangle,$$

is an interval-valued fuzzy set on $\mathbb{R}$ with the lower trapezoidal fuzzy number $\widetilde{A}^L$ expressed by:

$$\mu_{\widetilde{A}^L}(x) = \begin{cases} \gamma \dfrac{x - a_1^L}{a_2^L - a_1^L}, & \text{if } a_1^L \leq x \leq a_2^L, \\ \gamma, & \text{if } a_2^L \leq x \leq a_3^L, \\ \gamma \dfrac{a_4^L - x}{a_4^L - a_3^L}, & \text{if } a_3^L \leq x \leq a_4^L, \\ 0, & \text{otherwise,} \end{cases}$$

and the upper trapezoidal fuzzy number $\widetilde{A}^U$ expressed by:

$$\mu_{\widetilde{A}^U}(x) = \begin{cases} \delta \dfrac{x - a_1^U}{a_2^U - a_1^U}, & \text{if } a_1^U \leq x \leq a_2^U, \\ \delta, & \text{if } a_2^U \leq x \leq a_3^U, \\ \delta \dfrac{a_4^U - x}{a_4^U - a_3^U}, & \text{if } a_3^U \leq x \leq a_4^U, \\ 0, & \text{otherwise,} \end{cases}$$

where

- $a_1^L \leq a_2^L \leq a_3^L \leq a_4^L$,
- $a_1^U \leq a_2^U \leq a_3^U \leq a_4^U$,
- $a_1^U \leq a_1^L$,
- $a_4^L \leq a_4^U$,
- $0 < \gamma \leq \delta \leq 1$.

Observe that

$$\mu_{\widetilde{A}^L}(x) \leq \mu_{\widetilde{A}^U}(x).$$

This means that the least and greatest grades of membership of x in the interval $\widetilde{\widetilde{A}} = [\mu_{\widetilde{A}^L}(x), \mu_{\widetilde{A}^U}(x)]$, are $\mu_{\widetilde{A}^L}(x)$ and $\mu_{\widetilde{A}^U}(x)$ respectively.

Let $F_{IVTN}(\gamma, \delta)$ be the family of all level (γ, δ)-interval-valued trapezoidal fuzzy numbers, that is,
$F_{IVTN}(\gamma, \delta) = \{\widetilde{\widetilde{A}} = [\widetilde{A}^L, \widetilde{A}^U] = \langle (a_1^L, a_2^L, a_3^L, a_4^L; \gamma), (a_1^U, a_2^U, a_3^U, a_4^U; \delta) \rangle :$
$\widetilde{A}^L \in F_{TN}(\gamma), \widetilde{A}^U \in F_{TN}(\delta), a_1^U \leq a_1^L; a_4^L \leq a_4^U\}$, where $0 < \gamma \leq \delta \leq 1$.

Definition 3 (Ebrahimnejad [11]) A triangular fuzzy number denoted by:

$$\widetilde{\widetilde{A}} = [\widetilde{A}^L, \widetilde{A}^U] = \langle (a_1^L, a_2^L, a_3^L; \gamma), (a_1^U, a_2^U, a_3^U; \delta) \rangle$$

is an interval-valued fuzzy set on $\mathbb{R}$ with the lower triangular fuzzy number $\widetilde{A}^L$ expressed by :

$$\mu_{\widetilde{A}^L}(x) = \begin{cases} \gamma \dfrac{x - a_1^L}{a_2^L - a_1^L}, & \text{if } a_1^L \le x \le a_2^L, \\ \gamma \dfrac{a_3^L - x}{a_3^L - a_2^L}, & \text{if } a_2^L \le x \le a_3^L, \\ 0, & \text{otherwise.} \end{cases}$$

and the upper triangular fuzzy number $\widetilde{A}^U$ expressed by:

$$\mu_{\widetilde{A}^U}(x) = \begin{cases} \delta \dfrac{x - a_1^U}{a_2^U - a_1^U}, & \text{if } a_1^L \le x \le a_2^U, \\ \delta \dfrac{a_3^U - x}{a_3^U - a_2^L}, & \text{if } a_2^U \le x \le a_3^U, \\ 0, & \text{otherwise.} \end{cases}$$

where

- $a_1^U \le a_1^L \le a_2^U \le a_2^L \le a_3^L \le a_3^U$,
- $a_1^U \le a_1^L$,
- $a_4^L \le a_4^U$,
- $0 < \gamma \le \delta \le 1$.

Observe that $\mu_{\widetilde{A}^L}(x) \le \mu_{\widetilde{A}^U}(x)$.

Theorem 1 *[12] Let $\tilde{\tilde{A}} \in F_{IVTN}(\gamma, \delta)$. The signed distance of $\tilde{\tilde{A}}$ from O_1(y-axis) is given as follows:*

$$d(\widetilde{A}, O_1) = \frac{1}{4}[a_1 + a_2 + a_3 + a_4], \quad \widetilde{A}^L = \widetilde{A}^U = \widetilde{A}.$$

3 Time Variant Multi-objective Multi-item Solid Transportation Problem in Sustainable Development in Terms of (γ, δ) Interval Valued Fuzzy Numbers

Through transportation, vehicles usually consume energy and fuel. In any type of transportation system, some toxic gases are emitted into the atmosphere, directly or indirectly. This pollution factor is considered as a carbon emission. Consequently, the pollution factor is considered as one of the objective functions, for minimizing the carbon emission. Again, for a production system, energy is needed, so the pollution factor such as carbon emission is directly or indirectly involved in the system. In this aspect, we incorporate an objective function to minimize the pollution factor of the carbon emission whose penalties are considered as(γ, δ) interval valued fuzzy numbers.

The transport time is an important factor in sustainable development. In most public and private industrial systems the principle aim of goods transport is to reduce transport costs and maximize profits in a faster time. However, in many scenarios, no one wants to address the optimal situation of controlling pollution during fabrication and transportation. So, in a faster world, to maintain our nature in the best conditions, it is time to consider not only the benefits but also the sustainability of nature during the process. Since time is an essential factor, we construct an additional function objective function to minimize the transport time.

A solid transportation problem is a general case of transportation problem in which homogenous products will be transported from a set of sources or origins to a set of destinations via k, $(k > 2)$ different conveyances (trucks, cargo vans, good trains, etc.). Generally, we must carry together p, $(p > 2)$ different types of items and this defines the multi-item transportation problem. In our study we consider time variant Multi-Objective Multi-Item Solid Transportation Problem in Sustainable Development with parameters expressed as (γ, δ) interval valued fuzzy numbers.

Here, we use the following notations:

- x_{ijk}^p: the unknown quantity to be transported from i^{th} origin to j^{th} destination by k^{th} conveyance of pth item (decision variable);
- $\forall\, i$: means "for $i = 1, \ldots, m$";
- $\forall\, j$: means "for $j = 1, \ldots, n$";
- $\forall\, k$: means "for $k = 1, \ldots, K$";
- $\forall\, p$: means "for $p = 1, \ldots, P$";
- m : number of sources of the transportation problem;
- n : number of destinations;
- K : number of conveyances (modes of transportation);
- P: number of items;
- T_{ijk}^p:Time required for transporting from i^{th} origin to j^{th} destination by k^{th} conveyance of pth item;
- $\approx{A}_i^p = [\widetilde{A}_{L_i}^p, \widetilde{A}_{R_i}^p] = \langle(\widetilde{a}_{L_i}^p, \gamma), (\widetilde{a}_{R_i}^p, \delta)\rangle$: amount of product available at i^{th} origin for p^{th} item;
- $\approx{B}_j^p = [\widetilde{B}_{L_j}^p, \widetilde{B}_{R_j}^p] = \langle(\widetilde{b}_{L_j}^p, \gamma), (\widetilde{b}_{R_j}^p, \delta)\rangle$: demand of product at j^{th} destination for p^{th} item;
- $\approx{E}_k = [\widetilde{E}_{L_k}, \widetilde{E}_{R_k}] = \langle(\widetilde{e}_{L_k}, \gamma), (\widetilde{e}_{R_k}, \delta)\rangle$: amount of product which can be carried by the kth conveyance;
- $\approx{C}_{ijk}^{po} = [\widetilde{C}_{L_{ijk}}^{po}, \widetilde{C}_{R_{ijk}}^{po}] = \langle(\widetilde{C}_{L_{ijk}}^{po}, \gamma), (\widetilde{C}_{R_{ijk}}^{po} \delta)\rangle$: cost for the transportation problem from i^{th} origin to j^{th} destination by k^{th} conveyance of pth item for o objectives $o = 1, \ldots, O$;

The mathematical model is defined as follows:

$$(P_1)\begin{cases} \min Z^o = \sum_{p=1}^{P}\sum_{i=1}^{m}\sum_{j=1}^{n}\sum_{k=1}^{K} \widetilde{\widetilde{C}}{}_{ijk}^{po} x_{ijk}^{p} & \forall\, o \\ \min T = \sum_{p=1}^{P}\sum_{i=1}^{m}\sum_{j=1}^{n}\sum_{k=1}^{K} T_{ijk}^{p} y_{ijk}^{p} & \\ \sum_{j=1}^{n}\sum_{k=1}^{K} x_{ijk}^{p} \le \widetilde{\widetilde{A}}{}_i^{p}, & \forall\, i,\, p, \\ \sum_{i=1}^{m}\sum_{k=1}^{K} x_{ijk}^{p} \ge \widetilde{\widetilde{B}}{}_j^{p}, & \forall\, j,\, p, \\ \sum_{p=1}^{P}\sum_{i=1}^{m}\sum_{j=1}^{n} x_{ijk}^{p} \le \widetilde{\widetilde{E}}_k, & \forall\, k, \\ x_{ijk}^{p} \ge 0, & \forall\, i, j, k, p, \\ y_{ijk}^{p} = \begin{cases} 0, \text{ if } x_{ijk}^{p} = 0 \\ 1, \text{ if } x_{ijk}^{p} \ne 0 \end{cases} & \forall\, i, j, k, p. \end{cases}$$

4 Solution Procedure

The mathematical model described above cannot be solved directly because of the presence of (γ, δ) interval valued fuzzy numbers, For these raison we use two steps .

4.1 *Reduction of (γ, δ) Interval Valued Fuzzy Numbers into Interval Number*

Using the signed distance, we obtain the following model:

$$(P_2)\begin{cases} \min Z^o = \sum_{p=1}^{P}\sum_{i=1}^{m}\sum_{j=1}^{n}\sum_{k=1}^{K} \\ \left[\left(\frac{C_{ijk_1}^{po}+C_{ijk_2}^{po}+C_{ijk_3}^{po}+C_{ijk_4}^{po}}{4}\right)^L, \left(\frac{C_{ijk_1}^{po}+C_{ijk_2}^{po}+C_{ijk_3}^{po}+C_{ijk_4}^{po}}{4}\right)^U\right] x_{ijk}^{p} \quad \forall\, o \\ \min T = \sum_{p=1}^{P}\sum_{i=1}^{m}\sum_{j=1}^{n}\sum_{k=1}^{K} T_{ijk}^{p} y_{ijk}^{p} \\ \sum_{j=1}^{n}\sum_{k=1}^{K} x_{ijk}^{p} \le \left[\left(\frac{A_{i\,1}^{p}+A_{i\,2}^{p}+A_{i\,3}^{p}+A_{i\,4}^{p}}{4}\right)^L, \left(\frac{A_{i\,1}^{p}+A_{i\,2}^{p}+A_{i\,3}^{p}+A_{i\,4}^{p}}{4}\right)^U\right], \quad \forall\, i, p, \\ \sum_{i=1}^{m}\sum_{k=1}^{K} x_{ijk}^{p} \ge \left[\left(\frac{B_{j\,1}^{p}+B_{j\,2}^{p}+B_{j\,3}^{p}+B_{j\,4}^{p}}{4}\right)^L, \left(\frac{B_{j\,1}^{p}+B_{j\,2}^{p}+B_{j\,3}^{p}+B_{j\,4}^{p}}{4}\right)^U\right], \quad \forall\, j, p, \\ \sum_{p=1}^{P}\sum_{i=1}^{m}\sum_{j=1}^{n} x_{ijk}^{p} \le \left[\left(\frac{E_{k1}+E_{k2}+E_{k3}+E_{k4}}{4}\right)^L, \left(\frac{E_{k1}+E_{k2}+E_{k3}+E_{k4}}{4}\right)^U\right], \quad \forall\, k, \\ x_{ijk}^{p} \ge 0, \quad \forall\, i, j, k, p, \\ y_{ijk}^{p} = \begin{cases} 0, \text{ if } x_{ijk}^{p} = 0 \\ 1, \text{ if } x_{ijk}^{p} \ne 0 \end{cases} \quad \forall\, i, j, k, p. \end{cases}$$

Here we define a function that depends on time and takes values in the interval [0,1] that we will use to reduce the interval-valued cost parameter to the real-valued parameter (cf. Maity et al. [20])

4.2 Reduction of Interval into Real Number

As defined by Maity et al. [20], in Model 2, the transportation cost is of the interval type:

$$\Big[\Big(\frac{C^{po}_{ijk_1}+C^{po}_{ijk_2}+C^{po}_{ijk_3}+C^{po}_{ijk_4}}{4}\Big)^L,\Big(\frac{C^{po}_{ijk_1}+C^{po}_{ijk_2}+C^{po}_{ijk_3}+C^{po}_{ijk_4}}{4}\Big)^U\Big].$$

Let us consider a parameter that depends on time. Let t_0 be the time assigned by the DM as the minimum range period. If delivery occurs during the minimum range period, the minimum transportation cost:

$$\Big(\frac{C^{in}_{ijk_1}+C^{in}_{ijk_2}+C^{in}_{ijk_3}+C^{in}_{ijk_4}}{4}\Big)^L$$

must be paid. Due to the delay in delivery of the product, the cost becomes

$$C^{in}_{ijk}=(1-\lambda^{C^{po}_{ijk}})\Big(\frac{C^{po}_{ijk_1}+C^{po}_{ijk_2}+C^{po}_{ijk_3}+C^{po}_{ijk_4}}{4}\Big)^L+\lambda^{C^{po}_{ijk}}\Big(\frac{C^{po}_{ijk_1}+C^{po}_{ijk_2}+C^{po}_{ijk_3}+C^{po}_{ijk_4}}{4}\Big)^U.$$

Here, $\lambda^{C\ in}_{ijk}$ is a parameter for each p such that;

$$\lambda^{C^{po}_{ijk}}=\begin{cases}0 & \text{if } T^p_{ijk}<t_0\\ \frac{T^p_{ijk}-t_0}{T^p_{ijk}} & \text{if } T^p_{ijk}\geq t_0\end{cases}$$

Where T^p_{ijk} is the time of transporting p item from i origin to j destination by k conveyance, then $\lambda^{C^{po}_{ijk}}$ is an increasing function of time.

In addition, all objective functions in the MOMISTP model are not the kind to ship goods an a given time. It can also be the profit of delivering the goods. Basically, when goods are urgently required, overtime for all the the entire staff working in the production system, production cost becomes higher; therefore, DM expects a greater profit since the need is met on an urgent basis. In this case, if the goods are demanded in a minimum time t_0, then DM makes a maximum profit. Here, it is assumed that if the order duration and the purchase time are t_0, then profit is

$$\Big(\frac{C^{po}_{ijk_1}+C^{po}_{ijk_2}+C^{po}_{ijk_3}+C^{po}_{ijk_4}}{4}\Big)^U,$$

which is the maximum value, otherwise, the value of production cost becomes

$$C_{ijk}^{po} = \lambda^{C_{ijk}^{po}}\Big(\frac{C_{ijk_1}^{po} + C_{ijk_2}^{po} + C_{ijk_3}^{po} + C_{ijk_4}^{po}}{4}\Big)^L + (1 - \lambda^{C_{ijk}^{po}})\Big(\frac{C_{ijk_1}^{po} + C_{ijk_2}^{po} + C_{ijk_3}^{po} + C_{ijk_4}^{po}}{4}\Big)^U.$$

Pollution factor of carbon emission is incorporated into one of the objective functions in our formulated MOMISTP. It is observed that pollution factor increases when using more power required on an emergency basis. As the pollution factor is not a crisp quantity, we choose here the interval valued pollution factor by the following way:

$$C_{ijk}^{po} = \lambda^{C_{ijk}^{po}}\Big(\frac{C_{ijk_1}^{po} + C_{ijk_2}^{po} + C_{ijk_3}^{po} + C_{ijk_4}^{po}}{4}\Big)^L + (1 - \lambda^{C_{ijk}^{po}})\Big(\frac{C_{ijk_1}^{po} + C_{ijk_2}^{po} + C_{ijk_3}^{po} + C_{ijk_4}^{po}}{4}\Big)^U.$$

That indicates that, if $T_{ijk}^{p} = t_0$ then the pollution factor takes maximum value and as much as T_{ijk}^{p} increases and, then, the pollution factor decreases.

Proposition 1 *Maity et al. [20] The transportation cost C_{ijk}^{po} with minimization type attains minimum value when transportation time in the o objective function for the i origin to the j destination of p items by k conveyance.tends to the minimum value and conversely.*

Proposition 2 *Maity et al. [20] The profit C_{ijk}^{po} with maximization type attains maximum value when profit in the o objective function for the i origin to the j destination of p items by k conveyance tends to the minimum value and conversely.*

4.3 Hybrid Method for Solving Multi-objective Transportation Problem

Step 1: Solve the multi-objective transportation problem as a single objective transportation problem, taking each time only one objective as objective function and ignoring all others.

Step 2: Calculate:

$$L^o = \min Z^o, \quad \forall\, o$$

$$U^o = \max Z^o, \quad \forall\, o$$

$$L = \min T,$$

$$U = \max T,$$

Step 3: Define and choose an appropriate membership functions $\mu_o(Z^o(x), \mu(T))$ (linear, hyperbolic, exponential, quadratic) .
Step 4: Define the weights for each deviations.
Step 5: Develop the proposed model as follows:

(a) For linear membership function, the model may be written as:

$$\begin{cases} \min \sum_{o=1}^{O} (1 - W_{o1}) r_{o1} + (1 - W_{o2}) r_{o2} + (1 - W_o) r_o + (1 - H) r \\ \qquad +(1 - H_1) r_1 + (1 - H_2) r_2, & \\ \sum_{p=1}^{P} \sum_{i=1}^{m} \sum_{j=1}^{n} \sum_{k=1}^{K} C_{ijk}^{po} x_{ijk}^{p} - (1 - W_o) r_o = L_o, & \forall\, o, \\ \sum_{p=1}^{P} \sum_{i=1}^{m} \sum_{j=1}^{n} \sum_{k=1}^{K} T_{ijk}^{p} y_{ijk}^{p} - (1 - H) r = L, & \\ \frac{U_o - Z_o}{U_o - L_o} - (1 - W_{o1}) r_{o1} + (1 - W_{o2}) r_{o2} = 1, & \forall\, o, \\ \frac{U - T}{U - L} - (1 - H_1) r_1 + (1 - H_2) r_2 = 1, & \forall\, o, \\ \sum_{j=1}^{n} \sum_{k=1}^{K} x_{ijk}^{p} \leq a.\left(\frac{A_{i\,1}^{p} + A_{i\,2}^{p} + A_{i\,3}^{p} + A_{i\,4}^{p}}{4}\right)^{L} & \\ \qquad +(1 - a).\left(\frac{A_{i\,1}^{p} + A_{i\,2}^{p} + A_{i\,3}^{p} + A_{i\,4}^{p}}{4}\right)^{U}, & \forall\, i, p, \\ \sum_{i=1}^{m} \sum_{k=1}^{K} x_{ijk}^{p} \geq b.\left(\frac{B_{j\,1}^{p} + B_{j\,2}^{p} + B_{j\,3}^{p} + B_{j\,4}^{p}}{4}\right)^{L} & \\ \qquad +(1 - b).\left(\frac{B_{j\,1}^{p} + B_{j\,2}^{p} + B_{j\,3}^{p} + B_{j\,4}^{p}}{4}\right)^{U}, & \forall\, j, p, \\ \sum_{p=1}^{P} \sum_{i=1}^{m} \sum_{j=1}^{n} x_{ijk}^{p} \leq c.\left(\frac{E_{k1} + E_{k2} + E_{k3} + E_{k4}}{4}\right)^{L} & \\ \qquad +(1 - c).\left(\frac{E_{k1} + E_{k2} + E_{k3} + E_{k4}}{4}\right)^{U}, & \forall\, k, \\ x_{ijk}^{p} \geq 0, & \forall\, i, j, k, p, \\ y_{ijk}^{p} = \begin{cases} 0, \text{ if } x_{ijk}^{p} = 0 \\ 1, \text{ if } x_{ijk}^{p} \neq 0 \end{cases} & \forall\, i, j, k, p. \end{cases}$$

(b) For quadratic membership function, the model may be written as:

$$\begin{cases} \min \sum_{o=1}^{O} (1 - W_{o1})r_{o1} + (1 - W_{o2})r_{o2} + (1 - W_o)r_o + (1 - H)r \\ \quad +(1 - H_1)r_1 + (1 - H_2)r_2, \\ \sum_{p=1}^{P}\sum_{i=1}^{m}\sum_{j=1}^{n}\sum_{k=1}^{K} C_{ijk}^{po} x_{ijk}^{p} - (1 - W_o)r_o = L_o, & \forall\, o, \\ \sum_{p=1}^{P}\sum_{i=1}^{m}\sum_{j=1}^{n}\sum_{k=1}^{K} T_{ijk}^{p} y_{ijk}^{p} - (1 - H)r = L, \\ \frac{U_o - Z_o}{U_o - L_o} + a_{o1}Z_o^2 - a_{o1}(L_o + U_o)Z_o + a_{o1}L_oU_o \\ \quad -(1 - W_{o1})t_{o1} + (1 - W_{o2})t_{o2} = 1 & \forall\, o, \\ \frac{U - T}{U - L} + a_1T^2 - a_1(L + U)T + a_1LU \\ \quad -(1 - H_1)t_1 + (1 - H_2)t_2 = 1 & \forall\, o, \\ \sum_{j=1}^{n}\sum_{k=1}^{K} x_{ijk}^{p} \le a.\Big(\frac{A_{i\,1}^p + A_{i\,2}^p + A_{i\,3}^p + A_{i\,4}^p}{4}\Big)^L \\ \quad +(1 - a).\Big(\frac{A_{i\,1}^p + A_{i\,2}^p + A_{i\,3}^p + A_{i\,4}^p}{4}\Big)^U, & \forall\, i, p, \\ \sum_{i=1}^{m}\sum_{k=1}^{K} x_{ijk}^{p} \ge b.\Big(\frac{B_{j\,1}^p + B_{j\,2}^p + B_{j\,3}^p + B_{j\,4}^p}{4}\Big)^L \\ \quad +(1 - b).\Big(\frac{B_{j\,1}^p + B_{j\,2}^p + B_{j\,3}^p + B_{j\,4}^p}{4}\Big)^U, & \forall\, j, p, \\ \sum_{p=1}^{P}\sum_{i=1}^{m}\sum_{j=1}^{n} x_{ijk}^{p} \le c.\Big(\frac{E_{k1} + E_{k2} + E_{k3} + E_{k4}}{4}\Big)^L \\ \quad +(1 - c).\Big(\frac{E_{k1} + E_{k2} + E_{k3} + E_{k4}}{4}\Big)^U, & \forall\, k, \\ x_{ijk}^{p} \ge 0, & \forall\, i, j, k, p, \\ y_{ijk}^{p} = \begin{cases} 0, \text{ if } x_{ijk}^{p} = 0 \\ 1, \text{ if } x_{ijk}^{p} \ne 0 \end{cases} & \forall\, i, j, k, p. \end{cases}$$

(c) For hyperbolic membership function, the model may be written as:

$$\begin{cases} \min \sum_{o=1}^{O} (1 - W_{o1})r_{o1} + (1 - W_{o2})r_{o2} + (1 - U_o)r_o + (1 - H)r \\ \quad +(1 - H_1)r_1 + (1 - H_2)r_2, \\ \sum_{p=1}^{P}\sum_{i=1}^{m}\sum_{j=1}^{n}\sum_{k=1}^{K} C_{ijk}^{po} x_{ijk}^{p} - (1 - W_o)r_o = L_o, & \forall\, o, \\ \sum_{p=1}^{P}\sum_{i=1}^{m}\sum_{j=1}^{n}\sum_{k=1}^{K} T_{ijk}^{p} y_{ijk}^{p} - (1 - H)r = L \\ \alpha_o Z_o(x) - \alpha_o \frac{(L_o + U_o)}{2} - (1 - W_{o1})r_{o1} + (1 - W_{o2})r_{o2} = 1, \\ \alpha T - \alpha \frac{(L + U)}{2} - (1 - H_1)r_1 + (1 - H_2)r_2 = 1, \\ \sum_{j=1}^{n}\sum_{k=1}^{K} x_{ijk}^{p} \le a.\Big(\frac{A_{i\,1}^p + A_{i\,2}^p + A_{i\,3}^p + A_{i\,4}^p}{4}\Big)^L \\ \quad +(1 - a).\Big(\frac{A_{i\,1}^p + A_{i\,2}^p + A_{i\,3}^p + A_{i\,4}^p}{4}\Big)^U, & \forall\, i, p, \\ \sum_{i=1}^{m}\sum_{k=1}^{K} x_{ijk}^{p} \ge b.\Big(\frac{B_{j\,1}^p + B_{j\,2}^p + B_{j\,3}^p + B_{j\,4}^p}{4}\Big)^L \\ \quad +(1 - b).\Big(\frac{B_{j\,1}^p + B_{j\,2}^p + B_{j\,3}^p + B_{j\,4}^p}{4}\Big)^U, & \forall\, j, p, \\ \sum_{p=1}^{P}\sum_{i=1}^{m}\sum_{j=1}^{n} x_{ijk}^{p} \le c.\Big(\frac{E_{k1} + E_{k2} + E_{k3} + E_{k4}}{4}\Big)^L \\ \quad +(1 - c).\Big(\frac{E_{k1} + E_{k2} + E_{k3} + E_{k4}}{4}\Big)^U, & \forall\, k, \\ x_{ijk}^{p} \ge 0, & \forall\, i, j, k, p, \\ y_{ijk}^{p} = \begin{cases} 0, \text{ if } x_{ijk}^{p} = 0 \\ 1, \text{ if } x_{ijk}^{p} \ne 0 \end{cases} & \forall\, i, j, k, p. \end{cases}$$

(d) For the Exponential membership function, the model way be written as

$$\begin{cases} \min \sum_{o=1}^{O} (1-W_{o1})r_{o1} + (1-W_{o2})r_{o2} + (1-W_o)r_o + (1-H)r \\ \quad +(1-H_1)r_1 + (1-H_2)r_2, \\ \sum_{p=1}^{P}\sum_{i=1}^{m}\sum_{j=1}^{n}\sum_{k=1}^{K} C_{ijk}^{po} x_{ijk}^{p} - (1-W_o)r_o = L_o, & \forall\, o, \\ \sum_{p=1}^{P}\sum_{i=1}^{m}\sum_{j=1}^{n}\sum_{k=1}^{K} T_{ijk}^{p} y_{ijk}^{p} - (1-H)r = L, \\ \dfrac{\exp^{(-e\phi_o(x))} - \exp^{-e}}{1-\exp^{-e}} - (1-W_{o1})r_{o1} + (1-W_{o2})r_{o2} = 1, & \forall\, o \\ \dfrac{\exp^{(-f\phi)} - \exp^{-f}}{1-\exp^{-f}} - (1-H_1)r_1 + (1-H_2)r_2 = 1, & \forall\, o, \\ \sum_{j=1}^{n}\sum_{k=1}^{K} x_{ijk}^{p} \le a.\Big(\dfrac{A_{i\,1}^{p} + A_{i\,2}^{p} + A_{i\,3}^{p} + A_{i\,4}^{p}}{4}\Big)^{L} \\ \quad +(1-a).\Big(\frac{A_{i\,1}^{p} + A_{i\,2}^{p} + A_{i\,3}^{p} + A_{i\,4}^{p}}{4}\Big)^{U}, & \forall\, i, p, \\ \sum_{i=1}^{m}\sum_{k=1}^{K} x_{ijk}^{p} \ge b.\Big(\dfrac{B_{j\,1}^{p} + B_{j\,2}^{p} + B_{j\,3}^{p} + B_{j\,4}^{p}}{4}\Big)^{L} \\ \quad +(1-b).\Big(\frac{B_{j\,1}^{p} + B_{j\,2}^{p} + B_{j\,3}^{p} + B_{j\,4}^{p}}{4}\Big)^{U}, & \forall\, j, p, \\ \sum_{p=1}^{P}\sum_{i=1}^{m}\sum_{j=1}^{n} x_{ijk}^{p} \le c.\Big(\dfrac{E_{k1} + E_{k2} + E_{k3} + E_{k4}}{4}\Big)^{L} \\ \quad +(1-c).\Big(\frac{E_{k1} + E_{k2} + E_{k3} + E_{k4}}{4}\Big)^{U}, & \forall\, k, \\ x_{ijk}^{p} \ge 0, & \forall\, i, j, k, p, \\ y_{ijk}^{p} = \begin{cases} 0, \text{ if } x_{ijk}^{p} = 0 \\ 1, \text{ if } x_{ijk}^{p} \ne 0 \end{cases} & \forall\, i, j, k, p. \end{cases}$$

Step 6: Evaluate each objective functions for the obtained solution.
Step 7: Give the solution to the DM. If the DM satisfied with the obtained solution go to the step 8, else step 4.
Step 8: Stop.
Where

- $(1-W_{o1})r_{o1} + 1 - W_{o2})r_{o2}$, $(1-H_1)r_1 + (1-H_2)r_2$: represent the deviational functions
  from the optimum value of the membership function.
- $(1-W_o)r_o$, $(1-H)r$: represent the deviational function from the optimum value of the objective .
- $W_{o1}, W_{o2}, W_1, W_2, W_o, U$: represent the weights assigned by the decision maker

$$0 \le W_{o1}, W_{o2}, H_1, H_2, W_o, U \le 1$$

$$\sum_{o=1}^{O} W_{o1} + W_{o2} = 1$$

$$H_1 + H_2 = 1$$

$$\sum_{o=1}^{O} W_o = 1$$

$$U = 1$$

- $r_{o1}, r_{o2}, r_1, r_2, r_o, r$: represent the deviation from the optimum value of the objective
- $$0 \leq a, b, c \leq 1$$
- a_{o1}, a_1: represent a parameter associated with quadratic membership function.
- $$\alpha_o = \frac{6}{U_o - L_o}$$
- $$\alpha = \frac{6}{U - L}$$
- $$\phi_o = \frac{Z^o - L^o}{U^o - L^o}$$
- $$\phi = \frac{Z - L}{U - L}$$
- e, f: is a non-zero parameter prescribed by the decision maker.

5 Numerical Illustration

To demonstrate the applicability of this article, consider a farmer has two rice fields to produce two kind of rice at two different locations $\overset{\approx 1}{A}_1, \overset{\approx 1}{A}_2, \overset{\approx 1}{A}_2, \overset{\approx 2}{A}_2$, the rice produced is sold to two dealers located in different cities $\overset{\approx 1}{B}_1, \overset{\approx 1}{B}_2, \overset{\approx 2}{B}_1, \overset{\approx 2}{B}_2$, using two conveyances to transport the rice from location to dealers $\overset{\approx}{E}_1, \overset{\approx}{E}_2$. The problem is treated with three conflicting objective functions, namely the transport cost, profit and pollution factor. and the time corresponding to each of them. Here, the transport cost, the profit and the pollution factor are all considered as (γ, δ) interval valued fuzzy numbers. Here, there is a time associated with each of these intervals.
The dealer expects schedule time for transportation $t_0 = 4$ days.

Tables 1, 2, 3, 4, 5, 6, 7 and 8 present transportation costs, profits and pollution factor consecutively.

$$a_1^1 = \left\langle \begin{array}{c} (70, 90, 90, 100), \\ (65, 90, 100, 125) \end{array} \right\rangle a_2^1 = \left\langle \begin{array}{c} (72, 92, 97, 100), \\ (75, 95, 100, 150) \end{array} \right\rangle$$

$$a_1^2 = \left\langle \begin{array}{c} (30, 50, 70, 80), \\ (35, 55, 85, 90), \end{array} \right\rangle a_2^2 = \left\langle \begin{array}{c} (30, 40, 62, 70), \\ (35, 57, 75, 88) \end{array} \right\rangle,$$

$$b_1^1 = \left\langle \begin{array}{c} (20, 25, 30, 60), \\ (25, 35, 55, 75) \end{array} \right\rangle b_2^1 = \left\langle \begin{array}{c} (20, 30, 40, 50), \\ (25, 45, 55, 85) \end{array} \right\rangle$$

$$b_1^2 = \left\langle \begin{array}{c} (20, 30, 35, 50), \\ (35, 45, 55, 85) \end{array} \right\rangle b_2^2 = \left\langle \begin{array}{c} (25, 32, 32, 82), \\ (35, 57, 75, 88) \end{array} \right\rangle,$$

$$e_1 = \left\langle \begin{array}{c} (140, 149, 157, 159), \\ (140, 150, 160, 190) \end{array} \right\rangle$$

$$e_2 = \left\langle \begin{array}{c} (130, 157, 115, 182), \\ (130, 159, 168, 220) \end{array} \right\rangle.$$

Table 1 Cost C_{ijk}^{11}

i	j		j	
	1	2	1	2
1	⟨(10, 20, 30, 40), (5, 15, 35, 45)⟩	⟨(50, 60, 70, 90), (45, 55, 75, 95)⟩	⟨(25, 30, 40, 50), (25, 35, 45, 75)⟩	⟨(5, 15, 35, 80), (25, 26, 40, 70)⟩
2	⟨(60, 70, 80, 90), (55, 65, 85, 120)⟩	⟨(40, 50, 60, 100), (45, 65, 75, 105)⟩	⟨(25, 30, 40, 50), (25, 35, 45, 75)⟩	⟨(1, 5, 30, 60), (5, 15, 35, 80)⟩
k	1		2	

Table 2 Cost C_{ijk}^{12}

i	j		j	
	1	2	1	2
1	⟨(1, 10, 28, 40), (3, 12, 32, 43)⟩	⟨(40, 55, 65, 85), (40, 55, 75, 95)⟩	⟨(20, 30, 40, 50), (20, 35, 45, 75)⟩	⟨(10, 25, 30, 60), (10, 25, 35, 85)⟩
2	⟨(50, 60, 80, 90), (55, 65, 85, 95)⟩	⟨(45, 60, 70, 100), (45, 65, 75, 105)⟩	⟨(10, 30, 40, 50), (15, 35, 45, 75)⟩	⟨(7, 15, 30, 60), 9, 18, 32, 85)⟩
k	1		2	

Table 3 Profit C_{ijk}^{21}

i	*j*		*j*	
	1	2	1	2
1	⟨(6, 16, 32, 42), (7, 17, 37, 47)⟩	⟨(40, 52, 72, 92), (47, 57, 77, 97)⟩	⟨(27, 32, 42, 52), (27, 37, 47, 77)⟩	⟨(7, 15, 32, 62), (7, 17, 37, 82)⟩
2 2	⟨(52, 62, 82, 92), (57, 67, 87, 97)⟩	⟨(42, 62, 68, 102), (47, 67, 77, 107)⟩	⟨(27, 32, 42, 52), (27, 37, 47, 77)⟩	⟨(5, 15, 32, 62), (7, 17, 37, 82)⟩
k	1		2	

Table 4 Profit C_{ijk}^{22}

i	*j*		*j*	
	1	2	1	2
1	⟨(5, 12, 31, 43), (6, 15, 35, 48)⟩	⟨(40, 50, 68, 88), (43, 58, 78, 98)⟩	⟨(20, 33, 43, 53), (23, 38, 48, 78)⟩	⟨(10, 28, 33, 63), (13, 28, 38, 88)⟩
2	⟨(43, 53, 83, 93), (58, 68, 88, 98)⟩	⟨(38, 43, 93, 103), (48, 68, 78, 108)⟩	⟨(10, 33, 43, 53), (18, 38, 48, 78)⟩	⟨(8, 10, 33, 63), (12, 12, 35, 88)⟩
k	1		2	

Table 5 Pollution C_{ijk}^{31}

i	*j*		*j*	
	1	2	1	2
1	⟨(4, 12, 17, 22), (5, 15, 25, 32)⟩	⟨(10, 15, 20, 30), (15, 20, 25, 35)⟩	⟨(15, 20, 25, 30), (15, 27, 25, 35)⟩	⟨(1, 10, 28, 32), (5, 15, 32, 68)⟩
2	⟨(15, 40, 45, 60), (25, 45, 50, 60)⟩	⟨(30, 40, 45, 60), (45, 55, 65, 90)⟩	⟨(15, 20, 30, 35), (15, 25, 40, 50)⟩	⟨(8, 10, 20, 55), (10, 12, 32, 70)⟩
k	1		2	

Table 6 Pollution C_{ijk}^{32}

i	*j*		*j*	
	1	2	1	2
1 1	⟨(1, 10, 28, 40), (3, 12, 32, 43)⟩	⟨(30, 40, 52, 70), (40, 55, 65, 85)⟩	⟨(4, 8, 40, 50), (5, 10, 40, 70)⟩	⟨(5, 15, 30, 60), (10, 25, 35, 65)⟩
2	⟨(10, 20, 25, 40), (25, 45, 50, 65)⟩	⟨(35, 45, 50, 62), (40, 60, 70, 80)⟩	⟨(2, 5, 20, 45), (5, 15, 25, 55)⟩	⟨(3, 10, 15, 50), (9, 18, 32, 70)⟩
k	1		2	

Table 7 Time (in day) T^1_{ijk}

i	j		j	
	1	2	1	2
1	5	2	8	6
2	3	4	5	2
k	1		2	

Table 8 Time (in day) T^2_{ijk}

i	j		j	
	1	2	1	2
1	4	6	2	1
2	6	2	3	5
k	1		2	

The results of the numerical illustration for different types of membership function are presented in Tables 9, 10, 11 and 12.

Table 9 Optimal solution of the model for linear membership function

Weights	Optimal Solution	Optimal Value
$W_{11} = 0.16$, $W_{21} = 0.16$, $W_{31} = 0.16$ $W_{12} = 0.16$, $W_{22} = 0.16$, $W_{32} = 0.16$ $W_1 = 0.16$, $W_2 = 0.16$, $W_3 = 0.16$ $H = 1$ $H_1 = 0.5$, $H_2 = 0.5$	$x^1_{122} = 12.05$ $x^2_{111} = 11.95$ $x^2_{222} = 14.25$ and other variables are zero	$Z^1 = 90$ $Z^2 = 60$ $Z^3 = 120$ $T = 15$
$W_{11} = 0.16$, $W_{21} = 0.16$, $W_{31} = 0.16$ $W_{12} = 0.16$, $W_{22} = 0.16$, $W_{32} = 0.16$ $W_1 = 0.5$, $W_2 = 0.25$, $W_3 = 0.25$ $H = 1$	$x^1_{121} = 18$ $x^2_{111} = 15.25$ $x^2_{222} = 17$	$Z^1 = 94$ $Z^2 = 65$ $Z^3 = 100$ $T = 13$
$W_{11} = 0.16$, $W_{21} = 0.16$, $W_{31} = 0.16$ $W_{12} = 0.16$, $W_{22} = 0.16$, $W_{32} = 0.16$ $W_1 = 0.6$, $W_2 = 0.2$, $W_3 = 0.2$ $H = 1$ $H_1 = 0.8$, $H_2 = 0.2$	$x^1_{111} = 15.12$ $x^1_{122} = 6.05$ $x^2_{111} = 15$ $x^2_{222} = 12$	$Z^1 = 80$ $Z^2 = 50$ $Z^3 = 130$ $T = 16$
$W_{11} = 0.2$, $W_{21} = 0.1$, $W_{31} = 0.3$ $W_{12} = 0.1$, $W_{22} = 0.2$, $W_{32} = 0.1$ $W_1 = 0.4$, $W_2 = 0.3$, $W_3 = 0.3$ $H = 1$ $H_1 = 0.2$, $H_2 = 0.8$	$x^1_{111} = 11.95$ $x^1_{122} = 9.07$ $x^2_{111} = 13.53$ $x^2_{222} = 15.75$	$Z^1 = 78$ $Z^2 = 70$ $Z^3 = 95$ $T = 14$

Table 10 Optimal solution of the model for quadratic function

Weights	Optimal Solution	Optimal Value
$W_{11} = 0.16, W_{21} = 0.16, W_{31} = 0.16$	$x^1_{122} = 12.05$	$Z^1 = 90$
$W_{12} = 0.16, W_{22} = 0.16, W_{32} = 0.16$	$x^2_{111} = 11.95$	$Z^2 = 60$
$W_1 = 0.16, W_2 = 0.16, W_3 = 0.16$	$x^2_{222} = 14.25$	$Z^3 = 120$
$H = 1$	and other variables are zero	$T = 15$
$H_1 = 0.5, H_2 = 0.5$		
$W_{11} = 0.16, W_{21} = 0.16, W_{31} = 0.16$	$x^1_{121} = 18$	$Z^1 = 94$
$W_{12} = 0.16, W_{22} = 0.16, W_{32} = 0.16$	$x^2_{111} = 15.25$	$Z^2 = 65$
$W_1 = 0.5, W_2 = 0.25, W_3 = 0.25$	$x^2_{222} = 17$	$Z^3 = 100$
$H = 1$		$T = 13$
$W_{11} = 0.16, W_{21} = 0.16, W_{31} = 0.16$	$x^1_{111} = 15.12$	$Z^1 = 80$
$W_{12} = 0.16, W_{22} = 0.16, W_{32} = 0.16$	$x^1_{122} = 6.05$	$Z^2 = 50$
$W_1 = 0.6, W_2 = 0.2, W_3 = 0.2$	$x^2_{111} = 15$	$Z^3 = 130$
$H = 1$	$x^2_{222} = 12$	$T = 16$
$H_1 = 0.8, H_2 = 0.2$		
$W_{11} = 0.2, W_{21} = 0.1, W_{31} = 0.3$	$x^1_{111} = 11.95$	$Z^1 = 78$
$W_{12} = 0.1, W_{22} = 0.2, W_{32} = 0.1$	$x^1_{122} = 9.07$	$Z^2 = 70$
$W_1 = 0.4, W_2 = 0.3, W_3 = 0.3$	$x^2_{111} = 13.53$	$Z^3 = 95$
$H = 1$	$x^2_{222} = 15.75$	$T = 14$
$H_1 = 0.2, H_2 = 0.8$		

5.1 *Discussion of Results*

This paper presents Time Variant Multi-objective Multi-item Solid Transportation Problem in Sustainable Development in terms of (γ, δ) interval valued fuzzy numbers solved by hybrid method expressed by different membership functions linear,hyperbolic, exponential and quadratic

According to this example, we may conclude the following results.

- This method can be used to obtain multiple solutions points by varying the weights associated with the weights for each deviations.
- The optimal compromise solution does not change if we compare with the solution obtained by changing the kind of membership function
- This model treat a new model which ensure the sustainable development in transportation problem .
- Our proposed methodology is not comparable with any other existing techniques for solving multi objective multi-Item Solid Transportation Problem in terms of (γ, δ) interval valued fuzzy numbers

Table 11 Optimal solution of the model for hyperbolic function

Weights	Optimal Solution	Optimal Value
$W_{11} = 0.16$, $W_{21} = 0.16$, $W_{31} = 0.16$ $W_{12} = 0.16$, $W_{22} = 0.16$, $W_{32} = 0.16$ $W_1 = 0.16$, $W_2 = 0.16$, $W_3 = 0.16$ $H = 1$ $H_1 = 0.5$, $H_2 = 0.5$	$x^1_{122} = 12.05$ $x^2_{111} = 11.95$ $x^2_{222} = 14.25$ And other variables are zero	$Z^1 = 90$ $Z^2 = 60$ $Z^3 = 120$ $T = 15$
$W_{11} = 0.16$, $W_{21} = 0.16$, $W_{31} = 0.16$ $W_{12} = 0.16$, $W_{22} = 0.16$, $W_{32} = 0.16$ $W_1 = 0.5$, $W_2 = 0.25$, $W_3 = 0.25$ $H = 1$	$x^1_{121} = 18$ $x^2_{111} = 15.25$ $x^2_{222} = 17$	$Z^1 = 94$ $Z^2 = 65$ $Z^3 = 100$ $T = 13$
$W_{11} = 0.16$, $W_{21} = 0.16$, $W_{31} = 0.16$ $W_{12} = 0.16$, $W_{22} = 0.16$, $W_{32} = 0.16$ $W_1 = 0.6$, $W_2 = 0.2$, $W_3 = 0.2$ $H = 1$ $H_1 = 0.8$, $H_2 = 0.2$	$x^1_{111} = 15.12$ $x^1_{122} = 6.05$ $x^2_{111} = 15$ $x^2_{222} = 12$	$Z^1 = 80$ $Z^2 = 50$ $Z^3 = 130$ $T = 16$
$W_{11} = 0.2$, $W_{21} = 0.1$, $W_{31} = 0.3$ $W_{12} = 0.1$, $W_{22} = 0.2$, $W_{32} = 0.1$ $W_1 = 0.4$, $W_2 = 0.3$, $W_3 = 0.3$ $H = 1$ $H_1 = 0.2$, $H_2 = 0.8$	$x^1_{111} = 11.95$ $x^1_{122} = 9.07$ $x^2_{111} = 13.53$ $x^2_{222} = 15.75$	$Z^1 = 78$ $Z^2 = 70$ $Z^3 = 95$ $T = 14$

Table 12 Optimal solution of the model for Exponential function

Weights	Optimal solution	Optimal value
$W_{11} = 0.16$, $W_{21} = 0.16$, $W_{31} = 0.16$ $W_{12} = 0.16$, $W_{22} = 0.16$, $W_{32} = 0.16$ $W_1 = 0.16$, $W_2 = 0.16$, $W_3 = 0.16$ $H = 1$ $H_1 = 0.5$, $H_2 = 0.5$	$x^1_{122} = 12.05$ $x^2_{111} = 11.95$ $x^2_{222} = 14.25$ And other variables are zero	$Z^1 = 90$ $Z^2 = 60$ $Z^3 = 120$ $T = 15$
$W_{11} = 0.16$, $W_{21} = 0.16$, $W_{31} = 0.16$ $W_{12} = 0.16$, $W_{22} = 0.16$, $W_{32} = 0.16$ $W_1 = 0.5$, $W_2 = 0.25$, $W_3 = 0.25$ $H = 1$	$x^1_{121} = 18$ $x^2_{111} = 15.25$ $x^2_{222} = 17$	$Z^1 = 94$ $Z^2 = 65$ $Z^3 = 100$ $T = 13$
$W_{11} = 0.16$, $W_{21} = 0.16$, $W_{31} = 0.16$ $W_{12} = 0.16$, $W_{22} = 0.16$, $W_{32} = 0.16$ $W_1 = 0.6$, $W_2 = 0.2$, $W_3 = 0.2$ $H = 1$ $H_1 = 0.8$, $H_2 = 0.2$	$x^1_{111} = 15.12$ $x^1_{122} = 6.05$ $x^2_{111} = 15$ $x^2_{222} = 12$	$Z^1 = 80$ $Z^2 = 50$ $Z^3 = 130$ $T = 16$
$W_{11} = 0.2$, $W_{21} = 0.1$, $W_{31} = 0.3$ $W_{12} = 0.1$, $W_{22} = 0.2$, $W_{32} = 0.1$ $W_1 = 0.4$, $W_2 = 0.3$, $W_3 = 0.3$ $H = 1$ $H_1 = 0.2$, $H_2 = 0.8$	$x^1_{111} = 11.95$ $x^1_{122} = 9.07$ $x^2_{111} = 13.53$ $x^2_{222} = 15.75$	$Z^1 = 78$ $Z^2 = 70$ $Z^3 = 95$ $T = 14$

6 Conclusions

This work presents an evolutionary and efficient model for solving Time Variant Multi-Objective Multi-Item Solid Transportation Problem in Sustainable Development in terms of (γ, δ) interval valued fuzzy numbers. The problem studied has been modeled as multi objective Multi-Item Solid Transportation Problem and solved by hybrid method expressed by different membership functions.
For future

- There are other parameters to be taken into account in the model in term of Sustainable Development.
- For the numerical example with general instances, we can extend the results for a more detailed and elaborate numerical experiment.

References

1. Anukokila, P., Radhakrishnan, B., Anju, A.: Goal programming approach for solving multi-objective fractional transportation problem with fuzzy parameters. RAIRO-Oper. Res. **53**(1), 157–178 (2019)
2. Baidya, A., Bera, U.K., Maiti, M.: Breakable fuzzy multi-stage transportation problem. J. Oper. Res. Soc. China **3**(1), 53–67 (2015)
3. Bharati, S.K.:Trapezoidal intuitionistic fuzzy fractional transportation problem. In: Book: Soft Computing for Problem Solving, Advances in Intelligent Systems and Computing, pp. 833–842 (2018)
4. Bharati, S.K., Singh, S.R.: Transportation problem under interval-valued intuitionistic fuzzy environment. Prog. Artif. Intell. **20**(5), 1511–1522 (2018)
5. Chanas, S., Kolodziejezk, W., Machaj, A.: A fuzzy approach to the transportation problem. Fuzzy Sets Syst. **13**(3), 211–221 (1984)
6. Chanas, S., Delgado, M., Verdegay, J.L., Vila, M.A.: Interval and fuzzy extensions of classical transportation problems. Transp. Plan. Technol. **17**(2), 203–218 (1993)
7. Chiang, J.: Fuzzy linear programming based on statistical confidence interval and interval-valued fuzzy set. Eur. J. Operat. Res. **129**, 65–86 (2001)
8. Chiang, J.: The optimal solution of the transportation problem with fuzzy demand and fuzzy product. J. Inf. Sci. Eng. **21**, 439–451 (2005)
9. Dalman, H., Güzel, N., Sivri, M.: A fuzzy set-based approach to multi-objective multi-item solid transportation problem under uncertainty. Int. J. Fuzzy Syst. **18**(4), 716–729 (2016)
10. Dutta, D., Murthy, S.: Multi-choice goal programming approach for a fuzzy Transportation problem. IJRRAS **2**(2), 132–139 (2010)
11. Ebrahimnejad, A.: Fuzzy linear programming approach for solving transportation problems with interval-valued trapezoidal fuzzy numbers. Sadhana-Acad. Proc. Eng. Sci. **41**(3), 299–316 (2016)
12. Farhadinia, B.: Sensitivity analysis in interval-valued trapezoidal fuzzy number linear programming problems. Appl. Math. Model. **38**(1), 50–62 (2014)
13. Gupta, A., Kumar, A.: A new method for solving linear multi-objective transportation problems with fuzzy parameters. Appl. Math. Model. **36**(4), 1421–1430 (2012)
14. Hari, G.A., Suresh, M., Sivakumar, G.: On solving fuzzy transportation problem based on distance based defuzzification method of various fuzzy quantities using centroid. Malaya J. Matematik **S**(1), 410–426 (2020)

15. Jimenez, F., Verdegay, J.L.: Uncertain solid transportation problems. Fuzzy Sets Syst. **100**(1–3), 45–57 (1998)
16. Jiménez, F., Verdegay, J.: Solving fuzzy solid transportation problems by an evolutionary algorithm based parametric approach. Eur. J. Oper. Res. **117**(3), 485–510 (1999)
17. Kumar, R., Edalatpanah, S.A., Jha, S., Singh, R.: A Pythagorean fuzzy approach to the transportation problem. Complex Intel. Syst. **5**(2), 255–263 (2019)
18. Liu, S.T.: Fuzzy total transportation cost measures for fuzzy solid transportation problem. Appl. Math. Comput. **174**(2), 927–941 (2006)
19. Mahapatra, D.R., Roy, S.K., Biswal, M.P.: Stochastic based on multi-objective transportation problems involving normal randomness. Adv. Model. Optim. **12**(2), 205–223 (2010)
20. Maity, G., Roy, S.K., Verdegay, J.L.: Time variant multi-objective interval-valued transportation problem in sustainable development. Soft Comput. Sustain. **11**(21), 6161 (2019)
21. Malini, S.U., Kennedy, F.C.: An approach for solving fuzzy transportation problem using octagonal fuzzy numbers. Appl. Math. Sci. **7**(54), 2661–2673 (2013)
22. Nagarajan, A., Jeyaraman, K.: Solution of chance constrained programming problem for multi-objective interval solid transportation problem under Stochastic environment using fuzzy approach. Int. J. Comput. Appl. **10**(9), 19–29 (2010)
23. Oheigeartaigh, M.: A fuzzy transportation algorithm. Fuzzy Sets Syst. **8**(3), 235–243 (1982)
24. Pramanik, S., Roy, T.: Multi-objective transportation model with fuzzy parameters: priority based fuzzy goal programming approach. J. Transp. Syst. Eng. Inform. Technol. IT **8**(3), 40–48 (2008)
25. Rani, D., Gulati, T.R.: Uncertain multi objective multi product solid transportation problems. Sadhana **41**(5), 531–539 (2016)
26. Samuel, A.E., Venkatachalapathy, M.: Modified Vogel's approximation method for fuzzy transportation problems. Appl. Math. Sci. **5**(28), 1367–1372 (2011)
27. Sheng, Y., Yao, K.: A transportation model with uncertain costs and demands. Int. J. Inform. **15**(8), 1–7 (2012)
28. Shunmugapriya, S., Uthra, G.: Ranking interval valued fuzzy numbers. Int. J. Sci. Res. Rev. **7**(4), 1232–1242 (2018)
29. Sifaoui, T., Aïder, M.:Uncertain interval programming model for multi-objective multi-item solid transportation problem with budget constraint and safety measure. Soft Comput. 1–25 (2019)
30. Singh, S.K., Yadav, S.P.: Intuitionistic fuzzy multi-objective linear programming problem with various membership functions. Ann. Oper. Res. **269**(1–2), 693–7707 (2018)
31. Wei, S.H., Chen, S.M.: Fuzzy risk analysis based on interval-valued fuzzy numbers. Exp. Syst. Appl. **36**, 2285–2299 (2009)
32. Yao, J.S., Lin, F.T.: Constructing a fuzzy flow-shop sequencing model based on statistical data. Int. J. Approx. Reason. **29**, 215–234 (2002)
33. Zimmermann, H.I.: Fuzzy programming and linear programming with several objective functions. Fuzzy Sets Syst. **1**(1), 45–55 (1978)

On the Generation of Alternative Solutions for a Perishable Food Distribution Problem

Hanane El Raoui, David A. Pelta, Antonio Rufián-Lizana, Mustapha Oudani, and Ahmed El Hilali Alaoui

Abstract Solving a perishable food distribution problem in a real world setting is a very complex task. This is due to products characteristics, and the requirements of customers. To ensure a safe, quality product with a desired service level, a bunch of specifications should be included during the decision/optimization process. Generally, it's difficult to include all the parameters in the mathematical model. Thus, it's desired to have a set of alternative solutions to select from, that allows the decision maker to consider different perspectives. For this purpose, the current work outlines the application of a Modeling to Generate Alternatives approach, that can generate a set of near optimal solutions, but maximally different from the best one. A General Variable Neighborhood Search GVNS algorithm is applied to solve the problem. We show through computational experiments how the proposed procedure allows to generate a number of diverse solutions in a single run. We also show how the consideration of a fuzzy threshold constraint may allow to obtain interesting solutions over a set of criteria calculated after the optimization process, for an a posteriori analysis.

H. El Raoui · D. A. Pelta (✉)
Department of Computer Science and A.I., Universidad de Granada, Granada, Spain
e-mail: dpelta@ugr.es

H. El Raoui
e-mail: hanane.elraoui@usmba.ac.ma

H. El Raoui
Modeling and Mathematical Structures Laboratory, Sidi Mohamed Ben Abdellah University, Fez, Morocco

A. Rufián-Lizana
Dpto. Estadística e I.O., Universidad de Sevilla, Sevilla, Spain
e-mail: rufian@us.es

M. Oudani
TIC Lab, ESIN, International University of Rabat, Rabat, Morocco
e-mail: mustapha.oudani@uir.ac.ma

A. E. Alaoui
Université Euro Méditerranéenne, Fez, Morocco
e-mail: a.elhilali-alaoui@ueuromed.org

J. L. Verdegay et al. (eds.), *Computational Intelligence Methodologies Applied to Sustainable Development Goals*, Studies in Computational Intelligence 1036,
https://doi.org/10.1007/978-3-030-97344-5_17

Keywords Food distribution · Optimization · Generation of alternatives · A posterior analysis

1 Introduction

In recent years, the global food and grocery retail industry has seen a significant growth. This growth will result in an increasing number of delivery requests and therefore the freight movement growth will produce negative impact on costs and environment.

Nowadays, logistics systems in several sectors are struggling to enhance their sustainability performance, as a result of increasing environmental and societal problems such as population density, climate change, pollution. The 2030 Agenda for Sustainable Development, adopted by all United Nations Member States in 2015, provides a shared blueprint for peace and prosperity for people and the planet, now and into the future. At its heart are the 17 Sustainable Development Goals (SDGs). This research is related with some sustainability aspects (economic, environmental, social issues) associated to the movement of goods through the food supply chain.

The distribution of perishable food has become challenging due to the characteristics of these goods. Indeed, it is generally acknowledged that the handling of perishable food differs from the the processing of other products. Perishables has a short shelf life, and their quality decay over the time during the distribution process up to the consumption. Consequently, ensuring food quality has become a top priority that must be met in order to satisfy customers.

In addition to product quality, customer's satisfaction is often influenced by the punctuality of service delivery. Therefore, its becoming increasingly important that the customer's request arriving within the time frame he/she specified.

In real life, other specifications may need to be included in the decision making/optimization process, making of the latter a quite complex task. For instance, since we are dealing with a time sensitive products, customers such as restaurants may adopt the Just in Time JIT inventory management strategy to ensure the freshness of their meals. As a result, the customer may specify a target time to receive the product, and the fulfillment of this condition is a key factor to improve the service level. Another delivery constraint that may arises in real life is related to the priority level of customers. This issue of priority may be very significant when it comes to delivery to small restaurants. In fact, since their storage capacities are small, such restaurants must be cyclically supplied. Therefore, during the scheduling process, the DM need to prioritise this category of customers.

The distribution of perishable food in a real world setting involves a set of complex specifications that may be difficult to integrate into the mathematical models. Such parameters are often conflicting, and characterized by unquantifiable parameters [6, 9, 11].

The single best solutions provided by optimization techniques for a mathematical model can not be considered as optimal for the real world, since they may lack information about certain unmodelled features. Therefore, it's desirable to have a set of alternative solutions that allows the decision maker to analyse the problem from different perspectives, instead of having a single solution that focuses only on one objective, and restricts the reality. The approaches devoted to this purpose are generally known under the name *Modeling To Generate Alternatives* (MGA, in what follows).

The concept behind MGA approach is to generate a set of alternative solutions that are near to, but maximally different from the best solution. In such a way, these solutions need to perform well with respect to the modeled objective, while being interesting with respect to other criteria that can not be modelled.

Most of the MGA applications proposed in the literature use incremental approaches to generate alternatives, by iteratively re-running the optimization algorithm whenever a new solution is needed [1, 2, 7, 10, 18].

The aim of this work is to approach the solution of a perishable food distribution problems by means of the MGA approach. Initially, a transportation cost minimization problem is solved using a General Variable Neighborhood Search (GVNS) metaheuristic. Then, the obtained solution is used as a reference to generate (using again GVNS) a set of maximally different solutions for which, other criteria are taken into account.

Compared to the earlier MGA algorithms proposed in the literature, the proposed procedure can generate a number of diverse solutions in a single run. Another novelty, is the use of a fuzzy constraint in the admissible difference between the reference and alternatives transportation costs.

This chapter is structured as follows: In Sect. 2 we present the 'modeling to generate alternatives' approach. In Sect. 3 we state the problem of perishable food distribution, propose a mathematical formulation of the problem, and an heuristic to solve it. The applied modeling to generate alternatives approach used fro perishables distribution is detailed in Sect. 4. Section 5 provides the criteria used to analyse the alternative solutions a posteriori. A full example showing all the steps of the proposed GVNS-MGA approach is shown in Sect. 6. The last section wraps up the chapter.

2 The 'Modeling to Generate Alternatives' Approach

In most 'real world' decision-making and/or optimization problems, countless decision criteria and objectives can not be evidently included in the model formulation, and if included it may increase the complexity of the model up to a point where it can not be solved. Final decisions are often made based not only on modelled objectives, but also on more subjective decision makers goals and preferences [7].

Almost all of mathematical optimization algorithms have been devoted to finding single optimal solutions to single-objective problems or, in the best case, to determine sets of non-inferior solutions to multi-objective formulations [2, 9, 15]. We should always remember that we are, hopefully, obtaining the optimum solution for the model, and that it would be debatable whether it is the best for the real-world problem setting from different perspectives [2, 6, 9, 10].

Assuming the existence of unmodelled goals implies that non-conventional approaches are needed, not only to search the decision space for optimal solutions, but also to explore the decision region for alternative solutions. Such solutions can be sub-optimal for the model point of view, but valuable from additional perspectives.

The 'Modeling to generate alternatives' (MGA, in what follows) approach, proposed by [2, 10, 19], tries to generate a set of solutions that are quantifiable good across all the modelled goals while remaining as different as possible from each other.

Let's assume that the best solution from a mathematical model is X^*, withbreak $Z^* = F(X^*)$ as the objective value. To generate alternative solutions X that are maximally different from the best solution X^*, the following problem is proposed:

$$\mathrm{Max}\Delta(X, X^*) \tag{1}$$

Subject to:

$$X \in D \tag{2}$$

$$|F(X) - Z^*| \leq T \tag{3}$$

where Δ represents the difference between the alternative solution and the best one. The parameter T is the tolerance threshold related to the optimal objective value Z^* and should be defined by the decision maker.

3 A Model for Perishable Foods Distribution

3.1 *Problem Statement*

The sustainability of perishable food distribution is difficult to optimize owing to temperature control requirements. A perishable product's shelf life is heavily influenced by temperature. The latter rises during the transportation process, and when the vehicle stops to make a delivery. As a result, an estimated damage of 8–23% may occurs [13]. This deterioration in quality increases the risk of food waste.

A well designed distribution plan that emphasizes short transit times and distances can preserve the quality of products, and reduce the amount of wasted products. However, improving the sustainability of perishables distribution network necessitates balancing several competing goals, such as minimizing the travel costs (e.g., fuel consumption costs, refrigeration costs, the damage cost), maximize the average freshness, meeting consumers requirements to ensure their satisfaction (e.g., on-time delivery, good service level), and reducing environmental effect.

These trade-offs highlight the need of including multiple objectives when studying the issue of perishables distribution. However, it is not evident how to include all these conflicting goals in a single model, due to the complexity of the problem when considering a large number of parameters. To deal with this issue, we propose an initial model that focus on the travel cost minimization. Afterwards, we propose the application if a Modeling to Generate Alternatives approach to obtain other solutions that can be assessed with respect to the unmodelled objectives.

3.2 *Model Formulation and Solution Procedure*

The problem is represented as a direct graph denoted by $G = (V, A)$, being $V = \{0, 1, \ldots, n+1\}$ the set of vertices, nodes 0 and $n+1$ represent the depot, and the remaining nodes $C = \{1, \ldots, n\}$ are the customers to serve. Each customer i should be served within a specific time slot defined by $[e_i, l_i]$.

The problem is modeled under the assumption of fleet homogeneity: all the refrigerated trucks have the same refrigeration characteristics, the same load capacity, and the same constant velocity.

Decision variables
$x^k_{(i,j)}$: a 0–1 decision variable, equal to 1 in case the truck k travels on arc (i, j), 0 otherwise.
y^k_i: a 0–1 decision variable, equal to 1 in case the customer i is served by the truck k, 0 otherwise.
t^k_i: a continuous decision variable representing the starting service time at customer i using the truck k.

Parameters
$C_{(i,j)}$: the transportation cost from node i to j.
F: fixed cost associated to a vehicle.
$T_{(i,j)}$: the travel time from i to j.
$d_{(i,j)}$: the distance between node i and j.
C_e: the cost per unit time for the refrigeration during the transportation process.
U_i: the unloading time at customer i.
C'_e: the unit refrigeration cost during the unloading.
S_i: the necessary time to serve customer i.
q_i: the demand of customer i.
Q: the loading capacity of trucks.
K: the fleet of trucks.
$[e_i, l_i]$: the time window of customer i.

The Objective Function
The objective is to minimize the travel cost which includes: the fixed costs for using a vehicle denoted C_1, transportation costs C_2 and the refrigeration costs C_3.

- **The fixed costs**

The fixed costs of a vehicle are not related to the mileage, and represent the maintenance, and depreciation costs. Assuming that the depot has k refrigerated trucks to provide distribution services for the set of customers.
F represents the fixed cost related to a vehicle. The fixed costs can be formulated as:

$$C_1 = F \sum_{k \in K} y_0^k \tag{4}$$

- **The transportation costs**

We denote these costs by C_2, They are proportional to the vehicle mileage. We consider only the fuel consumption cost and express it as:

$$C_2 = \sum_{k \in K} \sum_{i \in V} \sum_{j \in V} C_{(i,j)} x_{(i,j)}^k d_{(i,j)} \tag{5}$$

- **The refrigeration costs**

Include two types of costs: the costs incurred by the vehicle's energy usage to maintain a specific temperature in the process of transportation, in addition to the costs of extra energy during the unloading.
The refrigeration costs during transportation process denoted C_{31} are expressed as follows:

$$C_{31} = C_e \sum_{k \in K} \sum_{i \in V} \sum_{j \in V} x_{(i,j)}^k T_{(i,j)} \tag{6}$$

The cost of energy supplied during the unloading C_{32} is expressed as:

$$C_{32} = C_e' \sum_{k \in K} \sum_{i \in V} y_i^k U_i \tag{7}$$

the total refrigeration cost

$$C_3 = C_{31} + C_{32} \tag{8}$$

3.2.1 Mathematical Model

Considering the objectives described above, the problem can be formulated as a Mixed integer Program (MIP) giving by the following:

$$\text{Min } C_1 + C_2 + C_3 \tag{9}$$

subject to:

$$x_{(i,i)}^k = 0, \quad \forall i \in V, \quad \forall k \in K \tag{10}$$

$$x^k_{(0,n+1)} = 1, \quad \forall k \in K \tag{11}$$

$$\sum_{k \in K} y^k_i = 1, \quad \forall i \in C \tag{12}$$

$$\sum_{i \in V} x^k_{(i,j)} = y^k_j, \quad \forall j \in C, \quad \forall k \in K \tag{13}$$

$$\sum_{j \in V} x^k_{(i,j)} = y^k_i, \quad \forall i \in C, \quad \forall k \in K \tag{14}$$

$$\sum_{j \in v} x^k_{(0,j)} \leq 1, \quad \forall k \in K \tag{15}$$

$$\sum_{i \in v} x^k_{(i,n+1)} \leq 1, \quad \forall k \in K \tag{16}$$

$$\sum_{i \in C} D_i y^k_i \leq Q, \quad \forall k \in K \tag{17}$$

$$t_i + S_i + T_{(i,j)} x^k_{(i,j)} - M(1 - x^k_{(i,j)}) \leq t_j, \quad \forall i, j \in V, \quad \forall k \in K \tag{18}$$

$$e_i \leq t^k_i \leq l_i, \quad \forall i \in C, \quad \forall k \in K \tag{19}$$

$$\sum_{k \in K} y^k_0 \leq K \tag{20}$$

$$\sum_{j \in V} x^k_{(i,j)} \leq y^k_i, \quad \forall i \in V, \quad \forall k \in K \tag{21}$$

$$x^k_{(i,j)} \in \{0, 1\}, \quad \forall i, j \in V, \quad \forall k \in K \tag{22}$$

$$t^k_i \geq 0, \quad \forall i \in C, \quad \forall k \in K \tag{23}$$

$$y^k_i \in \{0, 1\}, \quad \forall i \in C, \quad \forall k \in K \tag{24}$$

The objective (9) corresponds to the total cost minimization. Constraint (10) avoids going from a point to itself. Constraint (11) avoids going from the depot to node $n+1$ which represent the depot. Constraint (12) states that only one vehicle visits each customer exactly once. Constraints (13) and (14) ensure that each vehicle that arrives at a customer must depart for another destination. The vehicle departs from the depot 0, and returns to the depot node $n+1$ according to the Constraints (15) and (16). Constraint (17) guarantees the respect of vehicle loading capacity.

The connection between the start service at a specific customer and the next one is established by Constraint (18). Constraint (19) is the time window constraint. Constraint (20) guarantees that the number of trucks departing from the depot do not exceed the available fleet of trucks K. Constraint (21) states that the vehicle must traverse the arc (i, j) in order to serve the customer j.

3.2.2 General Variable Neighborhood Search

To solve the proposed problem we use the General Variable Neighborhood Search GVNS Meta-heuristic. In GVNS, Variable Neighborhood Descent (VND) is used for local search. When designing a GVNS, one should specify the following choices: the number of neighborhood structures, the order in which they will be explored, how the initial feasible solution is generated, the acceptance criterion and the stopping conditions. These elements define the configuration of the GVNS.

The main steps of GVNS are shown in Algorithm 1. GVNS starts by an initialization phase (line 0) in which a feasible solution is generated (we detail in the next subsection the heuristic used for initialization). The best solution is set as the first feasible solution (line 1). After selecting the neighborhood structures for the shaking stage N_s, and for local search N_k (line 2), the stopping condition is then chosen (line 3). The stopping condition corresponds to a number of iterations M that will be set in the computational experience. The loop corresponding to lines 4–25 is repeated M times. The shaking process (line 7), the local search (line 8), and the move decision (line 18) are repeated until $S = S_{max}$. In the shaking phase a solution X' is generated randomly at the Sth neighborhood of $X^*(X' \in N_s(X^*))$. Then the local search is performed to find a better solution X'' from X' using the N_K neighborhood structures.

4 MGA for Perishable Food Distribution

In order to generate a set of alternative solutions to the problem described, a MGA approach is adopted. As explained in Sect. 2 the objective in a MGA is to maximize the difference between the best (or a reference) solution, and the obtained alternatives (Eq. 1).

Therefore, we first need a way to measure the difference between the routing solutions. In this work we use the Jaccard's coefficient to measure the dissimilarities between routes. This method is detailed in Sect. 4.1.

The good alternatives are the ones that minimize the Jaccard's coefficient while respecting the threshold Constraint 3. So in second place, we describe how such constraint is managed as a fuzzy constraint and we propose to deal with it using the parametric approach.

Finally, the whole solving procedure to generate alternative routes using GVNS is described.

Algorithm 1: GVNS

```
Initialization;
Set the current best solution X* ← X0;
Select the set of neighborhood structures to be used in shaking phase N_S(s = 1, ..., s_max)
  and the set of structures for local search N_k (k = 1, ..., k_max);
Choose a stopping condition M;
while the stopping condition M is not meet do
    S ← 1;
    repeat
        Shaking : randomly create a solution X' from the S^th neighborhood of N_s(X*), the
        current best solution;
        Perform Local search using Variable Neighborhood Descent;
        set K ← 1 ;
        repeat
            explore the K^th neighborhood searching for the best neighbor X" of X' in N_k(X');
            if f(X") ≤ f(X') then
                Update x' ← X" and K ← 1;
            else
                K ← K + 1
            end
        until K = K_max;
        Move or not;
        if the local optima is better that the current best solution then
            Update x* ← X' and S ← 1
        else
            S ← S + 1
        end
    until S = S_max;
end
```

4.1 *Measuring Similarity of Solutions*

Several methods and indices have been proposed in the literature on vehicle routing problems in order to quantify the dissimilarities between solutions. The Hamming distance is the most common metric. In our study, we use the Jaccard's similarity coefficient. The Jaccard's similarity coefficient is a statistic metric used for comparing the similarity of two sets. It is defined as the cardinality of the intersection of the sets divided by the cardinality of the union of them, i.e.

$$J(A, B) = \frac{|A \cap B|}{|A \cup B|} \tag{25}$$

It is easy to see that if sets A and B contain the same elements, $A = B = A \cap B = A \cup B$, then Jaccard's similarity coefficient $J(A, B) = 1$. On the other hand, if sets A and B do not share any element at all, $|A \cap B| = 0$, so $J(A, B) = 0$.

We can now define the similarity between two routing solutions according to the Jaccard's similarity coefficient as the ratio of the number of shared arcs to the number of total arcs used in both solutions. Let $y_{ijk} = 1$ if arc (i, j) from vertex i to vertex j is used by any vehicle in solution r_k, and $y_{ijk} = 0$ otherwise. Then the similarity ζ_{pq} between solutions p and q is

$$\zeta_{pq} = \frac{\sum_{i=0}^{n}\sum_{j=0}^{n} y_{ijp} \cdot y_{ijq}}{\sum_{i=0}^{n}\sum_{j=0}^{n} sign(y_{ijp} + y_{ijq})} \tag{26}$$

where $y_{ijp} \cdot y_{ijq} = 1$ iff arc (i, j) is used by both solutions, and $sign\left(y_{ijp} + y_{ijq}\right) = 1$ if any of the solutions use it. If solutions p and q are the same, the sum in the numerator will equal the sum in the denominator, and therefore $\zeta_{pq} = 1$. On the other hand, if they are two completely different solutions with no arc in common, the numerator will equal 0, and then $\zeta_{pq} = 0$

4.2 *Fuzzy Constraint for the Transportation Cost Difference*

As stated before, besides the maximization of the solutions differences, we need to deal with the following constraint:

$$|F(X) - Z^*| \leq T \tag{27}$$

where T is the allowed increase in the transportation cost for the alternative solutions.

As the solutions represents routes on a map, it is not possible to assess if other alternatives are available so it has sense to define:

$$|F(X) - Z^*| \leq_f T \tag{28}$$

where $\leq_f$ implies that restriction in the threshold could be partially satisfied.

Then the membership function that represents the satisfaction degree of the fuzzy constraint is shown in Fig. 1 and it is defined as:

$$\mu(\tau) = \begin{cases} 1 & ; \tau < T \\ f(\tau) & ; T \leq \tau < T' \\ 0 & ; \tau \geq T' \end{cases} \tag{29}$$

where $T' = T + \Delta^t$ is the endurable distance threshold, $\tau = |F(X) - Z^*|$ is the difference of distance between the reference solution and the obtained alternative. $f(\tau)$ is a non-increasing function that is defined as:

$$f(\tau) = 1 - \frac{\tau - T}{\Delta^t} \tag{30}$$

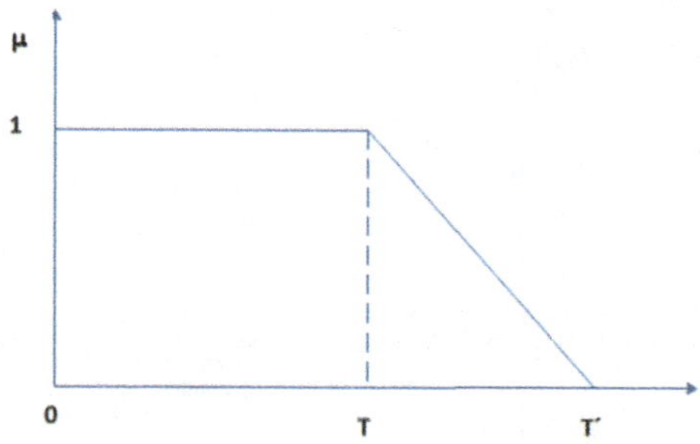

Fig. 1 The satisfaction degree of the fuzzy constraint

To our knowledge, no approach can solve the problem in its fuzzy form directly. The original fuzzy problem, on the other hand, can be transformed into a series of crisp problems.The Parametric Approach [14] can be used for this purpose, transforming the fuzzy problem into a set of crisp problems using the concept of $\alpha - cuts$. The problem is solved afterwards for each value of α using an optimization technique. This approach has been widely used in fuzzy optimization problems as in [3–5, 12]. In the context of routing problems, the approach was applied in [8, 14], just to cite some examples.

After some basic transformations, the fuzzy constraint is transformed into:

$$|F(X) - Z^*| \leq T + \Delta^t(1 - \alpha) \tag{31}$$

4.3 Solving Strategy

Up to this point, all the elements of the proposal presented. The problem to be solved is

$$\text{Min: Jaccard}(X, X^*) \tag{32}$$

$$|F(X) - F(X^*)| \leq T + \Delta^t(1 - \alpha) \tag{33}$$

The whole approach is depicted in Fig. 2.

At the beginning, it starts solving the initial problem using the GVNS. After several runs, the best solution obtained will be used as the reference solution X^*. Then, the new problem is solved by an adapted GVNS using the new objective function and constraint.

The generated alternatives (the solutions obtained by GVNS) are then evaluated according to new criteria not covered in the original problem. These criteria are: the damage cost, the average freshness, the service level, and the tardiness. Then, an a posteriori analysis will allow the decision maker to study the problem from different perspectives, in order to select an appropriate solution according to his/her preferences.

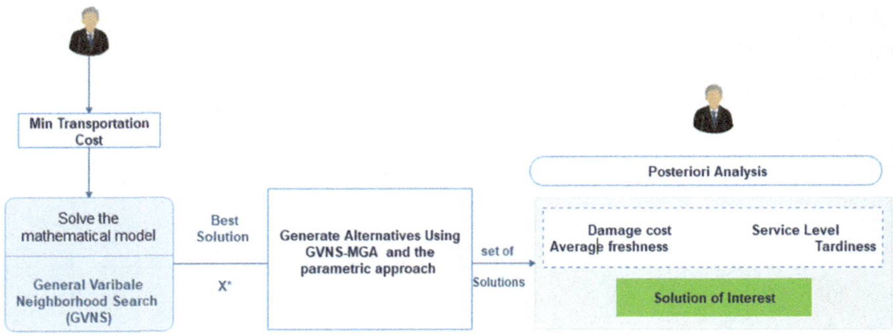

Fig. 2 The adopted solving strategy to generate alternative solutions

5 Additional Quality Criteria

The alternative solutions generated will be evaluated for a set of criteria relevant to perishable food distribution. It is expected that they can help the decision maker in selecting the most appropriate solution according to his/her preferences. In this work we consider the four criteria described below:

5.1 *Damage Cost*

The quality of perishable foods degrades with the progression of time and temperature fluctuations during transportation and handling. Damage costs are incurred when product quality falls below a certain threshold. The following feature can be used to express the quality of refrigerated products [16]:

$$D_t = D_0 e^{-\partial t}$$

where D_t, and D_0 are respectively the quality of product at time t and from the depot 0. The parameter ∂ is the product's spoilage rate, which is assumed to increase as the temperature rises. As a result, we distinguish between damage cost during distribution C_{dmg}, and the damage cost during the unloading C'_{dmg} due to the temperature changes (∂ varies also). The damage cost C_{dmg} is expressed as follows:

$$C_{dmg} = \sum_{k \in K} \sum_{i \in C} y_i^k P q_i \left(1 - e^{-\partial_1 (t_i^k - t_0^k)}\right) \quad (34)$$

where the coefficient ∂_1 reflects the spoilage rate of product when the vehicle is locked, t_i^k the arrival time of vehicle k at customer i, t_0^k the departure time of vehicle k from the depot,and y_i^k is 0–1 variable taking the value 1 in case the vehicle k is servicing the customer i, and 0 otherwise.

The damage cost during the unloading C_{42} is defined as:

$$C'_{dmg} = \sum_{k\in K}\sum_{i\in C} y_i^k P q_{in}\left(1 - e^{-\partial_2 S_i}\right) \tag{35}$$

With q_{in} the remaining quantity of product after servicing the customer i, the necessary time to serve is customer i is S_i , and ∂_2 is the spoilage rate when the vehicle is opened.

5.2 Average Freshness

The average freshness denoted by AVGF can be defined as in [17] by the following formula:

$$AVGF = \frac{\sum_{k\in K}\sum_{i\in C} y_i^k q_i e^{-\partial_1 (t_i^k - t_0^k)}}{\sum_{i\in C} q_i} \tag{36}$$

5.3 Service Level

This criterion reflects the fact that customers may want to obtain their products at a certain time so that they can begin planning meals, for example. Furthermore, restaurants often use the Just-in-Time inventory management approach to reduce storage costs. As a result, rather than servicing consumers within a time frame, we put as much emphasis on completing customer requests as quickly as possible. We use the target time as an indicator to determine the service level using the following function:

$$SL_i(t_i) = \begin{cases} 0 & t_i < e_i \\ f(t_i) & e_i \le t_i < T_g^i \\ 1 & t_i = T_g^i \\ g(t_i) & T_g^i < t_i \le l_i \\ 0 & t_i > l_i \end{cases} \tag{37}$$

$SL_i(t_i)$ represents the service level ensured for customer i if we deliver his demand at time t_i. T_g^i is the target time, e_i and l_i are respectively the lower, and upper bounds of the time window.

The function f is a non-decreasing function, while g is a decreasing function shown in Fig. 3, and defined as follows:

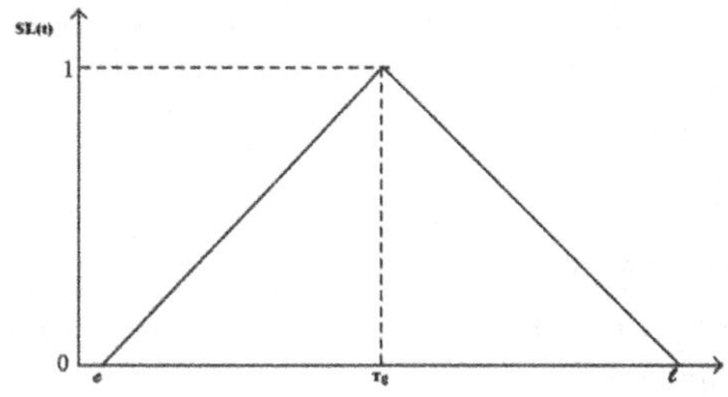

Fig. 3 Service level measuring function

$$f(t_i) = \frac{t_i - e_i}{T_g^i - e_i} \tag{38}$$

$$g(t_i) = \frac{l_i - t_i}{l_i - T_g^i} \tag{39}$$

A good alternative solution is the one that maximizes the following function:

$$SL = \sum_{i \in C} SL_i(t_i) \tag{40}$$

5.4 *Total Tardiness*

With the aim of increasing customer satisfaction levels. we consider that customers are served according to their priority level. When consumers have varying levels of interest, an issue like this occurs. The use of priority indexes in our problem is motivated by the fact that a customer may have a variety of locations to service and expectations for serving each node. Customers' priorities may be set by the decision-maker. We assume that we have a pre-defined set of priorities. To consider priority indexes, we define a precedence matrix P, where $P_{ij} = 1$ indicates that the customer i should be supplied before the customer j, and $P_{ij} = 0$ if customer i might be supplied after customer j.

A good solution is the one that minimize the tardiness as much as possible. Tardiness is increased when a lower-priority customer is served before a higher-priority customer, and can be expressed using the following formula:

$$\sum_{i \in C} \sum_{j \in C} t_j - t_i \tag{41}$$

with t_j, t_i are the arrival times at customer j and i respectively.

6 Application Example

6.1 Data and Parameters

The proposed approach is illustrated in a problem with 50 customers. Data is taken from instance RC101, from Solomon's data set. For customer's target time, we assume that it's the midpoint of the corresponding time window. The refrigerated vehicles used to deliver products have a fixed cost of 25 €, and the fuel consumption is estimated to 3 €/km. For the GVNS parameters, we use the following values: $\delta_1 = 0, \delta_2 = 1, \Lambda = 2, \mu = 1$. The stopping criteria M which correspond to the number of iterations, is fixed to 10. The other parameters are: $P = 20$ €/Unit, $\partial_1 = 0.002$, $\partial_2 = 0.003$, $C_e = 0.03$ €/unit, $C'_e = 0.04$ €/unit and $\delta = 0.8$

The experiments were carried out on an Intel Core i7 processor with 1.99 GHz speed and 8 GB RAM.

6.2 Computational Results

To generate a set of alternative solutions, we start by finding the best solution. For so, we solve the proposed mathematical model using GVNS. The best solution is obtained by performing 10 runs, and it is used to initialize the GVNS in order to obtain other solutions that are maximally different from it.

Then, the new problem is solved for different values of α, with $\alpha \in [0, 0.1, \ldots, 1]$. For each value of α, we run the algorithm 10 times. The algorithm provides the routing solutions with their corresponding objective values (the total travel cost). We start by running the algorithm for $\alpha = 1$ and report the alternative solutions, Afterwards we decrease the value of α by 0.1 and re-run the algorithm.

Table 1 shows the results obtained. For each α, just different solutions are reported. Every solution is compared against the reference solution in terms of the Jaccard's coefficient and all the other criteria: the damage cost, the average freshness, the service level, and the tardiness. A positive value refers to an increment, and negative one to a decrement. The interpretation of the percentages depends on the criteria evaluated: a positive percentage in the transportation cost, damage cost and tardiness is not desirable. However, in the average freshness and service level positive values means improvements.

We can observe that the proposed approach allows us to generate a set of diverse solution that shows a different behaviour for each criterion. Thus, the decision maker (DM) can analyse the alternatives from different perspectives and select the appropriate solution depending on his preference. Indeed, if the DM has an economic-centric preference he would select the cheapest solution considering both the transportation and damage cost. Otherwise, if the DM is more focused on the quality of product and service, he would opt for the routing solution that maximize the average freshness, the service level, and minimize the tardiness.

Table 1 Experiments results of 10 runs for each value of α

		Jaccard coefficient	Travel cost	Damage cost	Average freshness	Service level	Tardiness
	Ref Solution	1	3158.98	6297.407	31.99	14.66	226698.47
$\alpha = 1$	s_1^1	0.78	0.08	−0.09	−0.15	5.51	1.39
	s_1^2	0.51	0.75	0.41	−1.68	19.15	5.07
	s_1^3	0.81	0.31	0.22	0.05	14.49	−15.44
	s_1^4	0.84	0.87	0.26	−2.09	2.12	2.33
	s_1^5	0.87	0.64	2.07	−1.75	−5.05	3.06
	s_1^6	0.81	0.35	0.27	−1.59	−0.86	−4.79
	s_1^7	0.81	0.56	1.82	−2.07	−8.25	−3.62
$\alpha = 0.9$	–	–	–	–	–	–	–
$\alpha = 0.8$	s_3^1	0.63	0.30	−1.07	0.02	−0.57	−11.93
	s_3^2	0.49	0.53	1.37	−2.04	−3.44	3.36
	s_3^3	0.66	0.95	0.17	−2.22	7.65	3.74
$\alpha = 0.7$	s_4^1	0.81	0.21	−0.99	0.15	−6.10	−13.43
	s_4^2	0.63	0.65	1.73	−2.19	−2.71	4.24
$\alpha = 0.6$	s_5^1	0.68	1.08	−0.73	−1.94	−3.98	−11.16
$\alpha = 0.5$	s_6^1	0.66	0.66	0.49	−1.54	13.63	−20.08
$\alpha = 0.4$	s_7^1	0.66	0.78	0.83	−1.91	−14.35	−10.46
$\alpha = 0.3$	s_8^1	0.63	0.40	0.14	−0.08	20.01	−13.89
	s_8^2	0.63	0.43	0.18	−1.72	4.67	−3.40
$\alpha = 0.2$	–	–	–	–	–	–	–
$\alpha = 0.1$	–	–	–	–	–	–	–
$\alpha = 0$	s_{11}^1	0.68	1.18	0.48	−2.04	16.60	10.88

By analysing the given solutions, we have found that the more we relax the problem (in terms of α), we get new interesting solutions. To illustrate let's consider the solution S_1^3 given by $\alpha = 1$ (when no flexibility in the threshold is allowed), compared to the reference solution (the best). The latter improves slightly the average freshness by 0.05%, but the service level is increased by 14.49%, and the tardiness is reduced by 15.44%. Which can be considered as significant enhancement. For $\alpha = 0.7$ a new solution is generated S_4^1, that has the same Jaccard coefficient but performs differently for the given criteria. S_4^1 reduce the damage cost by 1%, the tardiness by 13.43%, and improve slightly the average freshness by 0.15%. Furthermore, the obtained enhancement implies just an increment of 0.21% in the travel cost, which implies that this solution could be a compromise one for that has economic focus as well as customer satisfaction focus. For $\alpha = 1$ we have found solutions that improves only one criterion compared to the reference solution such as S_1^6 which only reduce the tardiness. Yet, when we relax the problem for $\alpha = 0.3$ we obtain solutions S_8^1 and S_8^2 that have almost the same travel cost as S_1^6, but too different from the best solution with a Jaccard equal to 0.63. S_8^1 and S_8^2 respectively improves the service

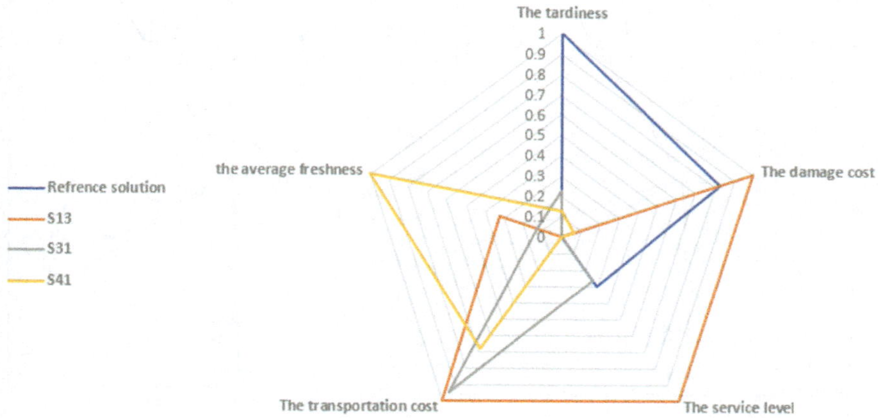

Fig. 4 The three compromised solutions compared to the reference solution for each criteria

level by 20% and 4.67%, and reduce the tardiness by 13.89% and 3.4%. We can conclude that by considering fuzzy threshold we increase our chances of getting diverse solutions. These solutions performs well for a set of criteria that are relevant for the studied problem, and help the decision maker in analysing different aspect instead of considering only the cost criteria.

For the set of alternatives we can observe that there are some solutions producing improvements in only one criterion, other for two criteria, and best ones allows an improvement in three criteria. By focusing on the latter, we have found three solutions S_1^3, S_3^1 and S_4^1 that are shown in Fig. 4 in order to analyse them in terms of the evaluation criteria, and help the DM in selecting the compromised one. From the plot we can see that the solution S_1^3 is the best one in terms of service level improvement, and in terms of average freshness solution S_4^1 is the best. However, solution S_3^1 is the one that minimize at most the damage cost. Thus, we have three different good solution dependent to the DM preferences, but if the latter focus most on customer satisfaction then the solution S_1^3 is the best one because instead of the service level, this solution is the one that minimize at most the tardiness.

Another interesting solution for the decision maker is S_1^1 shown in Fig. 5. This solution is quite different from the reference solution (Jaccard coefficient 0.78), but interestingly allows an improvement in the service level, and reduce the damage cost with almost the same transportation cost. S_1^1 lead to a slight decrement in the average freshness, and an increase of the tardiness, but it can be considered as a good alternative if the DM has an economic and customer-centric perspective.

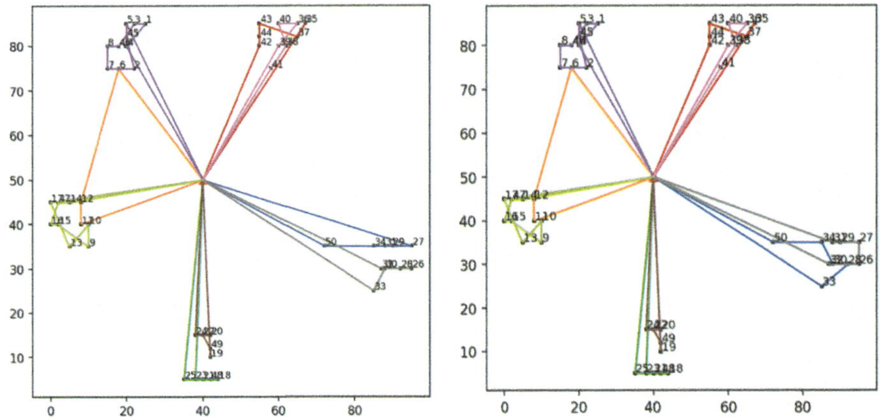

Fig. 5 Left: the reference solution. Right: an alternative one S_1^1

7 Conclusion

Real world decision making for perishable food distribution is a very complicated process as it's impacted by a set of parameters that exists in real life, and need to be considered, but it can not be included in a model due to its complexity. Moreover, these specifications can be unacquainted at the modelling stage. In this case, it's doubtful that a single solution would be able to meet all the real specifications. Therefore, it's desirable for the decision maker to have a set of solutions that are good enough for the modeled objective, but can perform also well when it comes to other unmodelled criteria. This ability to consider different perspectives will grant a decision adapted to the real context.

This work propose a MGA approach with a fuzzy constraint and a GVNS metaheuristic algorithm that can generate a set of near-optimal but maximally different solutions. Experimental results demonstrated the ability of the proposed approach to generate alternatives that provides different perspectives for the studied problem.

Acknowledgements D. Pelta acknowledges support from projects PID2020-112754GB-I0, MCIN/AEI /10.13039/501100011033 and B-TIC-640-UGR20 from UGR-FEDER-2020 Research Programme.

References

1. Baugh Jr, J.W., Caldwell, S.C., Brill Jr., E.D.: A mathematical programming approach for generating alternatives in discrete structural optimization. Eng. Optim. **28**(1-2):1–31 (1997)
2. Brill, E.D., Jr., Chang, S.-Y., Hopkins, L.D.: Modeling to generate alternatives: The HSJ approach and an illustration using a problem in land use planning. Manag. Sci. **28**(3), 221–235 (1982)

3. Brito J, Martínez, F.J., Moreno, J., Verdegay, J.L.: A grasp–vns hybrid for the fuzzy vehicle routing problem with time windows. In: International Conference on Computer Aided Systems Theory, pp. 825–832. Springer (2009)
4. Brito, J., Martínez, F.J., Moreno-Pérez, J.A., Verdegay, J.L.: Aco-grasp-vns metaheuristic for vrp with fuzzy windows time constraints. In: International Conference on Computer Aided Systems Theory, pp. 440–447. Springer (2011)
5. Brito, J., Moreno, J.A., Verdegay, J.L.: Transport route planning models based on fuzzy approach. Iranian J. Fuzzy Syst. **9**(1), 141–158 (2012)
6. Brugnach, M., Tagg, A., Keil, F., de Lange, W.J.: Uncertainty matters: computer models at the science-policy interface. Water Resour. Manag. **21**(7), 1075–1090 (2007)
7. Gunalay, Y., Yeomans, J.S.: Simulation-optimization techniques for modelling to generate alternatives in waste management planning. J. Appl. Oper. Res. **3**(1), 23–35 (2011)
8. Guzmán, V.C., Pelta, D.A., Verdegay, J.L.; Fuzzy maximal covering location models for fighting dengue. In: 2016 IEEE Symposium Series on Computational Intelligence (SSCI), pp. 1–7. IEEE (2016)
9. Janssen, J., Krol, M., Schielen, R., Hoekstra, A.: The effect of modelling quantified expert knowledge and uncertainty information on model based decision making. Environ. Sci. and Policy **13**(3), 229–238 (2010)
10. Loughlin, D.H., Ranjithan, S.R., Brill, E.D., Jr., Baugh, J.W., Jr.: Genetic algorithm approaches for addressing unmodeled objectives in optimization problems. Eng. Optim. **33**(5), 549–569 (2001)
11. Matthies, M., Giupponi, C., Ostendorf, B.: Environmental decision support systems: current issues. In: Methods and Tools (2007)
12. Melian, B., Verdegay, J.L.: Using fuzzy numbers in network design optimization problems. IEEE Trans. Fuzzy Syst. **19**(5), 797–806 (2011)
13. Osvald, A., Stirn, L.Z.: A vehicle routing algorithm for the distribution of fresh vegetables and similar perishable food. J. Food Eng. **85**(2), 285–295 (2008)
14. Verdegay, J.: Fuzzy mathematical programming. In: Fuzzy information and decision processes. mm gupta and e. sanchez (1982)
15. Walker, W.E., Harremoës, P., Rotmans, J., Van Der Sluijs, J.P., Van Asselt, M.B., Janssen, P., Krayer von Krauss, M.P.: Defining uncertainty: a conceptual basis for uncertainty management in model-based decision support. Integr. Assess. **4**(1), 5–17 (2003)
16. Wang, S., Tao, F., Shi, Y., Wen, H.: Optimization of vehicle routing problem with time windows for cold chain logistics based on carbon tax. Sustainability **9**(5), 694 (2017)
17. Wang, X., Wang, M., Ruan, J., Li, Y.: Multi-objective optimization for delivering perishable products with mixed time windows. Adv. Prod. Eng. Manag. **13**(3), 321–332 (2018)
18. Zechman, E.M., Ranjithan, R.S.: Generating alternatives using evolutionary algorithms for water resources and environmental management problems. J. Water Resour. Plann. Manag. **133**(2), 156–165 (2007)
19. Zechman, E.M., Ranjithan, S.R.: An evolutionary algorithm to generate alternatives (EAGA) for engineering optimization problems. Eng. Optim. **36**(5), 539–553 (2004)

Solving Two-Stage Multi-objective Transportation Problem Using Goal Programming and Its Application to Sustainable Development

Gurupada Maity and Sankar Kumar Roy

Abstract In modern civilization, people are very fast and busy, and they try to achieve their goals in a short time. Most of the time, they do not consider the damage of nature to access their profits. In this regard, corporate world is moving based on use of technologies and machines. As a consequence, common people are losing their jobs in offices, factories, corn fields, etc. and living their life in anxiety and poverty. Every technology or machine is running by power of fuel, which produces pollution directly or indirectly to the nature. Misusing of technologies and machines for better profit cause high rate of pollution day by day. Now it is the time to recover the nature for sustaining human life in earth along with the consideration of removing among unemployment of youths. In this study, we incorporate a *two-stage Multi-Objective Transportation Problem (MOTP)* and solve it through goal programming. Considering goals corresponding to the objective functions of both stages, we solve *two-stage MOTP* and then derive Pareto-optimal solution. Thereafter, a numerical example based on a real life scenario is presented to show the effectiveness of the study in sustainable development. Finally, conclusions and future directions are presented regarding the present study.

Keywords Sustainable development · Transportation problem · Multi-objective programming · Pareto-optimal solution · Goal programming

1 Introduction

Sustainable development can be defined as the development that meets the needs of the present without compromising the ability of future generations to meet their own needs [34]. To achieve sustainable development, we need to harmonize economic growth, social inclusion and environmental protection. In fact, ensuring energy secu-

G. Maity · S. K. Roy (✉)
Department of Applied Mathematics with Oceanology and Computer Programming, Vidyasagar University, Midnapore 721102, West Bengal, India

J. L. Verdegay et al. (eds.), *Computational Intelligence Methodologies Applied to Sustainable Development Goals*, Studies in Computational Intelligence 1036,
https://doi.org/10.1007/978-3-030-97344-5_18

rity, mitigating climate change, and improving air quality in the most populated areas (urban areas) have become the main concerns of governments. The main challenge for a transportation system to be sustainable is how to design it so that it is economically viable, benefits of all people especially those whose livelihoods depend on a good transportation system and is eco-friendly. Sustainable transportation plays a fundamental role in the socio-economic development of a country and considers in several dimensions: economic development, environmental preservation, and social development. More specifically, a sustainable transportation system is essential to guarantee the facts: (i) mobility and efficient access for all users; (ii) a safe and eco-friendly mode of transportation; (iii) an economically sustainable system; and (iv) public health, since high levels of pollution have been associated with serious health problems, i.e., cardio-respiratory morbidity, mortality, cancer etc.

The classical Transportation Problem (TP) can be described to a special case of linear programming problem and its model is applied to determine how many units of commodity of goods to be shipped from each origin to various destinations, satisfying supply and demand constraints, while it optimizes total cost of transportation. In broad sense TP is used in economic development, industrial management, both passenger and freight transportation modes etc. and their critical role for the economic development of numerous countries.

In a TP, sometimes it is also to be considered that before transporting the goods in the destinations from the sources, goods are to be stored at warehouses from sources. So, the Decision Maker (DM) utilizes the concept for managing *two-stage transportation* in his position which maximizes the profit. A TP is called a *two-stage transportation problem*, if it consists of transporting the goods by two stages, namely, *stage 1 transportation problem* and *stage 2 transportation problem*. The TP which is considered in two stages of collecting the goods in warehouses is called *stage 1 transportation problem*; the TP is chosen in another stage of distributing the goods is referred to as *stage 2 transportation problem*. In Fig. 1, a network is designed to illustrate a *two-stage transportation problem*.

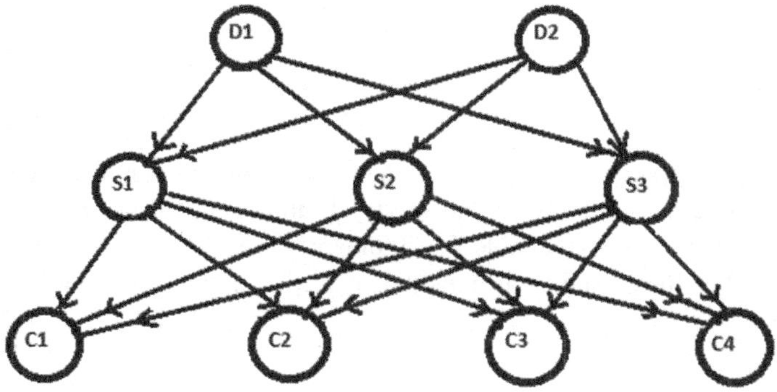

Fig. 1 Network represents a *two-stage transportation problem*

Here, the DM has three warehouses such as S1, S2 and S3 which collect the goods from two cities: D1 and D2 (*stage 1 transportation problem*). Again the goods are supplied to four cities: C1, C2, C3 and C4 (*stage 2 transportation problem* by the help of the DM. So, the goods are in and out at the warehouses S1, S2 and S3 by two-times transportation, and so it is a *two-stage transportation problem*.

Furthermore, single objective TP is not adequate to solve complicated scenarios in real-life situations, therefore we consider MOTP. Multi-objective optimization has been applied in many fields of science, including engineering, economics and logistics where optimal decisions need to be taken in the presence of trade-off between two or more conflicting objectives. Minimizing cost while maximizing comfort while buying a car, and maximizing performance whilst minimizing fuel consumption and emission of pollutants of a vehicle are examples of multi-objective optimization problems involving two and three objectives, respectively. In most of the real-life transportation systems, transportation cost is paid by the purchasers and selling the items get benefits to provide in favor of the supplier. So, supplier would like to optimize benefit and at the same time the purchasers would like to minimize cost of objective functions. So, a conflicting situation occurs in the system. Furthermore in a transportation system, if we consider the production at the supplying origins, then power is required to produce items, again at the time of delivering the items, vehicles are used, as a whole during the energy consumption or transportation environment is polluted. So, there is one more objective function to be added in the transportation, namely, minimizing the pollution. So, it is required to consider a TP in multi-objective environment.

Again, there may be more than one objective function in a *two-stage TP* which makes it a *multi-objective two-stage TP*. In this situation, taking decision is not an easy task for the DM unless he or she selects some proper goals to the objective functions. So, we make a connection between the goals and the objective functions of a *two-stage MOTP*.

For an MOTP, no single solution exists that simultaneously optimizes each objective function. In that case, the objective functions are said to be conflicting, and there exists a number of (possibly infinite) Pareto optimal solutions. A solution is called Pareto optimal, if none of the objective functions can be improved in value without degrading some of the other objective values. In the absence of additional subjective preference information, all Pareto optimal solutions are treated equally good because vectors cannot be ordered completely. Researchers studied MOTP from different viewpoints, and thus, there exist several solution philosophies and goals when setting and solving the problems. The goal may be to find a representative set of Pareto optimal solutions, and/or quantify the trade-off in satisfying the different objectives, and/or finding a single solution that fulfils the subjective preferences of DM. There are several approaches to solve MOTP such as goal programming, fuzzy programming, utility function approach, conic scalarization approach, weighting methods, etc. In particular here we use goal programming technique to achieve Pareto optimal solution.

Goal Programming (GP), an analytical approach, is considered to solve the decision making problem where targets have been assigned to all objective functions which are conflicting and non-commensurable to each other and the DM interests to maximize the achievement level of the corresponding goals. The main concept of GP is that 'minimize the deviation between the achievement goals and the achievement levels'. The mathematical model of multi-objective decision making can be considered in the following form:

$$\begin{aligned} &\text{minimize} \sum_{t=1}^{K} w_t |Z^t(x) - g_t| \\ &\text{subject to } x \in F, \end{aligned}$$

where F is the feasible set and w_t $(t = 1, 2, \ldots, K)$ is the weight attached to the deviation of the achievement function, $Z^t(x)$ is the t-th objective function of the t-th goal and g_t is the aspiration level of the t-th goal, and $|Z^t(x) - g_t|$ represents the deviation of the t-th goal. Let us take $d_t^+ = Z^t(x) - g_t$ if $Z^t(x) \geq g_t$, otherwise $d_t^+ = 0$ $(t = 1, 2, \ldots, K)$. Also, we put $d_t^- = g_t - Z^t(x)$ if $Z^t(x) \leq g_t$, otherwise $d_t^- = 0$. Then, $Z^t(x) - g_t = d_t^+ - d_t^-$, which implies that $Z^t(x) - d_t^+ + d_t^- = g_t$. Furthermore, $|Z^t(x) - g_t| = d_t^+ + d_t^-$. Thus, the GP model reduces to

$$\begin{aligned} \textbf{GP: } &\text{minimize} \sum_{t=1}^{K} w_t (d_t^+ + d_t^-) \\ &\text{subject to } Z^t(x) - d_t^+ + d_t^- = g_t, \\ &\qquad d_t^+ \geq 0, d_t^- \geq 0 \ (t = 1, 2, \ldots, K), \\ &\qquad x \in F, \end{aligned}$$

where d_t^+ and d_t^- are over- and under- achievements of the t-th goal, respectively.

In most of the real-life situations DM fails to recognize proper goals, then DM considers goals by intervals $[g_t^l, g_t^u]$ based on his experience. In this situation, consider $g_t = g_t^l$ for minimization problem and $g_t = g_t^u$ for maximization problem in model GP. Another important factor is that, if values $Z^t(x)$ are in different scales and there is a certain gap between the upper bound and lower bound of goals then we reduce the weights w_t by $w_t/(g_t^u - g_t^l)$.

Sustainability is not about a threat analysis; it is a system analysis. Specifically, it is about how environmental, economic and social systems interact to their mutual advantages or disadvantages at various space-based scales of operations. In the present study, we present how the economic conditions of people can be improved, environmental and health would be cured satisfying the profit goals. In every system where machines are running by energy. For that energy consumption, there are several types of gases, exerted by the machines that create pollution in the atmosphere. Most of the cases, the fact of pollution is not considered by the agency or person or governing body, they just focus on maximizing profit or minimizing expenditure. But, in most of the situations everyone forget to keep the atmosphere pollution free for upcoming future. So, it is the time to think about minimizing pollution and to optimize the other objectives as much as possible by considering our valuable time in the system.

The following are the main contributions of our study:

- Formulate the mathematical model of a class of TP, namely, *two-stage MOTP*,
- Selection of goal and solution by goal programming are incorporated,
- Sustainable development is incorporated by providing a numerical example.

The rest of the chapter is organized as follows: In the next section (Sect. 2), the review of related research is presented. The problem environment of *two-stage MOTP* is defined in Sect. 3. Section 4 presents the mathematical model formulation of *two-stage MOTP*. The solution procedure is described in the Sect. 4.1. To show the application of the proposed mathematical model of the MOTP, a numerical example of MOTP is included in Sect. 5. The result and discussion regarding the numerical example is presented in Sect. 5.1. Analyzing sustainable development through present study is described in Sect. 6. The conclusions with the future study are described in Sect. 7.

2 Review of Related Research

The basic TP was originally initiated by Hitchcock [9] and later on by Kantorovich [13], who has given an algorithm for obtaining the solution of the TP. He developed a procedure to solve the TP, which has a close link with the simplex method for solving TP as the primal simplex transportation method. A large number of research works on TP has been done in different directions by several researchers such as Roy et al. [32], Vincent et al. [35], Ghosh and Roy [7], James et al. [12], Ghosh et al. [8]. Luathep et al. [19] solved an optimization problem for mixed transportation network design. Mardanya et al. [23] presented a study on bi-level multi-objective TP using fuzzy logic. Lei et al. [15] solved a problem on transportation cost allocation under the consideration of a fixed route. Midya et al. [25] solved fuzzy multiple objective fractional optimization in rough approximation and its application to the fixed-charge transportation problem. Reisach et al. [27] employed a study on special areas of operational research problems. Maity et al. [22] introduced a study on multimodal transportation problem and use it in neural networking area.

GP is used widely as a tool to solve the MOTP, and it is accepted by many researchers and industrialists due to its easiness of use, adaptability and efficiency to provide the most appropriate solutions in multi-objective decision making problems. Charnes and Cooper [2] developed a goal programming model for linear programming problems in which conflicting goals were incorporated as constraints. Later on, Ignizio and Perlis [10] introduced a solution approach of GP, known as sequential GP method. Tamiz et al. [33] and Romero [28] incorporated some useful studies in decision making problems under goal environment. Roy et al. [31] presented a study on goal programming to solve a both way MOTP. Liao and Kao [17] described a method based on quality function development, fuzzy-extended analytic hierarchy process and Multi-Segment GP. Roy and Maity [29] incorporated a study on time and cost minimizing TP using interval parameters. A study on single-sink fixed-charge, multi-objective, multi-index stochastic TP was developed by Midya and Roy [24]. Recently, a study on solving MOTP using conic scalarization approach has been introduced by Roy et al. [30].

A first approach to sustainable transportation consists in the selection and measurement of the externalities it generates. There are many transportation externalities, being the release of pollutant particles which are mostly common. A method by Dulebenets [6] introduced a study on green optimization problem involving carbon emission. A study on analyzing the trade-off between CO_2 emissions and passenger service level in the airline industry introduce by Jalalian et al. [11]. The way of environmental cost procedure has been revealed very popular in the study of Lera-Lopez et al. [16]. Liotta et al. [18] described a study on freight transportation and defines sustainable development on it. Das et el. [3] incorporated a study on green optimization under fuzzy and non-fuzzy techniques. Midya et al. [26] studied on green supply chain management problem under fuzzy environment. Karakikes and Nathanail [14] discussed on sustainable development on a logistics based on simulation. A multi-objective solid transportation-location problem with variable carbon emission in inventory management is presented by Das et al. [5]. A study on sustainable development in transportation problem introduced by Maity et al. [21]. Das and Roy [4] solved a multi-objective transportation problem and used it to reduce carbon emission.

Many researchers adopted various methodologies for analyzing sustainability by several decision making problems. Nevertheless, to the best of our knowledge, no one attempt to solve *two-stage MOTP* and use it for sustainable development. Here, we present a new approach for solving a *two-stage MOTP* by goal programming which could be of a great impact for sustainable development.

3 Problem Environment

Sustainable development is highly accessible through optimization problems related to transportation. Again, many real-life problems can be formulated by the concept of TP in which objection function is referred to sustainable development of nature or produce employment to the unemployment people. Here, we formulate a *two-stage MOTP* in which we consider an objective function to minimize carbon emission for reducing pollution and another objective function is considered in such a way that less use of machines can remove unemployment for the people.

Here, we first define the background of *two-stage MOTP*. A large number of investigations has been accommodated by different researchers for solving MOTP with goals through the objective functions. But, in many cases, we observe that the objective function is associated with the goals, and then solution procedures are developed for obtaining solutions to the best according to decision maker's choice. In this research, we formulate a mathematical model of *two-stage MOTP*. In a broad sense, a *two-stage TP* refers to a TP in which goods are collected at some warehouses, and then the goods are delivered to several destinations. So, the warehouses are the demand points for sources and the supply points for destinations. Goods are transported at two stages, i.e., for gathering and distributing, and the whole system is managed by only one DM. The DM wishes to maximize his benefit and to minimize the transportation costs. Another important notion for this MOTP is that if the DM wishes to maximize his profit without considering the transportation cost, then the customers will be affected. In this situation, the optimal goals for the objective functions and the corresponding solutions are not specified in the literature till now. Based on all these ideas, we propose a new way to select the optimal goals for the objective functions of considered MOTP.

Under the consideration of goal values to the objective functions, we analyze sustainable developments through two ways in this study. Selection of goals in proper forms reflects a better decision in favor of sustainable development.

4 Model Formulation

The mathematical formulation of classical TP is defined as follows:

Model 1

$$\text{minimize } Z = \sum_{i=1}^{m} \sum_{j=1}^{n} C_{ij} x_{ij}, \tag{1}$$

$$\text{subject to } \sum_{j=1}^{n} x_{ij} \leq a_i \quad (i = 1, 2, \ldots, m), \tag{2}$$

$$\sum_{i=1}^{m} x_{ij} \geq b_j \quad (j = 1, 2, \ldots, n), \tag{3}$$

$$x_{ij} \geq 0 \ \forall \ i \text{ and } j, \tag{4}$$

where x_{ij} and C_{ij} $(i = 1, 2, \ldots, m; \ j = 1, 2, \ldots, n)$ are decision variable and transportation cost per unit commodity from the i-th origin to the j-th destination respectively. a_i $(i = 1, 2, \ldots, m)$ and b_j $(j = 1, 2, \ldots, n)$ are availability and demand at the i-th origin and the j-th destination, respectively, and the feasibility condition is $\sum_{i=1}^{m} a_i \geq \sum_{j=1}^{n} b_j$.

The mathematical model of the MOTP can be posed as follows:
Model 2

$$\text{minimize/maximize} \quad Z^s = \sum_{i=1}^{m}\sum_{j=1}^{n} C_{ij}^s x_{ij} \quad (s = 1, 2, \ldots, K) \tag{5}$$

$$\text{subject to} \quad \sum_{j=1}^{n} x_{ij} \leq a_i \quad (i = 1, 2, \ldots, m), \tag{6}$$

$$\sum_{i=1}^{m} x_{ij} \geq b_j \quad (j = 1, 2, \ldots, n), \tag{7}$$

$$x_{ij} \geq 0 \quad \forall\ i \text{ and } j. \tag{8}$$

The feasibility condition is same as defined in Model 1. The objective functions in MOTP are conflicting to each other. We declare the following assumptions to construct mathematical model (for Model 3) of a *two-stage MOTP* as:

- Consider m sources D_i $(i = 1, 2, \ldots, m)$ with supply a_i $(i = 1, 2, \ldots, m)$ and p destinations $C_{j'}$ $(j' = 1, 2, \ldots, p)$ with demand $b_{j'}$ $(j' = 1, 2, \ldots, p)$, and there are no direct links between sources and destinations.
- Assume n nodes S_j $(j = 1, 2, \ldots, n)$ with equal supply and demand d_j, $(j = 1, 2, \ldots, n)$ to interact between sources D_i and destinations $C_{j'}$; this has the capacity of taking goods from sources D_i and delivering goods to the destinations $C_{j'}$.
- We treat T objective functions for stage 1 transportation, i.e., transportation from sources D_i to destination S_j; y_{ij} is taken as a decision variable and P_{ij}^t is the cost parameter for the t-th objective function.
- We describe K objective functions for stage 2 transportation, i.e., transportation from sources S_j to destination $C_{j'}$; here, $x_{i'j'}$ is taken as a decision variable and $C_{i'j'}^k$ is the penalty parameter for the k-th objective function. In fact, the penalty parameter may be the transportation cost per unit item, production cost per unit item of goods, selling cost per unit item of goods, etc.

Then, the mathematical model can be designed in the following form:
Model 3

$$\text{minimize/maximize} \quad W^t = \sum_{i=1}^{m}\sum_{j=1}^{n} P_{ij}^t y_{ij} \quad (t = 1, 2, \ldots, T), \tag{9}$$

$$\text{minimize/maximize} \quad Z^k = \sum_{i'=1}^{n}\sum_{j'=1}^{p} C_{i'j'}^k x_{i'j'} \quad (k = 1, 2, \ldots, K) \tag{10}$$

$$\text{subject to} \quad \sum_{j=1}^{n} y_{ij} \leq a_i \quad (i = 1, 2, \ldots, m), \tag{11}$$

$$\sum_{i=1}^{m} y_{ij} \geq d_j \quad (j = 1, 2, \ldots, n), \tag{12}$$

$$\sum_{j'=1}^{p} x_{i'j'} \leq d_{i'} \quad (i' = 1, 2, \ldots, n), \tag{13}$$

$$\sum_{i'=1}^{n} x_{i'j'} \geq b_{j'} \quad (j' = 1, 2, \ldots, p), \tag{14}$$

$$\sum_{i=1}^{m} y_{il} \geq \sum_{j'=1}^{p} x_{lj'} \quad (l = 1, 2, \ldots, n), \tag{15}$$

$$y_{ij} \geq 0 \;\forall\, i,\; j;\; x_{i'j'} \geq 0 \;\forall\, i',\; j'. \tag{16}$$

Here, $\sum_{i=1}^{m} a_i \geq \sum_{j'=1}^{p} b_{j'}$ is the feasibility condition. The set of constraints can be designed as per as the extra conditions are allowed to impose in the respective decision making problem.

The objective functions W^t $(t = 1, 2, \ldots, T)$ are of conflicting nature which we consider at the stage 1 of Model 3; so these functions are not summed up, even if they are of same type. A similar concept is adopted for the objective functions at the subsequent stage of Model 3. Also, there are different agents to pay the cost at different stages of transportation. Due to this fact, we consider the objective functions separately in Model 3.

Usually, the DM determines the goals into the objective functions and, thereafter, the solution is obtained according to the best fit of the goals. Here, our main interest is to constitute the way of assignment of goals to the objective functions. When there are more than one objective function, conflicting in nature, in many cases optimal solutions of the objective functions occur at different points. So, consideration of specific goals and the corresponding solution are a complex task for the DM.

Now, some useful definitions regarding our proposed study in the goal space are presented. The following definitions are given with respect to s objective functions.

Definition 1 (*Feasible goal*): Let Z^t be the t-th objective function of the MOTP. Then the feasible goal for the objective function Z^t refers to all the possible values of the objective function Z^t which is an interval $I^t = [\min\; Z^t, \max\; Z^t]$. Here, $\min\; Z^t$ and $\max\; Z^t$ are obtained under supply and demand constraints of the MOTP along with the nonnegativity conditions.

Definition 2 (*Feasible goal region*): Let I^t be the feasibility of goal for t-th objective function of the MOTP. Then the subset $I^1 \times I^2 \times \ldots \times I^s$ of $\mathbb{R}^s$ is the feasibility goal region for the MOTP with respect to s objective functions.

Definition 3 (*Optimum goal region*): In MOTP, an objective function is either maximization type or minimization type. It can be solved with respect to the constraints separately to get its optimal solution. Let $X_1^*, X_2^*, \ldots, X_s^*$, be the respective ideal solutions of s objective functions of the MOTP. Evaluating all these objective func-

Table 1 Pay-off matrix for ideal solutions of the objective functions

$$\begin{bmatrix} Z^1(X_1^*) & Z^1(X_2^*) & \cdots & Z^1(X_s^*) \\ Z^2(X_1^*) & Z^2(X_2^*) & \cdots & Z^2(X_s^*) \\ Z^3(X_1^*) & Z^3(X_2^*) & \cdots & Z^3(X_s^*) \\ \vdots & \vdots & \vdots & \vdots \\ Z^s(X_1^*) & Z^s(X_2^*) & \cdots & Z^s(X_s^*) \end{bmatrix}.$$

tions at all the ideal solutions, we formulate a pay-off matrix of format $s \times s$ as stated in Table 1.

Then, we find $I_o^i = [\min_{j=1,2,\ldots,s} Z^i(X_j^*), \max_{j=1,2,\ldots,s} Z^i(X_j^*)]$ $(i = 1, 2, \ldots, s)$.

The subset $I_o^1 \times I_o^2 \times \ldots \times I_o^n$ of $\mathbb{R}^s$ is the optimum feasibility goal region for the MOTP with respect to the given objective functions.

4.1 Algorithm for Selection of Goals to the Objective Functions In Two-Stage TP

Here, we present an algorithm for selection of goals to the objective functions in our *two-stage MOTP*.

- **Step 1**: First, we solve each of the objective functions corresponding to a stage 1 TP under the restriction of proposed problem. As a result, we obtain the solutions, $Y_1^*, Y_2^*, \ldots, Y_T^*$. Again, we solve each of the objective functions corresponding to a stage 2 TP under the restriction of our proposed problem and we calculate the solutions, $X_1^*, X_2^*, \ldots, X_K^*$. These solutions are called as ideal solutions; often, they are known as Pareto optimal solutions.
- **Step 2**: Using the ideal solutions obtained in Step 2, we formulate pay-off matrices, one is of format $T \times T$, and the other one is of format $K \times K$, according to Definition 3.
- **Step 3**: Obtaining $I_o^i = [\min_j W^i(Y_j^*), \max_j W^i(Y_j^*)]$ $(i = 1, 2, \ldots, T)$, we then find the optimum feasibility goal region $I_o^1 \times I_o^2 \times \ldots \times I_o^T$ of $\mathbb{R}^T$ for the stage 1 MOTP. In a similar way, we define $G_o^i = [\min_j Z^i(X_j^*), \max_j Z^i(X_j^*)]$ $(i = 1, 2, \ldots, K)$, and derive the optimum feasibility goal region $G_o^1 \times G_o^2 \times \ldots \times G_o^K$ in $\mathbb{R}^K$ for the stage 2 MOTP.
- **Step 4**: We consider p_t as the goal value to each of the objective functions W^t and g_k as the goal value to each of the objective functions Z^k. Based on the goal values, we construct a utility function $E = E(p_1, p_2, \ldots, p_T; g_1, g_2, \ldots, g_K)$ and, then,

we optimize the value of E within the feasible goal regions $I_o^1 \times I_o^2 \times \ldots \times I_o^T$ in $\mathbb{R}^T$ and $G_o^1 \times G_o^2 \times \ldots \times G_o^K$ in $\mathbb{R}^K$. If $p_1^*, p_2^*, \ldots, p_T^*; g_1^*, g_2^*, \ldots, g_K^*$ are the optimal solutions of E, then these are the optimal goals to the objective function of the MOTP.

- **Step 5**: We formulate a single objective function whose solution optimally accommodates the goals, and it also yields the solution of the MOTP.
- **Step 6**: Based on the results of optimal goals and possible optimized values of the objective functions, we solve the proposed problem using revised GP.
- **Step 7**: Stop.

5 Numerical Example

In real-life situations, sustainable development is well connected through many projects, but intension of excessive profits in a quicker time it is not considered by decision maker. In this regard, we consider a project on rice harvesting in corn field. Few decades ago, rice harvesting was a work of solely by men. In modern days, there are so many machines used in this work. Modern technology reduces the human hard works. But, machines consume fuel and emit carbon to the nature. In real-life situations, these technologies has more demands. Using a Rice Harvester Machine (RHM), a farmer can complete work of 50 labors in a day. Sometimes, machines are not usable in some fields, then farmer needs the hands of human labors. It is also noted that, overall expenditure is lesser in harvesting by machines rather than harvesting by human. So, farmer wishes to complete the works as much as possible by machines in a low cost. But, they do not consider pollution of carbon emission by machines. Nowadays, pollution of nature is growing highly. Again, using of excessive machines reduce the work of labors. Finally, we are in the situation that, labors are losing their works and nature is polluted due to excessive carbon emission. So, we need to sustain our nature and remove unemployment for the labors. In this regard, we choose a numerical example and solve it using goal programming.

A Business Man (BM) has two offices in two different locations O_1 and O_2, respectively, where the offices, RHM and human factors are collected from centers C_1 and C_2, respectively, they are appointed for harvesting in fields of locations D_1, D_2 and D_3 as per the requirements and work loads in the areas. The graphical network of the problem is shown in Fig. 2.

There is a manager independent of BM who controls the centers C_1 and C_2. A RHM (Greenfield Agro Impex type) consumes around 8 litres diesel per hour. So, it consumes 80 litres diesel in a day for 10 h of service. According to composition, the weights of one litre is 835 g. Diesel consists of 86.2% carbon, or 720 g of carbon per litre diesel. In order to combust this carbon produces to CO_2, 1920 grammes of oxygen are needed. The sum is then $(720 + 1920)$ g = 2640 g of CO_2/litre diesel. Thus a RHM emits $(2.64 \times 80)kg = 211.20$ kg. CO_2 in a day during harvesting. Therefore, a RHM emits approximately 0.2 ton CO_2 in a day during rice harvesting.

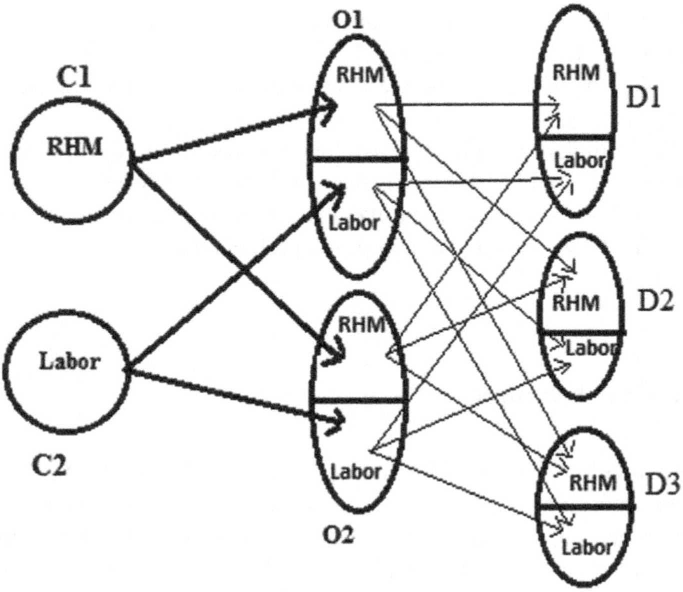

Fig. 2 Graphical representation of the numerical example.

Table 2 Transportation cost (in $) per RHM and per labor from C_1, C_2 to O_1, O_2

	O_1	O_2
C_1	40	38
C_2	2	1.9

This information may vary according to the geographical areas. Based on the above considerations, we formulate a mathematical model with the following decisions:

- W^1: minimizing transportation cost (in a day) for collecting RHM and labors from C_1 and C_2 (objective function for Stage 1),
- W^2: maximizing profit (in a day) for supplying RHM and labors to O_1 and O_2 (objective function for Stage 1),
- Z^1: maximizing profits for supplying RHM and labors to the destinations D_1, D_2 and D_3 (objective function of Stage 2),
- Z^2: minimizing pollution by minimizing carbon emission (objective function of Stage 2).

The supplied data of the problems are shown in Tables 2, 3, 4, 5 and 6.

In the aforementioned numerical example in Stage 1 transportation, the BM collects RHM and labors by proving minimum cost per day and in the Stage 2 transportation, he supplies RHM and labors to the destinations D_1, D_2 and D_3 by considering a profit margin. Considering the sustainable development, BM wishes to reduce the pollution, i.e., minimize the carbon emission. In this way the work of machine will

Table 3 Profit (in \$) per RHM and per labor for transporting C_1, C_2 to O_1, O_2.

	O_1	O_2
C_1	50	49
C_2	1	1.2

Table 4 Profit (in \$) for supplying RHM per day

	D_1	D_2	D_3
O_1	4.5	4.2	4.3
O_2	4.1	4.5	4.7

Table 5 Profit (in \$) for supplying labor per day

	D_1	D_2	D_3
O_1	0.11	0.15	0.10
O_2	0.12	0.10	0.14

Table 6 Carbon emission (in ton) of RHM in harvesting per day

	D_1	D_2	D_3
O_1	0.21	0.19	0.20
O_2	.22	0.23	0.20

be used lees in numbers whereas labors will be engaged more in numbers. Therefore, the aforementioned problem is a *two-stage MOTP*. Now, we solve the problem using our mathematical model.

The number of RHM and labors available at C_1 and C_2 are 50 and 2500 respectively. Also, it is noted that, work by a RHM in a day is equivalent to work of 50 human labors in a day. Minimum requirements of man power (altogether RHM and labor) at destinations D_1, D_2 and D_3 are 1000, 1500, 1100 respectively.

We formulate the mathematical model in which W^1 and W^2 are objective functions of minimization and maximization type for Stage 1 transportation and Z^1 and Z^2 are objective functions of maximization and minimization type in Stage 2 transportation, respectively. The decision variables for formulating the problem are as follows:

y_{1j}: number of RHMs transported from C_1 to O_j, $j = 1, 2$;
y_{2j}: number of labors transported from C_1 to O_j, $j = 1, 2$;
x_{ij}: number of RHMs supplied to D_j, $j = 1, 2, 3$;
z_{ij}: number of labors supplied to D_j, $j = 1, 2, 3$.

Based on the supplied data, the following mathematical model is formulated:
Model 4

$$\text{minimize } W^1 = 40y_{11} + 38y_{12} + 2y_{21} + 1.9y_{22}, \quad (17)$$
$$\text{maximize } W^2 = 50y_{11} + 49y_{12} + y_{21} + 1.2y_{22}, \quad (18)$$
$$\text{maximize} Z^1 = 4.5x_{11} + 4.2x_{12} + 4.3x_{13} + 4.1x_{21} + 4.5x_{22} + 4.7x_{23} + 0.11z_{11} + 0.15z_{12} + 0.10z_{13} + 0.12z_{21} + 0.10z_{22} + 0.14z_{23}, \quad (19)$$
$$\text{minimize } Z^2 = 0.21z_{11} + 0.19z_{12} + 0.20z_{13} + 0.22z_{21} + 0.23z_{22} + 0.20z_{23} \quad (20)$$
$$\text{subject to } y_{11} + y_{12} \leq 50, \quad (21)$$
$$y_{21} + y_{22} \leq 2600, \quad (22)$$
$$50y_{11} + y_{21} \geq 2000, \quad (23)$$
$$50y_{12} + y_{22} \geq 1800, \quad (24)$$
$$x_{11} + x_{12} + x_{13} \leq y_{11}, \quad (25)$$
$$x_{21} + x_{22} + x_{23} \leq y_{12}, \quad (26)$$
$$z_{11} + z_{12} + z_{13} \leq y_{21}, \quad (27)$$
$$z_{21} + z_{22} + z_{23} \leq y_{22}, \quad (28)$$
$$50(x_{11} + x_{21}) + z_{11} + z_{21} \geq 1000, \quad (29)$$
$$50(x_{12} + x_{22}) + z_{12} + z_{22} \geq 1500, \quad (30)$$
$$50(x_{13} + x_{23}) + z_{13} + z_{23} \geq 1100, \quad (31)$$
$$x_{i'j'} \geq 0,\ z_{i'j'} \geq 0,\ y_{ij} \geq 0,\quad \forall i', j = 1, 2; i = 1, 2; j' = 1, 2, 3. \quad (32)$$

Solving the objective functions W^1 and W^2 independently with respect to the Constraints (21)–(32), we construct the pay-off matrix (cf. Definition 3) which is shown in Table 7.

Table 8 represents the pay-off matrix (cf. Definition 3) which is designed by solving each objective function Z^1 and Z^2 with respect to the same constraints (21)–(32).

Utilizing the obtained solutions (cf. Tables 7 and 8), we determine interval goal:

[4450, 6940] for W^1, [4050, 5620] for W^2; [490, 618.6] for Z^1, and [3.8, 10.26] for Z^2.

Table 7 Pay-off matrix for W^1 and W^2

	W^1	W^2
Y_1^*	4450	4050
Y_2^*	6940	5620

Table 8 Pay-off matrix for Z^1 and Z^2

	Z^1	Z^2
X_1^*	618.6	10.26
X_2^*	490	3.8

We consider the normalized weights (p_t) corresponding to t-th objective function. Another important factor is that, the values $Z^t(x)$ and $W^t(x)$ are in different scales and there is a certain gap between the upper bound and lower bound of goals then we reduce the weights p_t by $p_t/(g_t^u - g_t^l)$. Using these weights, we construct the following revised GP model:

Model 5

$$\begin{aligned}
\text{minimize} \quad & \frac{p_1}{2490}(d_1^+ + d_1^-) + \frac{p_2}{1570}(d_2^+ + d_2^-) + \frac{p_3}{128.6}(d_3^+ + d_3^-) + \frac{p_4}{6.28}(d_4^+ + d_4^-) \\
& + \frac{p_1}{2490}(e_1^+ + e_1^-) + \frac{p_2}{1570}(e_2^+ + e_2^-) + \frac{p_3}{128.6}(e_3^+ + e_3^-) + \frac{p_4}{6.28}(e_4^+ + e_4^-) \\
\text{subject to} \quad & W^1 - d_1^+ + d_1^- = y_1;\ y_1 - e_1^+ + e_1^- = 4450; \\
& W^2 - d_2^+ + d_2^- = y_2;\ y_2 - e_2^+ + e_2^- = 5620; \\
& Z^1 - d_3^+ + d_3^- = y_3;\ y_3 - e_3^+ + e_3^- = 618.6; \\
& Z^2 - d_4^+ + d_4^- = y_4;\ y_4 - e_4^+ + e_4^- = 3.8; \\
& 4450 \le y_1 \le 6940;\ 4050 \le y_2 \le 5620; \\
& 490 \le y_3 \le 618.6;\ 3.8 \le y_4 \le 10.26; \\
& d_i^+ \ge 0,\ d_i^- \ge 0,\ e_i^+ \ge 0,\ e_i^- \ge 0\ (i = 1, 2, 3, 4); \\
& \text{and constraints } (17) - (32).
\end{aligned}$$

Model 5 is simply a linear programming problem which can be solved through simplex algorithm. As Model 5 contains the large number of variables, so we use LINGO iterative scheme to obtain the solution of Model 5.

5.1 *Results and Discussions*

Solving Model 5 using different normalized weights corresponding to objective functions, we obtain Pareto-optimal solutions, which are listed in Table 9.

Table 9 represents Pareto-optimal solutions in different weights, from the solution we find the following important facts:

- In the first row of Table 9, considering equal weights we get the solution in which the profit goals for BM and the manager of centers C_1 and C_2 achieve their profit goals around 70% and 90% respectively, whereas the goals of minimizing the transportation cost (W^1) and carbon emission (Z^2) are not satisfactory.
- In second row of Table 9, if the considered weight is completely biased in favor objective function, carbon emission (Z^2) then the goal is highly satisfactory for (Z^2), satisfactory level for transportation cost (W^1) is around 50%, but the profit goals for both BM and the manager of centers C_1 and C_2 are not satisfactory. As per as the business is concerned they would not like the decision in favor of sustainability of nature. In this case, the using of RHM will be diminished, so

Table 9 Solution of Model 5

Weight	Value of objective function	Value of decision variable
$p_1 = 0.25, p_2 = 0.25$ $p_3 = 0.25, p_4 = 0.25$	$W^1 = 6940, W^2 = 5220$ $Z^1 = 596.9, Z^2 = 9.7$	$y_{11} = 23, y_{12} = 12, y_{21} = 1345, y_{22} = 1255$; $x_{11} = 20, x_{12} = 3, x_{23} = 25, z_{12} = 1345$, $z_{23} = 1255$; Rest of all are zero
$p_1 = 0, p_2 = 0$ $p_3 = 0, p_4 = 1$	$W^1 = 5980, W^2 = 4160$ $Z^1 = 376, Z^2 = 3.8$	$y_{11} = 24, y_{12} = 0, y_{21} = 800, y_{22} = 1800$; $x_{12} = 20, z_{13} = 800, z_{21} = 1000, z_{22} = 500$, Rest of all are zero
$p_1 = 0, p_2 = 0.5$ $p_3 = 0.5, p_4 = 0$	$W^1 = 6940, W^2 = 5220$ $Z^1 = 603.5, Z^2 = 10.2$	$y_{11} = 20, y_{12} = 30, y_{21} = 1550, y_{22} = 1000$; $x_{12} = 20, x_{23} = 30, z_{12} = 1550, z_{23} = 1000$, Rest of all are zero
$p_1 = 0.1, p_2 = 0.1$ $p_3 = 0.1, p_4 = 0.7$	$W^1 = 6733, W^2 = 5220$ $Z^1 = 433, Z^2 = 3.8$	$y_{11} = 30, y_{12} = 14, y_{21} = 500, y_{22} = 2100$; $x_{12} = 20, z_{12} = 500, z_{21} = 1000, z_{23} = 1100$, Rest of all are zero

the labor will get the work in more number of days. Here the profits and carbon emission are calculated per day.

- From the third row, we see that the profit goals are satisfactory for both BM and the manager of centers C_1 and C_2, But the other goals, i.e., transportation cost and carbon emission are not satisfactory. Here, we see that RHM and labor both are used in high numbers significantly. As a result, the profits are achieved in high margin. In this case, the labor will get their work for lesser time as machines are used in a good numbers for harvesting.
- Finally, in last row again, we see that the pollution decreases which enhance that the profit margins are decreased. In this case, we see that the desired goal of profit diminishes to negative level for BM. As we solve the problem by revised GP, so we use positive as well as negative deviations of goals, so value of the objective functions in Pareto-optimal solution may fall outside of the selected goals in some cases. But, this case is well for labors as they can get maximum work in the harvesting process.
- In this problem, we consider the objective functions in two stages; Stage 1 beneficiary is the manager of centers C_1, C_2 and Stage 2 beneficiary is the BM who works as a broker of the system. The process is completely realistic as it happens almost

in daily life for implementing the project works such as building construction, farming, drainage system, etc.

- In the presented model, we establish the sustainable development by introducing objective function which minimizes carbon emission. Through the solution, it is observed that pollution can be reduced by selecting proper goals corresponding to the objective function.

6 Analyzing Sustainable Development by Two-Stage MOTP

The main contributions of this study can be broadly divided into two notions. Firstly, the study presents a new class of TP namely *two-stage MOTP*. Considering the brokery in the real-life scenarios, we employ a linear programming problem of a system where there are two decision makers independently engage to maximize their profit. The research presents a situation where, the *two-stage MOTP* is applicable in decision making problems. The utility of the *two-stage MOTP* is discussed through the numerical example. Now, we describe the sustainable development through our proposed model *two-stage MOTP*.

In almost every business sectors, private or government, labors and machines are working together in completion of any project. Today's life is almost impossible to step forward without using of machines. In modern civilization, if we do not use technologies, we are much behind than the progress in compare to others. But in the urban, using of technologies takes us in a challenging atmospheric situation, where global warming is utmost import factor to sustain life in earth. There are so many causes of global warming, among them carbon emission by fuel consumption is an important factor. So, we need to decrease carbon emission in every possible sector.

We should have to use technology but excessive using for more benefit is not acceptable. Here, we choose a numerical example which shows how better benefits are occurring without polluting the nature. In the presented numerical example, we show that how pollution is reduced by lesser use of RHM in harvesting. Consequently, labors will get more work by lesser use of RHM in the process.

Some people would like to say technology reduces human efforts. These should have to use in different weather conditions, unavailability of workers, etc. Of course this is true, but the pollution is increasing by the system after the harvesting also, by burning the rice sticks remains in the corn field. Who will take the responsibility to stop these? So governments, public, farmers have to make prompt decisions to sustain nature for future generation.

In this study, we present only one situation to establish our study, there are many areas such as banking sectors, road making, drainage system, etc. where the same works are done by machines for better profits as a result literate and illiterate both peoples are losing their works along with pollution in nature which is increasing in a high rate. So, better policies may sustain our nature as well as give a new life to work less people.

7 Conclusion and Outlook

In this study, we have proposed a *two-stage MOTP* with two interesting characteristics, that one being about the goal preferences of the DM under the preference of sustainable development. Solution of the MOTP under these goals through revised GP has been provided a technique for multi-objective decision making. But, the main drawback of revised GP consists in how the DM would select the goals for the objective functions. The core content of our proposed study has removed this difficulty by selecting the goals of the objective functions through an algorithm. Thereafter, we have solved the proposed problem by the revised GP approach. Another important factor of this study has been considered sustainable development by minimizing pollution factor. Furthermore, we have shown that lesser use of technology has been produced more works to the labors in rice harvesting. Realizing the global warming and the overall pollution, it is utmost essential to keep the nature neat and clean for us not looking maximum profit which creating harmful effects on the atmosphere along with removing unemployment situations in globe.

In the future scope of our study, we can address different uncertainty of data, such as fuzzy, stochastic, etc., and test several examples such as route selection problem with consideration of a real-life *two-stage TP*, technology selection problem, plant location selection problem, etc., using the proposed methodology. We especially provide the research direction of associating the approach with the method by Anderluh et al. [1] on scheduling traffic flow in a city. In addition to all the aforementioned, we can employ our study for selecting optimal goals in inventory optimization, supply chain management and other areas.

References

1. Anderluh, A., Hemmelmayr, V.C., Nolz, P.C.: Synchronizing vans and cargo bikes in a city distribution network. Central Eur. J. Oper. Research. (2016). https://doi.org/10.1007/s10100-016-0441-z
2. Charnes, A., Cooper, W.W.: Management Model and Industrial Application of Linear Programming, p. 1. Wiley, New York (1961)
3. Das, S.K., Roy, S.K., Weber, G.W.: Application of type-2 fuzzy logic to a multi-objective green solid transportation-location problem with dwell time under carbon tax, cap and offset policy: fuzzy vs. non-fuzzy techniques. IEEE Trans. Fuzzy Syst. 28(11), 2711–2725 (2020)
4. Das, S.K., Roy, S.K.: Effect of variable carbon emission in a multiobjective transportation-p-facility location problem under neutrosophic environment. Comput. Ind. Eng. **132**, 311–324 (2019)
5. Das, S.K., Pervin, M., Roy, S.K., Weber, G.W.: Multi-objective solid transportation-location problem with variable carbon emission in inventory management: a hybrid approach. Ann. Oper. Res. (2021). https://doi.org/10.1007/s10479-020-03809-z
6. Dulebenets, M.A.: Green vessel scheduling in liner shipping: Modeling carbon dioxide emission costs in sea and at ports of call. Int. J. Transp. Sci. Technol. **7**(1), 26–44 (2018)
7. Ghosh, S., Roy, S.K.: Fuzzy-rough multi-objective product blending fixed-charge transportation problem with truck load constraints through transfer station. Rairo-Oper. Res. **55**, S2923–S2952 (2021)

8. Ghosh, S., Roy, S.K., Ebrahimnejad, A., Verdegay, J.L.: Multi-objective fully intuitionistic fuzzy fixed-charge solid transportation problem. Complex Intell. Syst. **7**, 1009–1023 (2021)
9. Hitchcock, F.L.: The distribution of a product from several sources to numerous localities. J. Maths Phys. **20**, 224–230 (1941)
10. Ignizio, J.P., Perlis, J.H.: Sequential linear goal programming: Implementation via MPSX. Comp. Oper. Res. **6**, 141–145 (1979)
11. Jalalian, M., Gholami, S., Ramezanian, R.: Analyzing the trade-off between CO_2 emissions and passenger service level in the airline industry: mathematical modeling and constructive heuristic. J. Clean. Prod. **206**, 251–266 (2019)
12. James, J.H.L., Hsu, C.C., Chen, Y.S.: Improving transportation service quality based on information fusion. Transp. Res. Part A: Policy Pract. **67**, 225–239 (2014)
13. Kantorovich, L.V.: Mathematical methods of organizing and planning production. Manag. Sci. **6**(4), 366–422 (1960)
14. Karakikes, I., Nathanail, E.: Simulation techniques for evaluating smart logistics solutions for sustainable urban distribution. Procedia Eng. **178**, 569–578 (2017)
15. Lei, S., Rangarajan, A., Karwan, M.H., Pinto, J.M.: Transportation cost allocation on a fixed route. Comp. Indus. Eng. **83**, 61–73 (2015)
16. Lera-Lopez, F., Faulin, J., Sinchez, M.: Determinants of the willingness-to-pay for reducing the environmental impacts of road transportation. Transp. Res. Part D Transp. Environ. **17**, 215–220 (2012)
17. Liao, C.N., Kao, H.P.: Supplier selection model using Taguchi loss function, analytical hierarchy process and multi-choice goal programming. Comput. Ind. Eng. **58**, 571–577 (2010)
18. Liotta, G., Stecca, G., Kaihara, T.: Optimisation of freight flows and sourcing in sustainable production and transportation networks. Int. J. Prod. Econ. **164**, 351–365 (2015)
19. Luathep, P., Sumalee, A., Lam, W.H.K., Li, Z.C., Lo, H.K.: Global optimization method for mixed transportation network design problem: a mixed-integer linear programming approach. Transp. Res. Part B: Methodol. **45**(5), 808–827 (2011)
20. Maity, G., Roy, S.K.: Solving multi-objective transportation problem with interval goal using utility function approach. Int. J. Oper. Res. **27**(4), 513–529 (2015)
21. Maity, G., Roy, S.K., Verdegay, J.L.: Time variant multi-objective interval-valued transportation problem in sustainable development. Sustainability **11**(21), 6161 (2020)
22. Maity, G., Roy, S.K., Verdegay, J.L.: Analyzing multimodal transportation problem and its application to artificial intelligence. Neural Comput. Appl. **32**(4), 2243–2256 (2020)
23. Mardanya, D., Maity, G., Roy, S.K.: Solving bi-level multi-objective transportation problem under fuzziness. Int. J. Uncertain. Fuzz. Knowl. Based Syst. **29**(3), 411–433 (2021)
24. Midya, S., Roy, S.K.: Single-sink, fixed-charge, multi-objective, multi-index stochastic transportation problem. Am. J. Math. Manag. Sci. **33**, 300–314 (2014)
25. Midya, S., Roy, S.K., Weber, G.W.: Fuzzy multiple objective fractional optimization in rough approximation and its aptness to the fixed-charge transportation problem. RAIRO-Oper. Res. **55**(3), 1715–1741 (2021)
26. Midya, S., Roy, S.K., Vincent, F.Y.: Intuitionistic fuzzy multi-stage multi-objective fixed-charge solid transportation problem in a green supply chain. Int. J. Mach. Learn. Cybern. **12**, 699–717 (2021)
27. Reisach, U., DeTombe, D., Weber, G.W.: Special issue: International aspects of OR. Central Eur. J. Oper. Res. **24**(4), 797–922 (2016)
28. Romero, C.: Handbook of Critical Issues in Goal Programming. Pergamon Press (1991)
29. Roy, S.K., Maity, G.: Minimizing cost and time through single objective function in multi-choice interval valued transportation problem. J. Intell. Fuzzy Syst. **32**(3), 1697–1709 (2017)
30. Roy, S.K., Maity, G., Weber, G.W., Alparslan, Gök. S..Z..: Conic scalarization approach to solve multi-choice multi-objective transportation problem with interval goal. Ann. Oper. Res. **253**(1), 599–620 (2017)
31. Roy, S.K., Maity, G., Weber, G.W.: Multi-objective two-stage grey transportation problem using utility function with goals. Central Eur. J. Oper. Res. **25**(2), 417–439 (2017)

32. Roy, S.K., Midya, S., Weber, G.W.: Multi-objective multi-item fixed-charge solid transportation problem under twofold uncertainty. Neural Comput. Appl. **31**(12), 8593–8613 (2019)
33. Tamiz, M., Jones, D.F., Romero, C.: Goal programming for decision making: an overview of the current state-of-the-art. Eur. J. Oper. Res. **111**(3), 569–581 (1998)
34. United Nations: Transforming Our World: The 2030 Agenda for Sustainable Development. Division for Sustainable Development Goals, New York, NY, USA (2015)
35. Vincent, FYu., Kuo-Jen, H., Chang, A.Y.: An interactive approach for the multi-objective transportation problem with interval parameters. Int. J. Prod. Res. **53**(4), 1051–1064 (2015)

The manufacturer's authorised representative in the EU is Springer Nature Customer Service Centre GmbH, Europaplatz 3, 69115 Heidelberg, Germany. If you have any concerns regarding our products, please contact ProductSafety@springernature.com

Printed and bound by CPI Group (UK) Ltd, Croydon, CR0 4YY
28/09/2026
02224093-0007